**Corrosion Resistance
Against Hydrogen**

Edited by
Michael Schütze, Günter Schmitt
and Roman Bender

Corrosion Resistance
Against Hydrogen

DECHEMA WILEY
WILEY-VCH Verlag GmbH & Co. KGaA

Editors

Prof. Dr.-Ing. Michael Schütze
DECHEMA-Forschungsinstitut
Chairman of the Executive Board
Theodor-Heuss-Allee 25
60486 Frankfurt am Main
Germany

Prof. Dr. Günter Schmitt
Chief Executive of IFINKOR gGmbH
Institute for Maintenance and
Corrosion Protection Technology
Kalkofen 4
58638 Iserlohn
Germany

Dr. rer. nat. Roman Bender
Chief Executive of GfKORR e. V.
Society for Corrosion Protextion
Theodor-Heuss-Allee 25
60486 Frankfurt am Main
Germany

Cover Illustration
Source: DECHEMA – Forschungsinstitut,
Frankfurt (Main), Germany

Warranty Disclaimer
This book has been compiled from literature data with the greatest possible care and attention.
The statements made only provide general descriptions and information.
Even for the correct selection of materials and correct processing, corrosive attack cannot be excluded in a corrosion system as it may be caused by previously unknown critical conditions and influencing factors or subsequently modified operating conditions.
No guarantee can be given for the chemical stability of the plant or equipment. Therefore, the given information and recommendations do not include any statements, from which warranty claims can be derived with respect to DECHEMA e. V. or its employees or the authors.
The DECHEMA e. V. is liable to the customer, irrespective of the legal grounds, for intentional or grossly negligent damage caused by their legal representatives or vicarious agents.
For a case of slight negligence, liability is limited to the infringement of essential contractual obligations (cardinal obligations). DECHEMA e. V. is not liable in the case of slight negligence for collateral damage or consequential damage as well as for damage that results from interruptions in the operations or delays which may arise from the deployment of this book.

■ This book was carefully produced. Nevertheless, editors, authors and publisher do not warrant the information contained therein to be free of errors. Readers are advised to keep in mind that statements, data, illustrations, procedural details or other items may inadvertently be inaccurate.

Library of Congress Card No.: Applied for.

British Library Cataloguing-in-Publication Data:
A catalogue record for this book is available from the British Library.

Bibliographic information published by Die Deutsche Bibliothek
Die Deutsche Bibliothek lists this publication in the Deutsche Nationalbibliographie; detailed bibliographic data is available in the Internet at
<http://dnb.ddb.de>.

© 2014 DECHEMA e. V., Society for Chemical Engineering and Biotechnology, 60486 Frankfurt (Main), Germany

All rights reserved (including those of translation into other languages). No part of this book may be reproduced in any form – nor transmitted or translated into machine language without written permission from the publishers. Registered names, trademarks, etc. used in this book, even when not specifically marked as such, are not to be considered unprotected by law.

Typesetting Beltz Bad Langensalza GmbH, Bad Langensalza
Printing and Binding Strauss GmbH, Mörlenbach
Cover Design Graphik-Design Schulz, Fußgönheim

ISBN: 978-3-527-33712-5

Printed in the Federal Republic of Germany

Printed on acid-free paper

Contents

Preface *VII*

How to use the Handbook *IX*

Corrosion Resistance Against Hydrogen *1*
Authors: Dr. P. Drodten, Dr. D. Schedlitzki, Prof. Dr. E. Wendler-Kalsch

A   Metallic materials   *20*
    Aluminum and aluminum alloys, copper and copper alloys, iron, iron-based alloys and steels, nickel and nickel alloys, titanium and titanium alloys, zirconium and zirconium alloys

B   Nonmetallic inorganic materials   *348*
    Carbon and graphite, binders for building materials, glass, fused silica and silica glass, enamel, oxide ceramic materials, metal ceramic materials

C   Organic materials   *355*
    Thermoplastics, elastomers, thermoplastic elastomers, duroplastics

D   Materials with special properties   *369*
    Coatings and films, gaskets and packings, composite materials

E   Material recommendations   *384*

Bibliography *391*

Key to materials compositions *415*

Index of materials *457*

Subject index *465*

Preface

Practically all industries face the problem of corrosion – from the micro-scale of components for the electronics industries to the macro-scale of those for the chemical and construction industries. This explains why the overall costs of corrosion still amount to about 2 to 4% of the gross national product of industrialized countries despite the fact that billions of dollars have been spent on corrosion research during the last few decades.

Much of this research was necessary due to the development of new technologies, materials and products, but it is no secret that a considerable number of failures in technology nowadays could, to a significant extent, be avoided if existing knowledge were used properly. This fact is particularly true in the field of corrosion and corrosion protection. Here, a wealth of information exists, but unfortunately in most cases it is scattered over many different information sources. However, as far back as 1953, an initiative was launched in Germany to compile an information system from the existing knowledge of corrosion and to complement this information with commentaries and interpretations by corrosion experts. The information system, entitled "DECHEMA-WERKSTOFF-TABELLE" (DECHEMA Corrosion Data Sheets), grew rapidly in size and content during the following years and soon became an indispensable tool for all engineers and scientists dealing with corrosion problems. This tool is still a living system today: it is continuously revised and updated by corrosion experts and thus represents a unique source of information. Currently, it comprises more than 8,000 pages with approximately 110,000 corrosion systems (i.e., all relevant commercial materials and media), based on the evaluation of over 100,000 scientific and technical articles which are referenced in the database.

Increasing demand for an English version of the DECHEMA-WERKSTOFF-TABELLE arose in the 1980s; accordingly the DECHEMA Corrosion Handbook was published in 1987. This was a slightly condensed version of the German edition and comprised 12 volumes. Before long, this handbook had spread all over the world and become a standard tool in countless laboratories outside Germany. The second edition of the DECHEMA Corrosion Handbook was published in 2004. Together the two editions covered 24 volumes.

The present handbook compiles all information on the corrosion behavior of materials that are in contact with hydrogen or environments containing this gas. This compilation is an indispensable tool for all engineers and scientists dealing with corrosion problems in hydrogen containing environments of any industrial use.

About 90% of all hydrogen is commercially produced in petrochemical processes, e.g. by catalytic steam cracking (steam reforming) of natural gas (methane) or light crude oil fractions or by partial oxidation of heavy oil. Considerable amounts of hydrogen are also produced in numerous processes in refineries or coking plants as well as in electrochemical processes, such as chloralkali electrolysis. Hydrogen is used for numerous chemical processes, for example for the synthesis of ammonia and methanol as well as in hydrogenation processes such as for the production of gasoline or for fat hardening. In metal extraction processes (such as W, Mo, Co, etc) it is used as a reducing agent and is also used as a shielding gas for welding and metallurgical processes. Finally, it can be used as a fuel gas or can be liquefied and used as an aerospace fuel.

Corrosion is a complex phenomenon that depends on a number of parameters, related to both the environment and the metal. In this handbook the behavior of materials in contact with hydrogen containing gases and liquids is compiled.

The chapters are arranged by the agents leading to individual corrosion reactions, and a vast number of materials are presented in terms of their behavior in these agents. The key information consists of quantitative data on corrosion rates coupled with commentaries on the background and mechanisms of corrosion behind these data, together with the dependencies on secondary parameters, such as flow-rate, pH, temperature, etc. This information is complemented by more detailed annotations where necessary, and by an immense number of references listed at the end of the handbook.

An important feature of this handbook is that the data was compiled for industrial use. Therefore, particularly for those working in industrial laboratories or for industrial clients, the book will be an invaluable source of rapid information for day-to-day problem solving. The handbook will have fulfilled its task if it helps to avoid the failures and problems caused by corrosion simply by providing a comprehensive source of information summarizing the present state-of-the-art. Last but not least, in cases where this knowledge is applied, there is a good chance of decreasing the costs of corrosion significantly.

Finally the editors would like to express their appreciation to Dr. Rick Durham and Dr. Horst Massong for their admirable commitment and meticulous editing of a work that is encyclopedic in scope.

They are also indebted to Gudrun Walter of Wiley-VCH for her valuable assistance during all stages of the preparation of this book.

Michael Schütze, Günter Schmitt and Roman Bender

How to use the Handbook

The Handbook provides information on the chemical resistance and corrosion behavior of materials in hydrogen and hydrogen containing atmospheres.

The user is given information on the range of applications and corrosion protection measures for metallic, non-metallic inorganic, and organic materials, including plastics.

Research results and operating experience reported by experts allow recommendations to be made for the selection of materials and to provide assistance in the assessment of damage.

The objective is to offer a comprehensive and concise description of the behavior of the different materials in contact with the medium.

The book is subdivided according to four groups of materials A-D:

- **A Metallic materials**
- **B Non-metallic inorganic materials**
- **C Organic materials and plastics**
- **D Materials with special properties**

These material groups are each subdivided according to their chemical formula; the metals are classed according to different alloy groups. These groups are shown in the uniformly designed summary table at the start of each chapter.

The information on resistance is given as text, tables, and figures. The literature used by the authors is cited at the corresponding point. There is an index of materials as well as a subject index at the end of the book so that the user can quickly find the information given for a particular keyword.

Material recommendations are given for each of the four groups of materials and are summarized in the section:

- **E Material recommendations.**

The Handbook is thus a guide that leads the reader to materials that have already been used in certain cases, that can be used or that are not suitable owing to their lack of resistance.

In addition to the detailed descriptions, the corrosion resistance and chemical resistance of the materials is also summarized in a summary table at the start of the chapter. The resistance is labeled with three evaluation symbols in view of concise presentation. Uniform corrosion is evaluated according to the following criteria:

Symbol	Meaning	Area-related mass loss rate[1] x		Corrosion rate y
		g/(m²h)	g/(m²d)	mm/a
+	resistant	≤ 0.1	≤ 2.4	≤ 0.1[2]
⊕	moderately resistant	> 0.1 to ≤ 1.0	> 2.4 to ≤ 24.0	> 0.1 to ≤ 1.0
−	not resistant	> 1.0	> 24.0	> 1.0

[1] Data applies to steel; for Al, Mg and their alloys, 1/3 of the value must be used
[2] Values for Ta, Ti, and Zr are too high (*possible embrittlement due to hydrogen absorption in the event of corrosion!* Therefore, corrosion rate = 0.01 mm/a, see the individual cases)

The evaluation of the corrosion resistance of metallic materials is given

- for uniform corrosion or local penetration rate, in: mm/a
- or if the density of the material is not known, in: g/(m²h) or g/(m²d).

Pitting corrosion, crevice corrosion, and stress corrosion cracking or non-uniform attack are particularly highlighted.

The following equations are used to convert area-related mass loss rates, x, into the corrosion rate, y:

with x_1 in g/(m²h)

$$\frac{x_1 \cdot 365 \cdot 24}{\rho \cdot 1000} = y(mm/a)$$

with x_2 in g/(m²d)

$$\frac{x_2 \cdot 365}{\rho \cdot 1000} = y(mm/a)$$

where

x_1: value in g/(m²h)
y: value in (mm/a)
x_2: value in g/(m²d)
d: days
ρ: density of material in g/cm³
h: hours

In those media in which uniform corrosion can be expected, isocorrosion curves (corrosion rate y ≤ 0.1 mm/a) or resistance ranges for non-metallic materials are given where possible. The evaluation criteria for non-metallic inorganic materials are stated in the individual cases; depending on the material and medium, they may also be given as corrosion rates (mm/a).

If corrosion rate values are not given in the literature sources, then the resistance of the material is limited to the above mentioned resistance symbols. These resistance symbols are also used for the non-metallic inorganic materials and – if present in the literature – are supplemented by values for the corrosion rate.

The suitability of organic materials is generally evaluated by comparing property characteristics (e.g. mass, tensile strength, elasticity module or elongation at rupture) and other changes (e.g. cracking, swelling, shrinkage) after exposure to the medium with respect to these characteristics in the initial state before exposure. The extent of changes in the properties after exposure to the medium is decisive for the evaluation of the resistance to chemicals or the durability of the materials. The criteria listed below for the evaluation of the chemical resistance apply to thermoplastics used to manufacture pipes and are based on results from immersion tests with an immersion time of 112 days (see ISO 4433 Part 1 to 4). In principle, they are also applicable to other organic materials; however, they should be adapted to the individual material because, as the following table shows, the evaluation criteria are not consistent, even within a group of thermoplastics, but depend on the type of thermoplastic material.

Symbol	Meaning	Permissible limiting value[1]			
		of the mass change[2] %	of the tensile strength[3] %	of the elasticity module[3] %	of the elongation at rupture[3] %
+	resistant/durable	PE, PP, PB: −2 to 10 PVC, PVDF: −0.8 to 3.6	PE, PP, PB, PVC, PVDF: ≥ 80	PE, PP, PB: ≥ 38 PVC: ≥ 83 PVDF: ≥43	PE, PP, PB: ≥ 50 to 200 PVC, PVDF: 50 to 125
⊕	limited resistance/ limited durability	PE, PB, PB: > 10 to 15 or < −2 to −5 PVC, PVDF: < −0.8 to −2 or > 3.6 to 10	PE, PB, PB, PVC, PVDF: < 80 to 46	PE, PB, PB: < 38 to 31 PVC: < 83 to 46 PVDF: < 43 to 30	PE, PB, PB: < 50 to 30 or > 200 to 300 PVC, PVDF: < 50 to 30 or > 125 to 150
−	not resistant/not durable	PE, PP, PB: < −5 or > 15 PVC, PVDF: < −2 or > 10	PE, PP, PB, PVC, PVDF: < 46	PE, PP, PB: <31 PVC: < 46 PVDF: < 30	PE, PP, PB: < 30 or > 300 PVC, PVDF: < 30 or > 150

[1] The data applies to the values determined in the initial state without exposure to the medium which correspond to 100 %
[2] Relative mass change according to DIN EN ISO 175
[3] Tensile strength, elasticity module, and elongation to rupture according to DIN EN ISO 527-1
Scope of validity for PVC: PVC-U, PVC-HI, and PVC-C; for PE: PE-HD, PE-MD, PE-LD, and PE-X

Unless stated otherwise, the data was measured at atmospheric pressure and room temperature.

The resistance data should not be accepted by the user without question, and the materials for a particular purpose should not be regarded as the only ones that are suitable. To avoid incorrect conclusions being drawn, it must be always taken into account that the expected material behavior depends on a variety of factors that are often difficult to recognize individually and which may not have been taken into account deliberately in the investigations upon which the data is based. Under certain circumstances, even slight deviations in the chemical composition of the medium, in the pressure, in the temperature or, for example, in the flow rate are sufficient to have a significant effect on the behavior of the materials. Furthermore, impurities in the medium or mixed media can result in a considerable increase in corrosion.

The composition or the pretreatment of the material itself can also be of decisive importance for its behavior. In this respect, welding should be mentioned. The suitability of the component's design with respect to corrosion is a further point which must be taken into account. In case of doubt, the corrosion resistance should be investigated under operating conditions to decide on the suitability of the selected materials.

Corrosion Resistance Against Hydrogen

Authors: *Dr. P. Drodten, Dr. D. Schedlitzki, Prof. Dr. E. Wendler-Kalsch*

		Page			Page
Survey Table		1	A 20	Austenitic CrNi steels	131
			A 21	Austenitic CrNiMo steels	131
Preliminary Remarks		4	A 22	Austenitic CrNiMo(N) and CrNiMoCu(N) steels	131
A	**Metallic materials**	20			
			A 23	Special iron-based alloys	158
A 1	Silver and silver alloys	20	A 26	Nickel	166
A 2	Aluminum	20	A 27	Nickel-chromium alloys	171
A 3	Aluminum alloys	22	A 28	Nickel-chromium-iron alloys (without Mo)	171
A 5	Cobalt alloys	44			
A 6	Chromium and chromium alloys	44	A 29	Nickel-chromium-molybdenum alloys	176
A 7	Copper	49			
A 9	Copper-nickel alloys	53	A 30	Nickel-copper alloys	201
A 10	Copper-tin alloys (bronze)	55	A 32	Other nickel alloys	201
A 11	Copper-tin-zinc alloys (red brass)	55	A 35	Platinum metals (Ir, Os, Pd, Rh, Ru) and their alloys	207
A 12	Copper-zinc alloys (brass)	55	A 37	Tantalum, niobium and their alloys	214
A 13	Other copper alloys	56			
A 14	Unalloyed and low-alloy steels/cast steel	57	A 38	Titanium and titanium alloys	235
A 17	Ferritic chromium steels with < 13 % Cr	99	A 40	Zirconium and zirconium alloys	312
A 18	Ferritic chromium steels with ≥ 13 % Cr	103	A 41	Other metals and their alloys	325
A 19	Ferritic/pearlitic-martensitic steels	107	B	**Nonmetallic inorganic materials**	348
A 19.1	Martensitic steels	107			
A 19.2	Ferritic-austenitic steels/duplex steels	108	B 3	Carbon and graphite	348

		Page			Page
B 4	Binders for building materials (e.g. concrete, mortar)	348	D	Materials with special properties	369
B 6	Glass	349	D 1	Coatings and films	369
B 7	Fused silica and silica glass	349	D 2	Gaskets and packings	372
			D 3	Composite materials	380
B 8	Enamel	351	E	Material recommendations	384
B 12	Oxide ceramic materials	353			
B 13	Metal-ceramic materials	354			
			Bibliography		391
C	**Organic materials**	355	**Key to materials compositions**		415
Thermoplastics		363	**Index of materials**		457
Elastomers		366			
Thermoplastic elastomers		366	**Subject index**		465
Duroplastics		366			

Warranty disclaimer

This book has been compiled from literature data with the greatest possible care and attention. The statements made in this chapter only provide general descriptions and information.

Even for the correct selection of materials and correct processing, corrosive attack cannot be excluded in a corrosion system as it may be caused by previously unknown critical conditions and influencing factors or subsequently modified operating conditions.

No guarantee can be given for the chemical stability of the plant or equipment. Therefore, the given information and recommendations do not include any statements, from which warranty claims can be derived with respect to DECHEMA e.V. or its employees or the authors.

The DECHEMA e.V. is liable to the customer, irrespective of the legal grounds, for intentional or grossly negligent damage caused by their legal representatives or vicarious agents.

For a case of slight negligence, liability is limited to the infringement of essential contractual obligations (cardinal obligations). DECHEMA e.V. is not liable in the case of slight negligence for collateral damage or consequential damage as well as for damage that results from interruptions in the operations or delays which may arise from the deployment of this book.

Preliminary remarks
Index of preliminary remarks

V 1	Introduction	5
V 2	Physical and chemical properties	5
V 3	Production	7
V 4	Storage and transportation	8
V 5	Applications	10
V 6	Reactions between materials and hydrogen	10
V 6.1	Types of damage to metal materials and influencing parameters	11
V 6.1.1	Generation and effects of atomic hydrogen	11
V 6.1.2	Embrittlement and crack formation	12
V 6.1.3	Damage at temperatures below 200°C	12
V 6.1.4	Damage at temperatures above 200°C	14
V 6.2	Types of damage to organic materials	15
V 7	Corrosion testing	15
V 7.1	Metal materials	15
V 7.2	Examination methods in compressed hydrogen	16
V 8	Analytical determination of hydrogen	18
V 9	Corrosion protection	19
V 9.1	Cladding	19
V 9.2	Multilayer structures	19

V 1 Introduction

Hydrogen has the element symbol H derived from "hydrogenium" and is the smallest and lightest of all elements. Therefore, it occupies the first place with its atomic number 1 in the Periodic Table of the Elements. The hydrogen atom consists of one proton as the atomic nucleus and one electron. At room temperature hydrogen is a gas containing hydrogen not in atomic but in molecular form with 2 atoms being combined to form one molecule (H_2). In chemical compounds the oxidation number of hydrogen is usually + 1, but can be also – 1 in several compounds, such as metal hydrides.

Pure hydrogen gas is colorless, odorless, tasteless, not poisonous, and flammable.

Bonded with oxygen or air, hydrogen forms explosive mixtures, i.e. "oxyhydrogen" (for flammability range refer to Table 1). In these mixtures both gases react explosively according to

Equation 1 $H_2 + {}^1/_2\, O_2 \rightarrow H_2O$

V 2 Physical and chemical properties

Several physical properties are summarized in Table 1 [1–3]. As in all liquids, hydrogen is only slightly soluble in water. In contrast to highly soluble gases, its solubility may increase with rising temperature. Its solubility in water reaches the minimum value at 50 °C and increases with rising temperature (Table 2).

Parameter	Unit	Value
Atomic mass		1.00794
Molar mass	g/mole	2.016
Melting point	°C	–259.20
Boiling point	°C	–252.77
Density in liquid state	g/cm^3	70.811
Heat of evaporation	kJ/kg	454.3
Density at 0 °C and 1.013 bar	g/cm^3	0.0899
Specific heat at 0 °C	J/(mole K)	28.61
Coefficient of thermal conductivity at 15 °C and 1.013 bar	J/(s m K)	0.177
Critical temperature	°C	–240
Critical pressure	bar	13.0
Critical density	g/cm^3	30.1
Autoignition temperature	°C	560
Flammability range in air	vol%	4–75

Table 1: Physical properties of hydrogen [1]

°C	0	10	20	30	40	50	60	70	80
10^3 mole%	1.755	1.576	1.455	1.377	1.330	1.310	1.312	1.333	1.371

Table 2: Solubility of hydrogen in water at a hydrogen pressure of 1.013 bar depending on temperature [2]

The amount of a gas dissolved in a liquid is proportional to its partial pressure over the liquid. The following holds true

$$\text{dissolved amount of gas } \frac{N\,cm^3 \cdot gas}{g \cdot liquid} = \lambda p_{gas}$$

Here λ is the technical solubility coefficient depending on temperature. Table 3 indicates the value λ for hydrogen in water at various temperatures. The values in the table correspond to the amount of hydrogen in Ncm³ (cm³ at 0 °C, 1 bar) dissolved at saturation of 1 g of water at the relevant temperature if the partial pressure of hydrogen is 1 bar.

°C	0	5	10	15	20	25	30	40	50	60	70	90
$\lambda \times 10^2$	2.09	1.98	1.89	1.82	1.76	1.71	1.67	1.61	1.58	1.57	1.57	1.60

Table 3: Technical solubility coefficient λ of hydrogen in water for 1 bar and various temperatures [3]

Figure 1 displays the thermal conductivity of hydrogen gas as a function of pressure for various temperatures [3.]

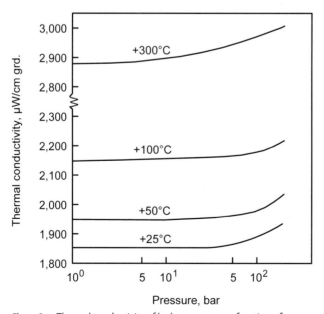

Figure 1: Thermal conductivity of hydrogen gas as a function of pressure [3]

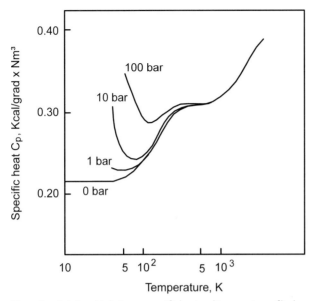

Figure 2: Relationship between specific heat and temperature of hydrogen at various pressures [3]

For information on the temperature dependence of the specific heat of gaseous hydrogen refer to Figure 2 [3].

Numerous metals with internal electron shells not completely filled (transition metals in the periodic table of the elements) are able to dissolve considerable amounts of hydrogen. Hydrogen is dissolved in the metals not in its molecular but in its atomic form (more specifically: as protons). It can be either dissolved in the lattice similar to an alloying element or chemically combined as hydride (refer to V 6).

V 3 Production

About 90% of all hydrogen is commercially produced in petrochemical processes, e.g. by catalytic steam cracking (steam reforming) from natural gas (methane) or light crude oil fractions according to Equation 2 or by partial oxidation of heavy oil according to Equation 3.

Equation 2 $\quad CH_4 + H_2O \rightarrow 3H_2 + CO$

Equation 3 $\quad 2\,C_nH_{2n+2} + n\,O_2 \rightarrow (2n+2)\,H_2 + 2n\,CO$

Other processes are coal gasification with subsequent conversion according to Equation 4 and Equation 5.

Equation 4 $\quad 3\,C + O_2 + H_2O \rightarrow H_2 + 3\,CO$

Equation 5 $\quad CO + H_2O \rightarrow H_2 + CO_2$

Also carbon monoxide formed in reactions according to Equation 2 and Equation 3 is reacted further according to Equation 5 to form hydrogen.

In technical terms, the electrolysis of water plays a role only if cheap electrical energy is available or a high purity degree of the produced hydrogen is required.

Considerable amounts of hydrogen are also produced in numerous processes in refineries or coking plants as well as in electrochemical processes, such as chlor-alkali electrolysis.

Presently, the production of hydrogen can be broken down as follows

- 77%   petrochemical processes
- 18%   coal gasification
- 4%   electrolysis of water
- 1%   other sources

Hydrogen produced from hydrocarbons in petrochemical processes has a purity of 97 to 99.5 vol%. The purity of electrolytically produced hydrogen is usually higher than 99.5 vol%. To obtain hydrogen with higher purity levels, impurities as a result of catalytic combustion (oxygen), drying or adsorption-diffusion processes need to be removed. Ultra-pure hydrogen required, for instance, for the semiconductor technology, is produced by diffusion through palladium-silver membranes or, since very recently, also by adsorption to hydride-forming materials. These processes can yield a purity degree of 7.0, i.e. 99.99999% corresponding to a residual share of impurities below 0.1 ppm (ml/m^3). Table 4 contains a list of purities of common commercial hydrogen grades.

Designation	2.5	3.0	3.8	5.0	5.3	5.6	6.0	6.0*
Purity, vol%	99.5	99.9	99.98	99.999	99.9993	99.9996	99.9999	99.9999
O$_2$/ppm (ml/m^3)		≤ 50	≤ 10	≤ 2	≤ 2	≤ 1	**	≤ 0.2
N$_2$/ppm (ml/m^3)		≤ 500	≤ 200	≤ 3	≤ 3	≤ 2		≤ 0.2
H$_2$O/ppm (ml/m^3)		≤ 100	≤ 20	≤ 5	≤ 2	≤ 1		≤ 0.5
CO/ppm (ml/m^3)								≤ 0.1
CO$_2$/ppm (ml/m^3)								≤ 0.1
HC/ppm (ml/m^3)				≤ 0.5	≤ 0.5	≤ 0.1		≤ 0.1

* = liquid hydrogen, ** = total content of impurities ≤ 1 ppm, HC = hydrocarbons

Table 4: Common purity degrees of commercially available hydrogen [2]

V 4   Storage and transportation

Hydrogen is stored as a gas in steel tanks under pressure (at 1.2 MPa to 1.6 MPa or at 20 MPa depending on the size of the tank) and is transported in steel cylinders and cylinder bundles (red color coding, valves with left-hand thread, 20 MPa). On

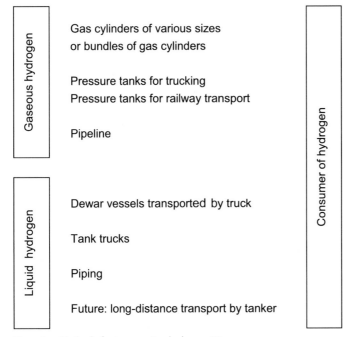

Figure 3: Methods for transporting hydrogen [2]

industrial estates with producing and consuming companies also pipeline networks are used to this end. In suitable low-temperature tanks hydrogen can be also stored and transported in its liquid state at a temperature of −253 °C. Figure 3 shows the various methods of transporting hydrogen.

To liquify hydrogen gas, energy must be removed until the condensation point is reached. This can be achieved either by cooling with external means, such as liquid nitrogen, or by magnetocaloric effects, or also by irreversible or isentropic decompression, i.e. without entropy change.

In contrast to most gases, the inversion temperature of hydrogen at which the Joule-Thomson coefficient changes from a negative value (heating at decompression) into a positive value (cooling at decompression) is very low as shown by the Joule-Thomson inversion curve in Figure 4.

A cooling effect is only obtained if decompression occurs at a point below the inversion curve. In the technical practice usually a temperature range from 77 K, i.e. the boiling point of liquid nitrogen at 0.1 MPa, to 25 K is used.

Liquid hydrogen is a colorless, very mobile liquid with low viscosity and low surface tension.

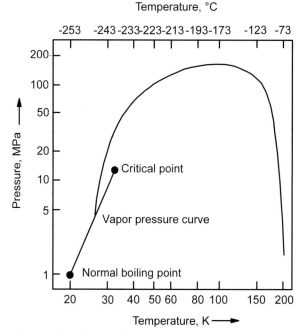

Figure 4: Joule-Thomson inversion curve [2]

V 5 Applications

Hydrogen is required for numerous chemical processes, e.g.

- for the synthesis of
  - ammonia
  - methanol
- in hydrogenation processes
  - for the production of gasoline
  - for fat hardening
- as a reducing agent for metal extraction (W, Mo, Co, etc.)
- as a shielding gas for welding and metallurgical processes
- as fuel gas
- liquified as aerospace fuel.

V 6 Reactions between materials and hydrogen

The suitability of constructional materials for the production and handling of gaseous hydrogen depends on the given process variables. Critical factors for the corrosive material stress include:

- Pressure
- Temperature
- The presence of other substances (impurities or additives)
- Residual stress in the component or operating stresses

V 6.1 Types of damage to metal materials and influencing parameters

When absorbing hydrogen, metal materials can become brittle and, in an extreme case, be damaged such that they fail. Manifestations of the embrittling effect of hydrogen on the mechanical properties of a metal material are:

- The ductility of the material in hydrogen is lower than in air or in another inert environment.
- Tensile stress or strain processes in hydrogen lead to crack formation.
- A subcritical crack growth occurs in hydrogen.
- The crack propagation velocity under static or alternating loads in hydrogen is higher than that in inert liquids.

Basic mechanical properties, such as the yield strength and the tensile strength, of the material remain almost unaffected. The most obvious manifestations of the influence of hydrogen on the strain behavior are, for instance, reduced reduction of area, notch tensile strength and crack behavior.

Basically, hydrogen can be absorbed from both gaseous and aqueous fluids. The decisive activity to this end of atomic hydrogen at the entry surface can be enhanced by promoters or reduced by inhibitors. The material can absorb hydrogen as follows

- by exposure to gaseous, molecular hydrogen under pressure at room temperature or at high temperatures
- by atomic hydrogen generated in corrosion processes directly on the metal surface
- by atomic hydrogen generated on the metal surface by cathodic polarization, e.g. cathodic protection or electrolytic metal deposition
- by atomic hydrogen generated at high temperatures, e.g. in welding processes from moisture or chemically combined water of the electrode coating or welding powder.

Damage to metal materials by gaseous hydrogen may be caused by various mechanisms. In any case hydrogen can diffuse into steel only in its atomic form. Molecular hydrogen needs to dissociate first on the metal surface.

V 6.1.1 Generation and effects of atomic hydrogen

Since damage to materials by hydrogen can only be caused by diffusible atomic hydrogen, dissociation of the molecular hydrogen in physical and chemical reactions is necessary.

Atomic hydrogen can be generated by

- hydrogen ions during electrolysis
- thermal dissociation
- chemisorption with subsequent dissociation on active metal surfaces.

The atomic hydrogen diffused into the metal material can either

- be interstitially dissolved in the lattice
- form a hydride with the metal atoms
- accumulate in so-called traps.

V 6.1.2   Embrittlement and crack formation

Hydrogen dissolved in the material within the solubility limits usually does not have any adverse effect on the material. Hydride formation leads to strong embrittlement of the material and can cause the formation of internal cracks, cracks or fractures under mechanical loading. Hydrogen accumulated in traps (e.g. dislocations, lattice vacancies, inclusions, harder microstructural regions) may locally build up high pressures and lead to internal separations, cracks and fractures. The precondition for crack initiation or the growth of existing cracks is that the material is subjected to static or dynamic tensile stress. Such tensile stress may occur as:

- static tensile stress
- tensile stress increasing over time
- alternating, low frequency tensile stress over time.

However, crack formation or crack growth only occurs if

- in case of static, i.e. constant over time, tensile stress, a critical value, i.e. a stress limit, is exceeded
- in case of increasing or alternating tensile stress, the strain rate is within a critical range.

The required tensile stress can be present in the component as residual stress or can result from operating stresses.

For more information regarding basic processes of hydrogen-induced damaging of metals and underlying reactions and mechanisms reference is made to [4].

V 6.1.3   Damage at temperatures below 200 °C

Metal materials are used for producing, transporting and storing hydrogen. Therefore, in terms of hydrogen technology special attention must be paid to interactions of metal materials – in particular steels – with molecular hydrogen at climatic ambient conditions.

The equilibrium of the reaction between molecular and atomic hydrogen

Equation 6 $H_2 \leftrightarrow 2H$

is completely on the side of molecular hydrogen at room temperature.
The equilibrium constant is reached

at 298 K: $K_p = \dfrac{p_H^2}{p_{H_2}} = 1.6 \cdot 10^{-71}$

where p_H and p_{H_2} are the partial pressures of the atomic and the molecular hydrogen.

If the temperature increases, the equilibrium of the reaction according to Equation 6 shifts to the right and the following equilibrium constants are obtained

at 1,000 K: $K_p = 1.3 \times 10^{-9}$
at 3,500 K: $K_p = 0.383$.

The equilibrium situation becomes even clearer by the ratio of dissociated and non-dissociated molecules, i.e. the number of the hydrogen atoms generated upon the establishment of the equilibrium divided by the number of the initial hydrogen molecules. In the state of equilibrium this value is called degree of dissociation α. At 298 K and 1 bar the value is

$\alpha = 2 \times 10^{-36}$

Hence, at room temperature there are 5×10^{35} hydrogen molecules for one hydrogen atom. Since under normal conditions 1 mole of hydrogen ($\approx$ 2 g) with a volume of 22.4 l contains 6×10^{23} molecules, the probability to find a hydrogen atom in 22 l hydrogen gas is very low.

Dissociation increases with increasing temperature. Table 5 contains a compilation of various values for degrees of dissociation at elevated temperatures [5].

K	500	1,000	1,500	2,000	3,000	3,500	4,000	5,000
α	3.5×10^{-24}	1.1×10^{-9}	8.7×10^{-6}	8.1×10^{-4}	0.0785	0.3180	0.6220	0.9546

Table 5: Degrees of dissociation α at 1 bar depending on temperature

According to Le Chatelier's Principle the degree of dissociation α also rises with decreasing pressure (Table 6).

Hence, in a state of equilibrium a noticeable concentration of hydrogen atoms in hydrogen gas can be expected only at high temperatures and low pressures.

Temperature	Pressure			
K	1 bar	0.1 bar	0.01 bar	0.001 bar
2,000	0.0008	0.0025	0.0081	0.0256
2,500	0.0126	0.0387	0.125	0.369
3,000	0.079	0.242	0.619	0.928
3,500	0.282	0.682	0.947	0.994
4,000	0.623	0.929	0.992	0.999

Table 6: Degree of dissociation α in hydrogen gas depending on total pressure [5]

The reaction according to Equation 6 in the direction of hydrogen atoms occurs with an extremely high dissociation enthalpy of 429.5 kJ/mole. Therefore, at low temperatures a considerable amount of energy is required to dissociate the amount of hydrogen molecules necessary for damaging the material.

The partial steps required to cause hydrogen-induced damage (embrittlement, crack formation) in molecular compressed gaseous hydrogen at climatic ambient temperatures are:

- adsorption on the metal surface
- dissociation of hydrogen in pure, active surface areas
- absorption of the resulting atomic hydrogen in to the metal lattice
- transportation of the atomic hydrogen in the matrix to sensitive sites, e.g. to zones in front of a notch or crack tip
- damage reaction (embrittlement, formation of internal cracks, crack growth).

The dissociation of the molecular, adsorbed hydrogen on the metal surface required to facilitate such damage processes necessitates a chemisorption process which can only occur on a clean active surface, e.g. a surface generated by plastic deformation. Plastic deformation in front of the crack tip releases energy in the form of heat of an amount sufficient to facilitate the dissociation of the H_2 molecule.

In practice this means that the dissociation of hydrogen is only possible at notches and crack tips where repeated or permanent plastic deformations occur as a result, for instance, of alternating loads. Factors increasing the probability of such hydrogen-induced crack growth are:

- amount and type of fluctuating stress
- frequency
- surface roughness (starting point for growable cracks)
- hydrogen pressure
- temperature
- strength of the material.

Moreover, the purity level of the hydrogen plays a role since impurities (e.g. oxygen or water) may act as inhibitors (competitive adsorption).

Therefore, hydrogen damage is not expected in plants operated under constant pressure since the preconditions for the dissociation of the hydrogen molecule into hydrogen atoms are not fulfilled. If, however, areas cannot be excluded with local plastic deformations as occurring as a result of filling and emptying operations in transport and storage tanks, which are alternately used, dissociation may occur on the generated active metal surfaces. Hydrogen diffusing into the material may lead to the damage mentioned above.

V 6.1.4 Damage at temperatures above 200 °C

The atomic hydrogen generated by dissociation according to Equation 6 at rising temperature is adsorbed on the surface in an adsorption process and by interacting with the electrons of the metal lattice. A typical phenomenon of chemical compressed hydrogen damage is the reaction of the diffused hydrogen with components of the material. In hydride-forming metals, e.g. titanium or tantalum, this can be the reaction with the atoms of the metal, e.g.:

Equation 7 $\quad\quad Ti + 4\,H \rightarrow TiH_4$

In metals not forming hydrides hydrogen can react with other microconstituents. In ferritic steels hydrogen reacts with the carbon of the steels at higher temperatures (T > 200 °C) according to Equation 8 and forms methane.

Equation 8 $\quad\quad Fe_3C + 4\,H \rightarrow 3\,Fe + CH_4$

Decarburization of the steel occurring in this process reduces the strength, and at the same time the formed methane gas leads to strong internal pressure increases. This causes internal separations, voids and cracks, which may result in failure of pressure parts. (refer to A 14)

Damage is also known to occur when annealing oxygen-containing copper in hydrogen gas (refer to A 7) according to the reaction

Equation 9 $\quad\quad CuO + 2\,H \rightarrow Cu + H_2O$

V 6.2 Types of damage to organic materials

A reaction with hydrogen is not expected at temperatures at which organic materials are used. However, also at room temperature constant permeation of hydrogen through the organic materials must be expected. Table 7 contains the permeation coefficient P of hydrogen for several plastics at 25 °C (P in 10^{-9} cm²/s mbar) [6].

Natural rubber	Polyethylene	Neoprene	Polystyrene	Perbunan	Mipolan
0.19–0.68	0.11–0.59	0.13–0.46	0.9	0.15	0.10

Table 7: Permeation coefficient P of hydrogen at 25 °C in various plastics [6]

V 7 Corrosion testing

V 7.1 Metal materials

Depending on the objective various examinations of the corrosion behavior of metallic materials can be carried out using the methods specified in test standards or test specifications. This concerns, for example, studies of

- stress corrosion cracking of bend, tensile and fracture-mechanical test specimens, e.g.:
  - DIN 50922 [7], DIN EN ISO 7539 Parts 1 to 7 [8]
  - ASTM G 38 (2001) [9], ASTM G 39 (1999) [10], ASTM G 49 (1985) [11], ASTM G 129 – 00 [12], ASTM G 142 – 98 [13], ASTM E 8 – 03 [14]

- corrosion fatigue cracking, e.g.:
  ISO 11782-1 [15], or ISO 11782-2 [16])
- behavior at higher temperatures, e.g.:
  DIN 50905-4 [17] or DIN EN ISO 2626 [18]).

However, it should be noted that such investigations in the laboratory only provide preliminary information on the general behavior of individual materials or on the combination of different materials and the results have only limited applicability to the behavior of the materials in practice.

Further information on testing procedures for examining the influence of hydrogen on the mechanical properties of metal materials is also provided in [162] and [163]. The results published there also show that often one test method is not sufficient to detect potential impacts on material properties by hydrogen.

V 7.2 Examination methods in compressed hydrogen

Since plastic deformation of the material is necessary to initiate damage caused by molecular hydrogen, three main examination methods are applied:

- slow strain rate tensile test (CERT test, CERT = Constant Extension Rate Tensile) in a hydrogen atmosphere
- slow burst test: disc test according to Fidelle [19] for testing the behavior of the material in compressed hydrogen
- fracture-mechanical test methods for examining hydrogen-induced crack growth.

The CERT test in a hydrogen atmosphere provides information on

- critical strain rates,
- critical temperature ranges,
- the influence of the surface condition,
- the influence of gas purity and gas pressure.

During this test a smooth or notched specimen is deformed at a constant strain rate over time until it breaks. The following criteria can be used for evaluating the susceptibility to stress corrosion cracking (SCC):

- reduction of area
- fracture elongation
- fracture energy.

The low-deformation sections of the fracture process caused by stress corrosion cracking are mainly reflected by a noticeable reduction of these material parameters compared to the values determined in an inert fluid.

Care should be exercised as to ensure a sufficiently slow strain rate since, if the strain rate is too high, the specimen breaks ductilely before hydrogen has been able to dissociate in a sufficient amount and diffuse into the material.

Whereas the slow strain rate tensile test simulates the behavior under uniaxial load, the specimen is tested under multiaxial load in the slow burst test. For this

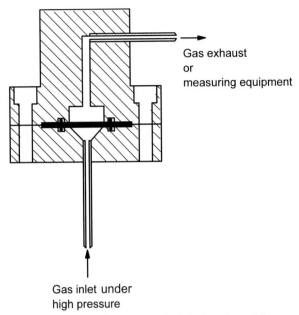

Figure 5: Test setup for examining the behavior of material in compressed hydrogen [19]

test a disk-type specimen with a diameter of 58 mm and a wall thickness of 0.75 mm is mounted with pressure-resistant seals between two flanges and loaded with compressed gas from one side until it bursts (Figure 5). This burst test is performed once with helium and once with hydrogen. The pressure is increased at a rate of about 65 bar/min. During the slow deformation process under hydrogen pressure atomic hydrogen can enter the material and impair deformability. The ratio of the bursting pressures under helium and under hydrogen P_{He}/P_{H2} serves to evaluate the susceptibility to hydrogen embrittlement [19].

The task of fracture mechanics is to provide information on the behavior of cracks existing or developing in a part under mechanical load. The relevant test methods yield quantitative results regarding the fracture stress and critical defects of the respective system. The mechanical load can be of a static or a dynamic type.

Fracture-mechanical methods describe the stress state of a crack subjected to tensile stress. The stress intensity factor K_I with the unit $N/mm^{3/2}$ or $MPa\sqrt{m}$ is used as a measure of stress concentration at the crack tip. K_I increases with increasing tensile stress σ perpendicular to the crack plane and with the square root of the crack depth a.

Therefore, in the field of linear elastic fracture mechanics the following holds true for statically loaded specimens with cracks:

Equation 10 $\qquad K_I = \sigma\sqrt{\pi a}f$

Here, "f" is a correction factor for the crack and component geometry.

In the validity range of linear elastic fracture mechanics, the parameters of materials are independent from the thickness of specimens. However, care must be exercised as to maintain the plane strain conditions during fracture-mechanical tests. This is achieved by sufficiently large specimens compared to the extension of the plastic zone in front of the crack tip. Experience shows that this is the case with specimens having a thickness b above 2.5 $(K_I/R_p)^2$.

In particular, fracture-mechanical parameters, such as the stress intensity limit value K_{IH} (stress intensity factor K_I under hydrogen atmosphere) causing incipient cracks to grow subcritically if it is reached or exceeded, or the crack propagation velocity of fatigue cracks da/dN as a function of the cyclic fracture intensity factor $\Delta K = K_{IHmax} - K_{IHmin}$ effective on the crack front and the frequency permit forecasts as to the behavior of the component in hydrogen.

Various types of specimens are available for determining fracture-mechanical parameters in corrosive fluids. For static test loads either DCB specimens (DCB = double cantilever beam) or WOL specimens (WOL = wedge opening load) are used to which stress is applied by means of screws, the specimens having been notched and provided with an incipient crack caused by alternating loads.

In CT specimens (CT = compact tension) the crack intensity is increased by gradually increasing the force over time until the initial crack growth is detected during the holding time at constant stress intensity. This type of specimens can be also used for determining the crack growth rate under alternating loads.

V 8 Analytical determination of hydrogen

The analytical determination of hydrogen absorbed by a metal is usually performed by stripping hydrogen from the metal specimen at higher temperatures. Depending on the applied temperature a difference is made between

- warm extraction
- hot extraction
- melting method.

When determining the hydrogen content a difference is made between the determination of the total dissolved hydrogen (total hydrogen content) and the determination of the diffusible hydrogen at certain temperatures below the melting point.

Total hydrogen is determined by applying the melting method, whereas for determining diffusible hydrogen the specimen is heated to a temperature suitable for the respective material and kept at this temperature for a sufficient period (warm or hot extraction). For unalloyed and low-alloy steels where the diffusion rate of hydrogen is high, heat extraction at 650 °C is usually sufficient to extract almost all hydrogen.

The melting method and the warm or hot extraction methods may be carried out either in vacuum or in an inert gas carrier flow. For vacuum extraction the gases released from the specimen are pumped down and the hydrogen is determined following its separation from the other gases by means of a pressure-volume measurement.

During the carrier gas method the gases released from the carrier gas are flushed out of the extraction furnace to separate foreign gases by means of absorption reagents and hydrogen is measured in a thermal conductivity cell [20].

V 9 Corrosion protection

Generally, corrosion protection can be provided by separating the sensitive material from the relevant corrosive medium by means of a protective film or a coating. In case of hydrogen, coating by using organic materials is not possible since hydrogen is able to diffuse, more or less quickly, into all organic systems. Therefore, corrosion protection is usually provided by applying a metal resistant to compressed hydrogen. To this end two different methods are primarily used.

V 9.1 Cladding

A material resistant to compressed hydrogen can be applied to the carrier material by:

- Roll cladding
- Explosive cladding
- Deposition welding
- Thermal spraying.

Roll cladded or explosively cladded blanks, e.g. sheets or bottoms, are jointed to form the desired unit. Deposition welding or thermal spraying may be also used to protect finished units.

V 9.2 Multilayer structures

For multilayer structures an internal core tube of a material resistant to compressed hydrogen is covered with one or more layers from which the pressure bearing part is manufactured. This external casing body is provided with degassing bores and tubes to drain any hydrogen diffusing through the core tube without applying pressure. These degassing tubes also facilitate detection of leaks of the core tube in that products leak from inside the reactor or the tank.

A
Metallic materials

A 1 Silver and silver alloys

Silver and silver alloys are able to dissolve considerable amounts of oxygen in the melt or at higher temperatures. If such oxygen-containing parts come into contact with hydrogen at high temperatures, hydrogen can dissociate on the surface and diffuse in atomic form into the material. There, hydrogen reacts with oxygen, forming water or water vapor. The high water vapor pressure causes the formation of blisters which may even link together to form bead-like cracks.

During the examinations described in [21] such flaws were produced by initially annealing silver specimens of purity degrees of 99.999%, 99.99% and 99.9% in air at 800 °C until the specimens were saturated with oxygen. According to Sievert's law $S = k\sqrt{p_{H2}}$ the solubility of oxygen in silver depends on the partial pressure of the gas. The value for oxygen at 800 °C is 1.2×10^{-5} mole/cm³ per bar. Then, the specimens were annealed in hydrogen gas for two hours. Depending on the structure of the material and the partial hydrogen pressure, the specimens exhibited blisters of various sizes as well as blisters linked together to form cracks at the grain boundaries. Blistering in hydrogen starts at a temperature of 800 °C after an aging time of 1 to 2 hours.

A 2 Aluminum

Usually, pure aluminum (99.99%) and super-purity aluminum (99.999%) do not absorb hydrogen under normal conditions in the presence of dry gaseous hydrogen [22]. The dry molecular hydrogen cannot dissociate at the natural Al_2O_3, layer which is always present, and hence the oxide film is a significant barrier for hydrogen absorption [22].

Hydrogen absorption is possible only under extremely critical conditions leading to the dissociation of hydrogen. Such conditions include, for instance:

Absorption of gaseous hydrogen by electron irradiation in the electron microscope [22, 23]. The influence of dry hydrogen on deformation and fracture processes of high-purity aluminum was examined in-situ in the electron microscope at ambient temperature. Hydrogen of a considerable amount was absorbed. The reason was strong dissociation and ionization of dry molecular hydrogen in the electron beam of the electron microscope. Of course, dissociated and partially ionized hydrogen can be absorbed.

If hydrogen absorption occurs under these critical conditions, the aluminum may soften since hydrogen increases the dislocation mobility and, hence, decreases the yield stress [22]. However, fractures were always of a ductile type, whether the aluminum was tested in vacuum or in dry or wet hydrogen gas.

The results of the in-situ electron-microscopic examinations show that they are suited for establishing the reasons of damage to a limited extent only since frequently they do not reflect the loads under real operating conditions.

Hydrogen solubility in liquid aluminum (molten metal)

Hydrogen absorption in liquid aluminum follows the reaction

Equation 11 $\quad \frac{1}{2} H_{2(gas)} \rightarrow \underline{H}_{(in\ Al-H)}.$

The solubility of atomic hydrogen is based on Sievert's law. This law describes the concentration of the dissociated solution of a biatomic gas as occurring, for instance, when H_2 is dissolved in liquid metals, the atoms dissolved in the molten metals being of a monoatomic type. The following holds true for the equilibrium solubility:

Equation 12 $\quad a_H = K \cdot \sqrt{p_{H2}}$

(p: partial pressure of the diatomic pressure, K: equilibrium constant).

As a result of numerous examinations an empirical equation for molten pure aluminum could be established for the hydrogen solubility (in % by weight) as a function of the temperature (K) for a partial hydrogen pressure of 1 atm ($\approx$ 1.013 bar) [24]:

Equation 13 $\quad \log_{10}(\% \text{ by weight } \underline{H})_{in\ Al-H} = -2{,}691.96/T - 1.32$

This empirical formula for the solubility of hydrogen in liquid aluminum is important when melting aluminum products since hydrogen in the smelt may have a detrimental effect on the properties of the material.

If, for instance, aluminum is remelted in a water vapor atmosphere at 900 °C, the absorption of hydrogen may be in a range from 0.25 to 0.28 cm³/100 g Al [25]. This content is sufficient to exert a negative effect on ductility. As illustrated in Figure 6, heat treatment of aluminum between 200 and 400 °C clearly reduces the ductility with increasing hydrogen content, i.e. including also the hydrogen content absorbed by remelting the aluminum (curve 4 in Figure 6). Only the hydrogen produced by electrolytic corrosion (0.52 to 1.23 cm³/100 g Al) leads to a further decrease of the relative elongation (curve 5 in Figure 6). Homogenization of aluminum at 435 ± 5 °C for 12 h yields a very low hydrogen content of 0.05 to 0.06 cm³/100 g Al.

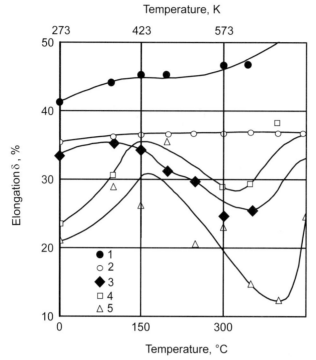

Figure 6: Influence of heat treatment on the relative elongation of Al 99 aluminum with different hydrogen contents [25]
1) 0.05–0.06 Ncm³/100 g Al
2) 0.10–0.12 Ncm³/100 g Al
3) 0.18–0.20 Ncm³/100 g Al
4) 0.25–0.28 Ncm³/100 g Al
5) 0.52–1.23 Ncm³/100 g Al

A 3 Aluminum alloys

Similar to pure aluminum face-centered cubic (fcc) aluminum alloys are immune to compressed hydrogen [26]. To prove this, tensile tests were performed with pure aluminum (99.993%) and the alloy EN AW-7039-T61 (EN AW-Al Zn4Mg3) in a compressed hydrogen atmosphere [27]. Changes were detected neither in the mechanical properties nor the fracture behavior.

Another study examined the influence of high-pressure hydrogen using notched and unnotched specimens from pure aluminum and the alloys EN AW-6061-T6 (EN AW-Al Mg1SiCu) and EN AW-7075-T73 (EN AW-Al Zn5,5MgCu) [28]. The tests were performed in air as well as in helium and hydrogen at a pressure of 35 and 70 MPa. After the desired gas pressure had been reached, the unnotched specimens were strained to their yield strength at a strain rate of 0.002/min and thereafter at a strain rate of 0.04/min until they failed. The notched specimens were loaded at 1.8×10^{-3} and 1.2×10^{-2} cm/min until they failed. Comparison of the results compiled in Table 8 and Table 9 does not reveal any noticeable reduction of the strength or ductility in the hydrogen atmosphere.

Material		Mechanical properties in air			
		R_p MPa	R_m MPa	A %	Z %
1100-O	u	49	98	36	90
	n		147		20
6061-T6	u	280	322	17	56
	n		630		6
7075-T73	u	462	518	12	32
	n		826		2.8

n: notched, u: unnotched

Table 8: Mechanical properties of notched and unnotched Al materials in air [28]

Material		R_m, MPa		A, %		Z, %	
		He	H_2	He	H_2	He	H_2
1100-O	u	110	110	42	39	93	93
	n	124	172				21
6061-T6	u	273	280	19	19	61	66
	n	504	546			9.5	11
7075-T73	u	455	455	15	12	37	35
	n	812	798			3.8	2.3

u: unnotched, n: notched

Table 9: Mechanical properties of unnotched and notched Al materials in helium and hydrogen at a gas pressure of 70 MPa [28]

In the literature the absence of embrittlement effects in precipitation hardened aluminum materials in dry gaseous hydrogen is justified by explaining that dry hydrogen is not able to penetrate the protective oxide film [29]. The thin oxide films prevent adsorption and absorption of hydrogen. Therefore, rupture of the oxide film is decisive for crack growth.

Dry hydrogen atmosphere are not able to initiate any subcritical crack growth in high-strength aluminum alloys [30]. This may only happen under the conditions of moisture. Crack propagation in aluminum alloys is decisively determined by the moisture content of hydrogen.

Aluminum-copper materials (alloy series 2xxx)

The aluminum alloy EN AW-2219-T6E46 (EN AW-Al Cu6Mn, Table 10) was checked for its suitability as a tank material for high-purity, high-pressure hydrogen [28]. To this end specimens were provided with an incipient crack, subjected

Si	Fe	Cu	Mn	Mg	Zn	V	Ti
0.20	0.30	5.8–6.8	0.20–0.40	0.02	0.10	0.05–0.15	0.02–0.10

Table 10: Chemical composition of the alloy EN AW-2219 [28]

to high-purity (99.999%), high-pressure hydrogen (36 MPa) and loaded at different stress intensity factors over different periods of time (up to 100 h). Subsequently, the specimens were ruptured in the tensile test. These tests were aimed at establishing the relationship between mechanical load and critical crack length. Figure 7 shows the results for the unwelded material 2219 and Figure 8 shows the results for the welded material. Considering the wall thickness of the hydrogen tank and the operating conditions, critical conditions for hydrogen-induced damage to the materials can be derived from these results. In general, these examinations revealed that the material EN AW-2219-T6E46 (AlCu6Mn) is better suited for high-pressure hydrogen gas tanks than, for instance, the nickel alloys Alloy 718 (2.4668, NiCr19NbMo) or TiAl6V4.

The material EN AW-2024-T351 (EN AW-Al Cu4Mg1) was used to examine by comparison the fatigue behavior under torsional stress (Mode III) in dry hydrogen,

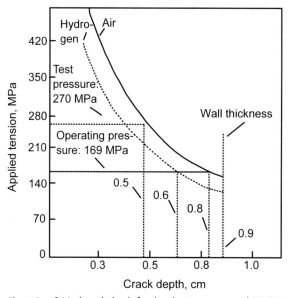

Figure 7: Critical crack depth for the aluminum material EN AW-2219-T6E46 in hydrogen gas at a gas pressure of 36 MPa [28]

dry argon and moist air [31]. The chemical composition and the mechanical properties of the examined material are summarized in Table 11 and Table 12. To remove the residual moisture in the technical Ar and H_2 gases, these gases were passed through liquid nitrogen. Five specimens each were tested in all ambient fluids under a load of 138 MPa and 50 Hz each.

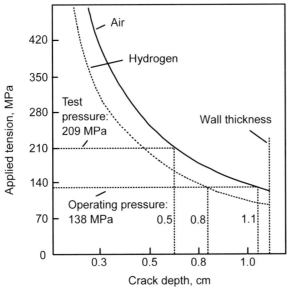

Figure 8: Critical crack depth for welded joints of sheets of EN AW-2219-T6E46 aluminum alloy with a thickness of 1.13 cm in hydrogen gas at a gas pressure of 36 MPa [28]

Alloy	Al	Cu	Mg	Zn	Mn	Cr	Fe	Si
2024	balance	3.8–4.0	1.2–1.8	0.25	0.3–0.9	0.1	0.5	0.5

Table 11: Chemical composition of the aluminum alloy EN AW-2024 (EN AW-Al Cu4Mg1) [31]

Alloy	Direction of test	R_p, MPa	R_m, MPa	A, %
2024-T351	L	365	470	17
	T	335	472	15
	S	298	414	5.8

L = longitudinal; T = transverse; S = thickness

Table 12: Mechanical properties of the aluminum alloy EN AW-2024-T351 [31]

The sequence of the increasing resistance of the material EN AW-2024-T351 under torsional stress (Mode III) in the tested fluids is as follows:

air with low moisture content – air with high moisture content – dry argon – dry hydrogen.

The results of the fatigue tests are summarized in Figure 9.

The higher number of cycles to failure in dry hydrogen and dry argon compared to moist air is attributable to the lack of moisture in the dry gases. The longer lifetime in air with high moisture content compared to air with low moisture content is attributable to the higher hydrogen activity and the resulting higher number of crack initiations [31]. The higher the number of crack nuclei, the lower is the prob-

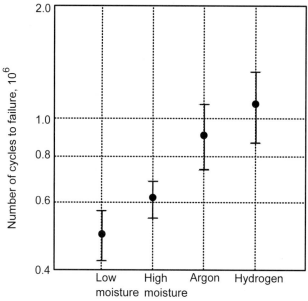

Figure 9: Influence of the atmosphere on the lifetime of the aluminum material EN AW-2024-T351, shear stress 138 MPa [31]

ability of growth of individual incipient cracks. In particular, this holds true for dry hydrogen. Due to the high number of crack initiations no incipient crack reaches a critical crack length in dry hydrogen, and hence lifetime is considerably increased. The number of cycles to failure in dry hydrogen is even higher than that in dry argon. To clarify this finding, the fracture surfaces were examined following the fracture of the specimens. Whereas the fracture in dry argon preferably took a path along the planes of maximum shear stress, "stepped" fracture surfaces were found in dry hydrogen, corresponding to a superposition of shear and tensile stresses and hence a great number of crack nuclei [31]. Comparing the fracture surfaces of the specimens failed in air with low moisture content with those in air with high moisture content, again a difference can be found regarding the mode of fracture. Whereas the fracture in air with low moisture content takes a path preferably along the planes of maximum shear stress, a high number of crack nuclei was found in air with high moisture content, corresponding to a superposition of shear and tensile stresses [31].

Another study [32] examined the fatigue behavior of the material EN AW-2024-T351 under torsional stress ($R = -1$). The maximum stress was 163 MPa and the frequency was 30 Hz. To specifically clarify the influence of hydrogen and moisture on corrosion fatigue, examinations were not only performed in gaseous dry, but also wet hydrogen gas (2% moisture). Again the atmospheres dry argon and moist air (15–20% and 90–95% RH) were used for comparison. Following the fracture of the specimens the crack propagation facets were counted under a stereo microscope. The results show that the number of crack propagation facets being

	Relative moisture %	Number of cycles to failure × 10³ cycles	Number of crack propagation facets
Argon	0	915	3
Hydrogen	0	991	48
Hydrogen	2	803	22
Air	15–20	531	9
Air	90–95	669	11

Mean value from 6 tests

Table 13: Fluid conditions, number of cycles to failure and number of crack propagation facets in the aluminum alloy EN AW-2024-T351

directly proportional to the number of the initiation sites are higher in dry hydrogen compared to dry argon and are considerably higher than in moist air (Table 13). In addition, it was established that the number of cycles to failure, i.e. the lifetime, also increases with an increasing number of crack propagation facets (Figure 10).

In the presence of moisture the lifetime of the examined aluminum material EN AW-2024-T351 (EN AW-Al Cu4Mg1) under torsional stress is generally reduced. The number of cycles to failure measured in wet hydrogen is clearly lower compared to dry hydrogen (cf. results in Figure 10). The specimens tested in hydrogen with a moisture content of 2–5% exhibit a shorter period for the initiation of cracks, a higher crack propagation and a lower lifetime compared to dry hydrogen [32].

The results in moist air [32] confirm the findings made in [31], i.e. a very high moisture content of 90–95% RH is less harmful than a lower moisture content

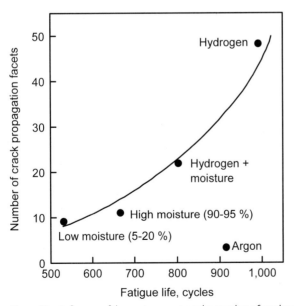

Figure 10: Influence of the environment on the number of crack propagation facets [32]

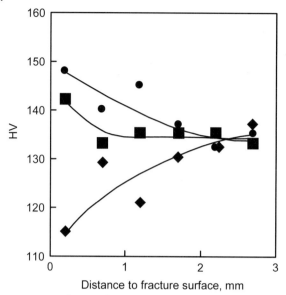

Figure 11: Influence of the atmosphere on the Vickers hardness in the fracture environment [32]. ● argon, ■ hydrogen, ◆ moist air

(15–20%). The reason is explained above. The fracture surfaces of the specimens tested in moist air exhibit black deposits and appear rough.

Measurements of the Vickers hardness revealed that, depending on the ambient fluid, the zones near the fracture surface may become harder or also softer [32].

The influence of high-purity, high-pressure hydrogen on the subcritical crack growth in the alloy EN AW-2219-T6 (EN AW-Al Cu6Mn) was examined using specimens with an incipient crack on the surface [33]. The results of such tests show that at a H_2 gas pressure of 36 MN/m² the threshold value K_0 for the initial crack growth amounts to 31 MN m$^{-3/2}$ (Figure 12). For welded specimens tested under the same test conditions a value K_0 = 29 MN m$^{-3/2}$ was established. Since these values amount to more than 80% of the fracture toughness, the material 2219-T6 is recommended as a material for pressure vessels for hydrogen gas [33].

The alloy 2021-T6 was tested to measure the influence of gaseous fluids on the crack propagation velocity of CT specimens (transverse specimens, TL direction) [34]. The alternating mechanical load was generated by a servo-hydraulic machine with a stress ratio R = K_{min}/K_{max} of 0.1 and 0.4 at frequencies of 0.5 and 9 Hz. The chemical composition of the material is indicated in Table 14. The tests were performed by comparison in the following fluids at room temperature:

– Dry, gaseous hydrogen, dried by passing through a cooling trap. The moisture content was < 10 ppm, corresponding to a partial pressure of water vapor of < 1 Pa.
– Lab air (moisture 40% RH)
– Distilled water
– Hydrazine

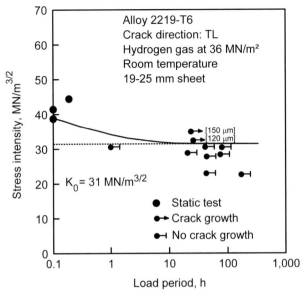

Figure 12: Determination of the threshold value K_0 for the subcritical crack growth in the material EN AW-2219-T6 (EN AW-AlCu6Mn) [33]

	Si	Fe	Cu	Mn	Mg	Sn	Zn	Ti	Zr	Cd	V
actual	0.10	0.17	6.4	0.32	0.005	0.06	0.03	0.05	0.18	0.15	0.09
Min	–	–	5.8	0.20	–	0.03	–	0.02	0.10	0.05	0.05
max	0.20	0.30	6.8	0.40	0.02	0.08	0.10	0.10	0.25	0.20	0.15

Table 14: Chemical analysis of the aluminum alloy 2021 [34]

The influence of the fluid on the crack propagation velocity da/dN depending on the amplitude of the stress intensity factor ΔK is summarized in Figure 13 for R = 0.1 and 9 Hz, in part also for 0.5 Hz.

Whereas a similar crack propagation behavior can be observed in dry hydrogen and moist air, the presence of distilled water or hydrazine leads to a sudden rise of the crack growth rate with increasing ΔK. The transition occurs at about $K = 7 MPa\sqrt{m}$ and is called ΔK^T. A reduction of the frequency from 9 Hz to 0.5 Hz has no effect in dry hydrogen gas, however leading to a clear increase in the crack growth rate per load cycle da/dN in moist air (cf. Figure 13). This finding proves that the crack propagation velocity decisively depends on the duration of exposure to moisture during one load cycle. It should be noted that the crack propagation above ΔK^T in dry hydrogen and moist air at 9 Hz is predominantly a transcrystalline process and is predominantly an intercrystalline process in distilled water and hydrazine. Higher crack rates when reducing the frequency or increasing the R value also lead to a transition from transcrystalline to intercrystalline crack propagation.

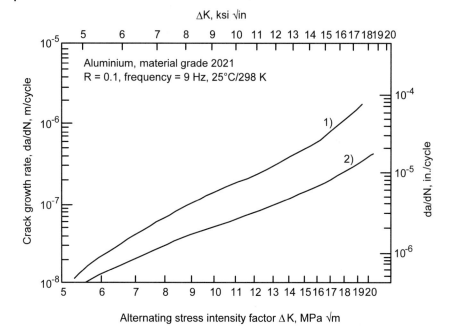

Figure 13: Crack growth rate in the aluminum material 2021-T6 under alternating loads depending on the amplitude of the stress intensity factor when exposed to dry hydrogen and hydrazine (1), moist air and distilled water (2) [34]

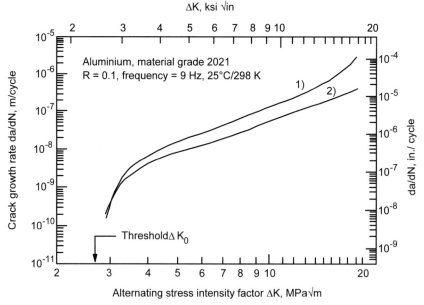

Figure 14: Fatigue behavior of aluminum alloy 2021-T6 under low load in distilled water (1) as well as dry hydrogen and moist air (2) [34]

The results for low mechanical loading in the range of the ΔK_0 value above which crack propagation is measurable, are indicated for the fluids dry hydrogen, moist air and distilled water in Figure 14. The ΔK_0 value is almost identical in all 3 atmospheres and the crack path is predominantly transcrystalline. Above ΔK_0 with the ΔK value increasing again, a clear difference is found between dry hydrogen as well as moist air and water. Crack propagation changes from transcrystalline to intercrystalline in distilled water.

Examinations of the influence of gaseous atmospheres on the fatigue behavior of the aluminum alloy 2021 have revealed that the presence of water vapor significantly accelerates the rate of the crack growth. The accelerated crack growth in distilled water with an intercrystalline crack mode is consistent with the examinations of the material EN AW-2219-T851 where an increase of the crack propagation velocity was observed with increasing water vapor pressure above a critical value of 2 Pa [34].

Aluminum-zinc materials (alloy series 7xxx)

According to current knowledge dry molecular hydrogen does not cause any embrittlement of aluminum alloys – neither under atmospheric pressure nor at high pressure and irrespective of the fact whether, or not, the parts are provided with an incipient crack on their surfaces. Neither does dry hydrogen accelerate the subcritical crack growth under static or cyclic loads [33]. In this connection Figure 15 shows by comparison the fatigue behavior of the material EN AW-7075-T651 (EN

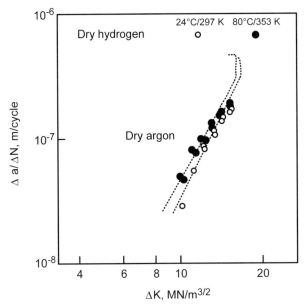

Figure 15: Crack propagation per load cycle $\Delta a/\Delta N$ as a function of the amplitude of the stress intensity ΔK for the high-strength material EN AW-7075-T651 in dry argon and dry hydrogen [33]

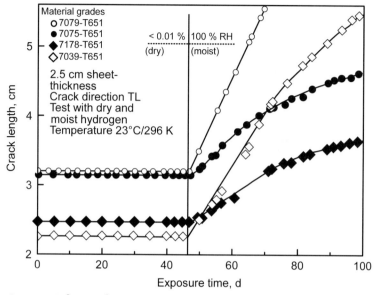

Figure 16: Influence of moisture on the subcritical crack growth of high-strength aluminum alloys in hydrogen gas [33]

AW-AlZn5.5MgCu) in dry hydrogen and in dry argon, not revealing any difference.

The resistance of high-strength aluminum alloys against subcritical crack growth in dry hydrogen gas is also illustrated in Figure 16. Four of the frequently used, high-strength Al alloys (7079-T651, EN AW-7075-T651 (AlZn5.5MgCu), EN AW-7178-T651 (AlZn7MgCu) and EN AW-7039-T651 (AlZn4Mg3)) were prenotched and mechanically loaded up to a level near the critical intensity factor K_{IC} and then exposed to dry hydrogen at RT at a pressure of 1 bar for 47 days [33]. Relative moisture was ≤ 0.01%. Crack growth did not occur; at least it was lower than the detection limit of 3×10^{-11} m/s. After measurable crack growth had not been detected after a period of 47 days, the relative moisture was increased to 100% with the effect that crack growth commenced immediately. This clearly proves that there is no crack growth in aluminum materials in dry hydrogen, but crack growth occurs in wet hydrogen.

In addition, examinations on the crack growth in hydrogen gas with saturated water vapor were performed by comparison with dry hydrogen [33]. Whereas crack propagation in dry hydrogen remained below the detection limit, crack propagation occurred in wet hydrogen (RH = 100%) depending on the composition of the alloy (Figure 17).

All findings show that for aluminum materials in gaseous fluids the content of water vapor is the decisive criterion for the embrittlement effect. To further confirm this, specifically the influence of the water vapor content on the fatigue behavior of the technical alloy AlZnMg 7017-T651 (Al, 5.0% Zn, 2.4% Mg) was tested [35]. The chemical composition and the mechanical properties of the examined material are

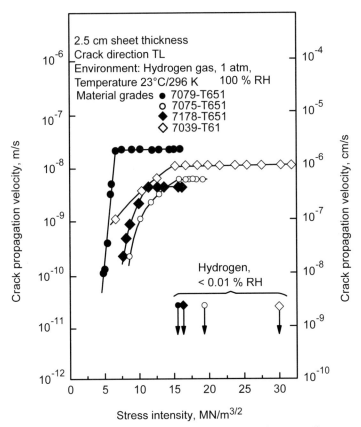

Figure 17: Crack propagation velocity of 4 high-strength aluminum alloys in wet hydrogen gas (RH = 100%) compared to dry hydrogen [33]

summarized in Table 15 and Table 16. The fatigue tests were performed in compliance with the ASTM E647 test method for the measurement of the crack propagation velocity using CT specimens (compact tension specimens). The mechanical load was applied sinusoidally with a stress ratio of R = 0.1 and frequencies of 1 Hz, 5 Hz and 10 Hz. The tests took place in high vacuum (10^{-5} Pa) as a reference fluid and in high-purity water vapor with different partial pressures (0.2–300 Pa) as well as in air

Zn	Mg	Fe	Si	Mn	Cr	Cu	Zr	Ti	Ni	Al
5.01	2.44	0.23	0.11	0.29	0.17	0.12	0.13	0.05	0.01	balance

Table 15: Chemical composition of the aluminum alloy 7017-T651 [35]

E, GPa	$R_{p0.2}$, MPa	R_m, MPa	K_{IC}, MPa $\sqrt{m}$
65.8	415	465	34.6

Table 16: Mechanical properties of the alloy 7017-T651 (LT direction) [35]

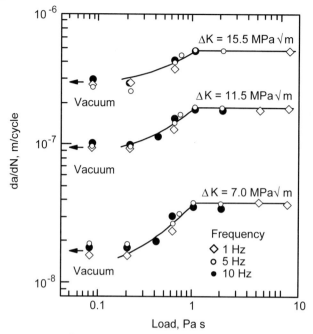

Figure 18: Influence of the atmosphere (water vapor pressure divided by frequency) on the fatigue crack growth rate of the aluminum alloy 7017-T651 under alternating loads (the continuous line reflects the calculated values) [35]

(50% RH). The results are summarized in Figure 18 [35]. Here, the influence of the fluid defined as the product of water vapor pressure and time during one load cycle is plotted against the crack propagation per load cycle. A linear correlation is found within a specific range. Hence, an increase of the water vapor pressure has the same effect on the crack propagation velocity as the reduction of frequency [35].

These results prove again that embrittlement of aluminum materials in gaseous fluids only occurs in the presence of water vapor. This embrittlement is caused by hydrogen generated by the reaction of water vapor with the material following

Equation 14 $\quad 2\,Al + 3\,H_2O \rightarrow Al_2O_3 + 6\,H^+ + 6e^-$

on fresh metal surfaces. Fresh metal surfaces are, for instance, obtained by alternating mechanical loading, in particular at the crack tip. Crack propagation is mainly determined by the amount of hydrogen absorbed by the material during one load cycle [35]. Hydrogen absorption can be influenced by lower alloy additions in the material [36]. For instance, the addition of chromium can reduce the absorption of hydrogen since it catalyzes recombination and hence the formation of molecular hydrogen, the hydrogen then escaping into the atmosphere.

Also the stress corrosion cracking behavior of aged aluminum materials in gaseous fluids, e.g. air, is decisively determined by the moisture content [37]. The mechanism of intercrystalline stress corrosion cracking of aluminum alloys involves –

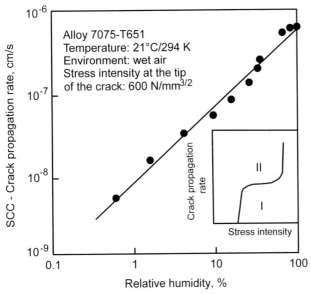

Figure 19: Influence of moisture on the crack propagation rate of a high-strength aluminum alloy (EN AW-7075) in area II (crack propagation rate independent of stress intensity) [37]

as shown by examinations performed at various research institutions [37] – hydrogen embrittlement of the grain boundaries, leading to the reduction of grain boundary cohesive strength. Figure 19 shows the influence of humidity on the crack propagation velocity of the high-strength AlZnMgCu alloy EN AW-7075-T651 (Table 17), the reaction according to Equation 13 being considered the hydrogen source. With an increasing offer of moisture also the amount of diffusible hydrogen increases at the crack tip. High concentrations of foreign atoms or also sensitive precipitation phases at the grain boundaries control the tendency to embrittlement, resulting in the gradual differences of susceptibility to stress corrosion cracking depending on the tempering conditions [37].

Summarizing the above, hydrogen can affect aluminum materials only in the presence of moisture. In this connection it should be noted that natural aging of aluminum alloys in moist air at room temperature (RT) can lead to noticeable hydrogen absorption. This is proven by the following test [38]. A rolled sheet of a high-purity AlZn5.63ZnMg2.4 alloy was first solution-annealed in dry argon (1 h 470 °C/H$_2$O) and subsequently aged in dry and moist air at room temperature for a period of 60 days. Thereafter, the specimens were deformed until fracture in dry argon at a strain rate of $1.3 \cdot 10^{-4}$ 1/s at RT. The results are summarized in Table

Al	Cu	Mg	Zn	Mn	Cr	Fe	Si
balance	1.2–2.0	2.1–2.9	5.1–6.1	0.3	0.18–0.28	0.5	0.4

Table 17: Chemical composition of high-strength alloy EN AW-7075 [37]

Aging atmosphere	Tensile strength, MN/m²	Fracture elongation, %
Dry air	285	42.4
	274	39.4
	261	40.5
	287	41.4
Moist air	238	22.1
	205	15.0
	222	19.0

Strain rate 1.3×10^{-4} 1/s

Table 18: Results from tensile tests on AlZn5.63Mg2.49 specimens aged in dry and moist air at RT [38]

18. Aging in moist air at RT led to a considerable reduction of the strength properties. Whereas the specimens aged in dry air exhibited only transcrystalline fractures, the share of intercrystalline fractures of sheets aged in moist air was up to 30%.

Since wet gases are considered already to belong to the moist atmospheres due to the reaction of water with aluminum, they are not dealt with in this "Compressed hydrogen" section.

Aluminum lithium materials (alloy series 8xxx)

AlLi-materials are a new generation of materials for the aerospace industry. Due to their specific weight of about 10% less than that of conventional aluminum alloys and their modulus of elasticity of up to 10% higher compared to conventional aluminum alloys, weight savings in a range of 15% are possible. A necessary precondition for the use of these materials in aircraft construction is good knowledge of the corrosion properties without and with mechanical loading, in particular also under the conditions of slowly and quickly alternating loads in aggressive gaseous fluids.

The material EN AW-8090-T8 (EN AW-Al Li2,5Cu1,5Mg1) was used to examine the fatigue behavior under the influence of dry hydrogen and dry helium as well as helium saturated with water vapor [39]. The Al-Cu-Mg material EN AW-2124-T851 (EN AW-Al Cu4Mg1) was used for comparison. The chemical composition of the alloy 8090-T8 was determined by cross-sectional analyses at 1/4 and 1/2 sheet thickness (Table 19). These analyses revealed that the Cu and Li values were slightly below the nominal values, which may have an impact on strength properties. The composition of the comparative alloy is indicated in Table 20; here, the nominal values are reached.

Both materials exhibited numerous precipitates and intermetallic phases along the grain and subgrain boundaries or former subgrain boundaries of the partially recrystallized material 2124.

The mechanical properties of both test materials were determined using 3 specimens each in ST direction (thickness direction) and are summarized in Table 21. As

	Li %	Cu %	Mg %	Zr %	Fe %	Si %	Ti %
1)	2.29	1.19	0.66	0.11	0.26	–	0.03
2)	2.25	1.06	0.60	0.12	0.23	0.04	0.05
nominal value	2.50	1.30	0.70	0.12	< 0.20	< 0.10	

1) Analysis at 1/4 sheet thickness
2) Analysis at 1/2 sheet thickness

Table 19: Chemical composition of the alloy EN AW-8090 [39]

	Cu %	Mn %	Mg %	Fe %	Si %
Specimen (2124)	4.30	0.68	1.56	0.12	0.06
nominal value	3.80–4.90	0.30–0.90	1.20–1.80	0.30 max	0.20 max

Table 20: Chemical composition of the alloy EN AW-2124 [39]

Material	Specimen No.	A %	Z %	$R_{p0.2}$ MPa	R_m MPa	E GPa
2124-T851	1	4.2	5.8	448	499	71.26
	3	4.7	6.2	442	484	74.39
	4	2.4	4.1	437	470	74.00
	average value	3.8	5.4	442	484	73.22
	deviation	± 1.21	± 1.13	± 5.3	± 14.7	± 1.7
8090-T8	L-1	1.7	3.5	345	456	78.94
	L-2	1.9	4.7	357	471	81.67
	L-3	1.5	3.2	346	454	79.32
	average value	1.7	3.8	349	460	79.98
	deviation	± 0.24	± 0.81	± 6.6	± 9.5	± 1.5

Table 21: Mechanical properties of the alloys EN AW-2124-T851 and EN AW-8090-T8

the results show, elongation A of the material EN AW-8090-T8 is only 1.7% and the reduction of area Z amounts to 3.8% only. The comparative values of the material EN AW-2124-T851 are A = 3.8 and Z = 5.4. The average modulus of elasticity E is 80.0 GPa for the alloy 8090 and 73.2 GPa for 2124. Hence, the stiffness of the Li-containing material is about 9.3% higher. Fractures of the 8090 specimens tested in the ST direction preferably occurred along the grain boundaries, whereas the 2124 specimens exhibited mixed transcrystalline and intercrystalline fractures. The lower ductility values of both materials in the ST direction are attributable to the precipitates.

Strain-controlled tests LCF tests (low cycle fatigue) according to ASTM E 606 and load-controlled tests HCF tests (high cycle fatigue) according to ASTM E 466 were performed on both alloy types (Al-Li and Al-Cu-Mg) in dry hydrogen and dry helium as well as helium saturated with water vapor.

HCF-tests

Crack initiation was determined by qualitative observation of slip bands, counting the incipient surface cracks and the number of cycles to failure in load-controlled HCF-tests. The following results were obtained:

- Considerably more slip bands were observed in the material 8090 compared to 2124. The slip band density increases with a higher stress amplitude.
- The slip band intensity increases with the aggressiveness of the atmosphere.
- In both materials the number of incipient surface cracks in wet helium is lower than in dry helium.
- In dry hydrogen the number of cracks clearly reduces at least in the material 2124.
- Crack initiation occurred at slip steps in the alloy 8090 and predominantly at major precipitates in the alloy 2124.

The influence of gaseous atmospheres and the stress amplitude on the finite life fatigue strength of both test materials under HCF-loading is summarized in Table 22.

Despite sometimes large scattering of the values, conclusions can be drawn from Table 22:

		Stress amplitude						
Gas	Material	118 MPa	138 MPa	152 MPa	166 MPa	186 MPa	207 MPa	276 MPa
He, dry	8090	700.5	54.5	34.7	22.7	9.1	8.2	1.3
	2124	–	54.7	73.5	31.5	20.4	23.4	3.9
H_2, dry	8090	419.9	68.2	60.1	41.9		8.4*	1.3
	2124	–	74.8	37.7	32.2	16.0	20.5	6.4
He, wet	8090	71.5	35.6	11.7	14.0	9.6	6.5	1.3
	2124	–	67.1	45.9	32.9	13.9	9.5	3.5

Number of cycles to failure: x 10^4; * load: 199 MPa

Table 22: Number of cycles to failure x 10^4 of the materials EN AW-8090-T8 and EN AW-2124-T851 depending on the fluid and the stress amplitude. Results from load controlled HCF-tests (ST direction)

As expected, the influence of the atmosphere is strongest at a lower load (118 MPa). Here, the resistance of the material 8090 significantly decreases in the sequence dry helium ≈ dry hydrogen ≈ wet helium.

At higher loads (276 MPa) the mechanical load plays the major role, whereas the atmosphere remains without almost any influence. The number of cycles to failure of the material 8090 is identical here.

Comparison of the materials shows that the resistance of the alloy 2124 in dry and wet helium is always higher than that of the alloy 8090. In dry hydrogen this only applies to lower and higher alternating mechanical loads. The influence of hydrogen on the material 2124 at medium stress amplitudes (150–170 MPa) seems to be a bit stronger.

LCF-tests

The influence of dry hydrogen and other gases (dry or wet helium) on the corrosion fatigue of the material EN AW-8090-T8 was also examined by performing

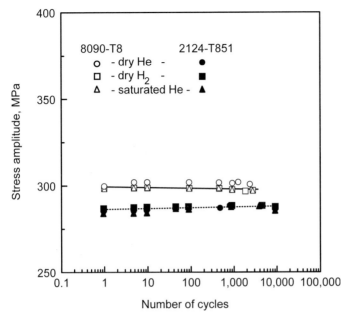

Figure 20: Dependence of the stress amplitude on the number of cycles as a result of the strain-controlled LCF-tests [39]

strain-controlled LCF-tests [39]. Again the material EN AW-2124-T851 was used for comparison. Six tests each were performed for all material + atmosphere corrosion systems. The results of these tests are stress amplitudes as a function of the number of load cycles. Figure 20 shows, for instance, the results of the materials 8090-T8 and 2124-T851 for a relatively low strain amplitude. The influence of the material, but no significant influence of the atmosphere can be observed here, which holds also true for all other strain-controlled LCF-results.

Hydrogen absorption in RSP-AlLi-alloys (Rapid Solidification Process)

Aluminum-lithium alloys are increasingly produced from rapidly solidifying powder (RSP) with subsequent extrusion. The manufacture and storage of the powders (40–100 µm) involves the problem of rapid oxidation by moisture. Hydrogen may be absorbed, which can cause H-induced embrittlement of the AlLi materials. To perform relevant tests to this end, RSP powders were produced from Al-3.4 Li-1.0 Mg-0.3 Cu-0.25 Zr [40]. Part of the powder was immediately degassed in vacuum (3 h, 450 °C) and extruded immediately thereafter (Specimen 2). Another part of the powder was first stored in contact with air for 15 days (Specimen 1) and then degassed and extruded. Both specimens were subsequently solution-annealed in a KNO_3 salt bath (1 h 530 °C/H_2O and 1 h 500 °C/H_2O). The hydrogen contents were determined and amounted to 4–6 ppm in Specimen 2 and 21 ppm in Specimen 1. Then, tensile tests were performed at different strain rates (1.7×10^{-2} to 10^{-5} 1/s) to detect an embrittlement effect, if any. Depending on the solution-annealing temperatures and the strain rate more or less distinctive fish eyes can be found on the fracture surfaces of the tested specimens, being indicative of hydrogen embrittlement. Summarizing the above:

- The hydrogen content of the tested RSP AlLi alloys is high enough under mechanical loading to form fish eyes as sites of initiating cracks.
- In the area near the fish eyes the fracture mode is of a mixed intercrystalline and transcrystalline type.
- The interaction of the RSP powder with air has a major influence on the hydrogen content in the alloy.
- Hydrogen embrittlement can be reduced at a lower solution-annealing temperature (1 h 500 °C/H_2O instead of 1 h 530 °C/ H_2O).
- The lower the strain rate, the more pronounced is the embrittlement.

Hydrogen solubility in liquid aluminum alloys (molten metals)

In contrast to pure aluminum (Al-H system) the question arises regarding aluminum alloys (Al-H-X system) how the alloying elements Cu, Si, Mg, Zn, Li, Fe and Ti influence the solubility of hydrogen in the molten metal.

Here, 3 interactions occur:

- aluminum with hydrogen
- aluminum with the alloying element
- hydrogen with the alloying element

Figure 21 proves that the individual alloying elements in the aluminum alloys exert different and strong influences on hydrogen solubility in molten metals [24].

Whereas hydrogen solubility in liquid aluminum materials decreases with increasing alloying contents of Cu, Si, Fe and Zn, it is increased by the alloying elements Ti, Mg and in particular by Li.

If the Al material contains several alloying elements, the activity coefficient f_H of hydrogen in molten metal results from:

Equation 15 $\quad f_H = f^H{}_H f^X{}_H$,

where $f^H{}_H$ considers the interaction between hydrogen and aluminum and $f^X{}_H$ is the interaction coefficient of hydrogen with the alloying element X.

The findings show that the alloying elements Cu, Si, Zn and Fe reduce the affinity of liquid aluminum towards hydrogen, attributable to the stronger bonds between these elements with Al [24]. It should be noted that, in particular, copper in aluminum alloys considerably reduces hydrogen solubility. On the other hand, the alloying elements Mg, Li, Ti [24, 41] and also Cr [41] increase hydrogen absorption in aluminum since these elements exhibit a high affinity towards hydrogen. Hence, it is also more difficult to remove the hydrogen. The interaction between H and Mg atoms can be modified by adding Cu [41]. Another paper [42] confirms that in solid and liquid aluminum alloys hydrogen solubility is considerably increased by the alloying element Li, being of importance in particular to AlLi materials.

There are various possibilities for the absorption of hydrogen, e.g. during the manufacturing process, for instance, when melting and casting of metals [43].

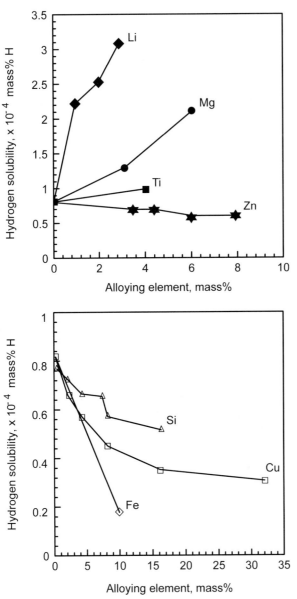

Figure 21: Influence of the alloying elements on hydrogen solubility in liquid Al at 700 °C and a partial hydrogen pressure of 1 atm (1.013 bar) [24]

During these processes the reaction of atmospheric water vapor with the metal (Equation 14) is considered the main source of hydrogen [42, 43]. The question is about the hydrogen content after the aluminum melt has solidified, since then the dissolved hydrogen may have a detrimental effect on the further processing of the aluminum products. First, it should be noted that the absorption of hydrogen in aluminum is fairly low compared to other metals. It amounts to about 1/100th of that in iron at the melting point of aluminum and is again reduced by a factor of 10 in the solid state [43].

Pore and blister formation

Due to the solubility of hydrogen strongly differing in the liquid and the solid state of aluminum materials (ratio 20:1) the molten metal may form blisters when solidifying [44]. When further processed, the pores filled with hydrogen lead to flakes in the forgings or to the formation of blisters in heat-treated or solution-annealed materials. Due to its lower strength the material EN AW-1199 (EN AW-Al 99.99) is more susceptible to blistering at higher temperatures. Whereas heat treatment at 343 °C causes low blistering or no blistering at all, heat treatment at higher temperatures, e.g. at 538 °C, clearly leads to the formation of blisters. The extent of damage depends on the hydrogen content and the thickness of the material (Table 23). In general, blistering decreases with decreasing thickness of the material and a decreasing content of hydrogen.

Thickness mm	Hydrogen content in ml/100 g		
	0.29 ml/100 g	0.17 ml/100 g	0.12 ml/100 g
	Hot-rolled from a thickness of 19 mm		
	blister formation		
3.2	extreme	strong	strong
1.6	strong	strong	moderate
1.0	considerable	low	very low
0.25	low	none	none
	Rolled from a thickness of 3.2 mm		
1.6	considerable	very low	very low
1.0	considerable	low	very low
0.25	low	none	none

Table 23: Influence of the hydrogen content on blistering of the material EN AW-Al 99.99 during heat treatment at 538 °C for 2 h [44]

Blistering is often found in inclusions [44]. The metallographic proof was provided elsewhere in this paper that there is a correlation between the hydrogen content and the number of inclusions per unit of area [45]. In particular in housing sheets blistering is not desired and gives rise to complaints.

The tests for determining hydrogen revealed that the average hydrogen content in plates is 50% lower compared to the molten metals, attributable to partial degassing [44]. The best way to avoid the formation of pores when molten metals solidify is to keep the hydrogen content as low as possible, the modern melting plants succeeding in doing so.

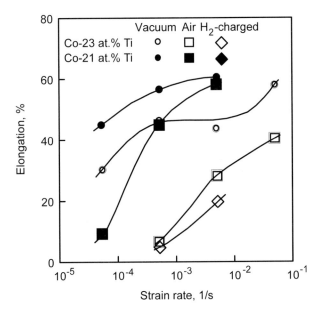

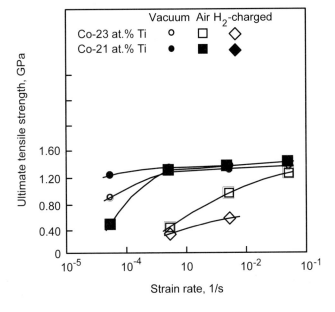

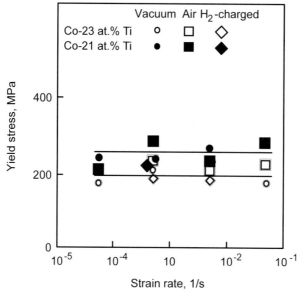

Figure 22: Strain rate-dependent fracture elongation (a), tensile strength (b) and yield stress (c) of the alloys Co-23 at.% Ti and Co-21 at.% Ti in vacuum, air and hydrogen [46]

A 5 Cobalt alloys

Among the cobalt alloys intermetallic compounds of a Co_3Ti type exhibit good high-temperature stability [46]. However, in a hydrogen atmosphere the ductility properties may be clearly reduced. In this connection Figure 22 shows, by comparison, the results of slow strain rate tensile tests of the alloys Co-23 at.% Ti and Co-21 at.% Ti in vacuum, in air and in hydrogen. Whereas the yield stress is hardly impaired, the tensile strength R_m and the fracture elongation of the hydrogen-charged specimens are strongly reduced depending on the strain rate. Also in air the stress and the toughness are impaired. Fractographically this impairment was identified to be hydrogen-induced grain boundary embrittlement with crack formation.

A 6 Chromium and chromium alloys

Although chromium and its alloys exhibit a good corrosion resistance in a corrosive environment as well as a good oxidation resistance at higher temperatures, their ductility at room temperature is low. However, the ductility of chromium can be improved by increasing its purity level. [47] reports on examinations of the influence of different ambient conditions on the toughness behavior of high-purity chromium. The toughness behavior was tested by slowly deforming specimens with a thickness of 0.5 mm in vacuum, air, hydrogen, nitrogen and oxygen. Deformation was achieved by slowly loading the specimens with an indentation stamp with a diameter of 2.5 mm. The rate of loading was varied between 0.01 mm/s and 1.0 mm/s. The deformation was measured depending on the load and the deformation energy was calculated from the area under the load-deformation curve and then selected to serve as the measure for the deformation capacity.

First, the specimens were kept in a vacuum chamber at 2.7×10^{-4} Pa for 180 min before the test gas was fed into the test chamber. The tests were carried out at a test pressure of the gases of 1.013 bar. The tests were carried out both in dry gases and wet gases having been passed through distilled water before hand. Prior to the test, the specimens were kept in the respective gas atmosphere for 12 h.

Figure 23 shows the load-deformation curves measured at a loading rate of 0.1 mm/s in vacuum, moist and dry air as well as in different dry gas atmospheres. It is noticeable that the toughness behavior is best in moist air and is even better than that in vacuum.

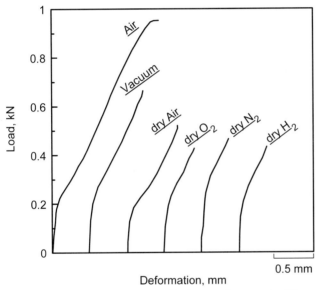

Figure 23: Load-deformation curve of high-purity chromium in different atmospheres at a loading rate of 0.1 mm/s (drawn offset for better understanding) [47]

The influence of moisture in the test atmospheres can be seen from Figure 24. Here the deformation energy is indicated as obtained at a loading rate of 0.1 mm/s from the relevant load-deformation curves in dry and moist atmospheres.

These results confirm the favorable influence of moisture in the test atmospheres on the deformation capacity of chromium as can be seen already from Figure 23. It can be assumed that also the oxide layer quickly formed at room temperature in the presence of moisture exerts a positive influence on the ductile behavior.

As shown in the diagrams in Figure 25, the loading rate has different effects on the deflection of the specimens in air and in hydrogen. Whereas the deflection

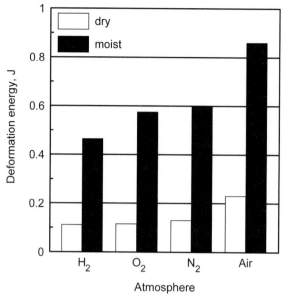

Figure 24: Deformation energy during the indentation test of high-purity chromium specimens in different dry and moist test atmospheres [47]

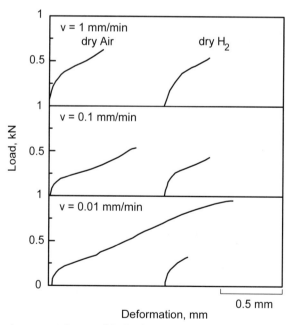

Figure 25: Influence of the loading rate on the ductile behavior of pure chromium in dry air and dry hydrogen [47]

and, hence, the toughness of the specimens increases with a decreasing loading rate in dry air, it is exactly the reverse behavior in dry hydrogen.

During the tests in hydrogen the specimens with the slowest loading rate of 0.01 mm/min exhibited a marked brittle behavior and broke without any plastic deformation. Hence, hydrogen has sufficient time to diffuse into the material and cause embrittlement.

To improve the ductility of chromium at room temperature, the same authors examined the influence of different alloying elements on the ductile behavior of chromium [48]. To this end, 0.5 mole% of the elements titanium, vanadium, manganese, iron, cobalt and nickel were alloyed with pure chromium (99.93%). In addition, vanadium was also added of amounts of 0.1 mole% to 5 mole%, since vanadium was expected to exert the strongest influence. The ductile behavior was assessed on the basis of the deformation of three-point bending test specimens performed either in air or in vacuum at loading rates of 0.1 mm/min.

Figure 26 summarizes the results obtained for specimens alloyed with 0.5 mole % from hardness measurements as well as the maximum bend angles determined for the bending specimens in air and in vacuum.

The bend angle obtained of about 120° shows that the ductility of the alloy with vanadium strongly increases, in particular, during the test in air. Higher or lower vanadium contents do not yield any improvement as can be seen from Figure 27.

Since among the examined alloys the Cr-0.5 mole% vanadium alloy had the strongest effect on the improvement of ductility, this material was additionally tested in oxygen, nitrogen and hydrogen atmospheres to identify potential embrittling influences of the atmosphere.

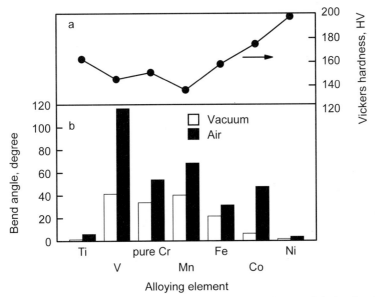

Figure 26: Influence of various alloying elements on the hardness and the bend angle of three-point bend specimens from chromium and chromium alloys

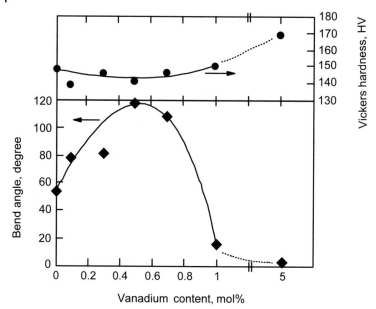

Figure 27: Hardness and bend angles of three-point bend test specimens from chromium-vanadium alloys with different vanadium contents [48]

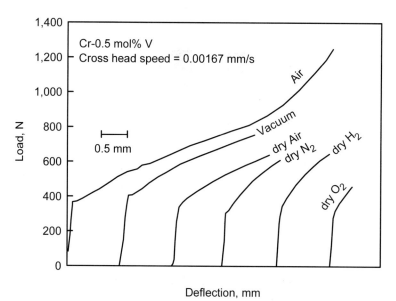

Figure 28: Stress-strain curves of a Cr-0.5 mole% vanadium alloy in different environments at room temperature [48]

As for the examinations described above the specimens were first kept in a vacuum chamber at 2.7×10^{-4} Pa for 180 min and thereafter the test gas was fed into the test chamber, the specimens were kept in the respective gas atmosphere for 12 h and then tested at a gas pressure of 1.013 bar. Again, the tests were performed in dry and wet gases. Figure 28 shows the stress-strain curves in the various gas atmospheres. Again, the ductility of the specimens in vacuum, dry air and the dry gases is obviously lower than in ambient air.

Figure 29 shows a comparison of the behavior in dry and moist gases using the maximum bend angles measured for the three-point bend test specimens.

Both figures show that also the chromium-vanadium alloy tends to hydrogen embrittlement as can be detected in pure chromium. The positive influence of moisture found already for pure chromium on the ductile behavior in different atmosphere is confirmed for this alloy.

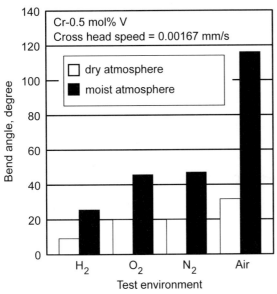

Figure 29: Bend angles of three-point bend test specimens obtained in various dry and moist gases [48]

A 7 Copper

For unalloyed copper grades a difference is made between oxygen-containing and oxygen-free grades. Liquid or also strongly heated copper absorbs oxygen from an oxygen-containing environment and forms copper(I)-oxide (Cu_2O). Refined copper not having been deoxidized (e.g. with phosphorus) always contains Cu_2O particles. The oxygen-containing copper grades listed in Table 24 with their designations according to DIN EN 1976 [49] and the former DIN 1708 [50] may contain up to

	DIN EN 1976		DIN 1708	
	Code	Number	Code	Number
	–	–	E-Cu57	2.0060
	Cu-ETP1	CR003A	–	–
	Cu-ETP	CR004A	E1-Cu58	2.0061
	Cu-FRHC	CR005A	E2-Cu58	2.0062
	Cu-FRTP	CR006A	F-Cu	2.0080

Table 24: Designations of oxygen-containing copper grades

0.040% oxygen according to the relevant standard and the material Cu-FRTP may contain up to 0.100% oxygen.

The oxygen-containing copper materials are sensitive to hydrogen at higher temperatures. If copper is annealed in a hydrogen-containing atmosphere, the hydrogen molecules are dissociated on the hot metal surface at temperatures above 400 °C and atomic hydrogen diffuses into the copper at a high diffusion rate. At high temperatures, the diffused hydrogen reduces the Cu_2O present in the microstructure and forms H_2O (refer to Equation 9). The resulting water vapor is not diffusible and at high temperatures the resulting high water vapor pressure leads to internal cracks and voids. If the reduced copper oxides occurred in the near subsurface region, also surface blisters with a diameter of several micrometers were formed.

This phenomenon of blistering is especially marked in a temperature range above 375 °C, in which the reduction of copper oxides and other oxides is preferred both thermodynamically and kinetically. Blistering may impair mechanical properties, in particular during deformation processes [51].

This process of copper embrittlement called "steam-embrittlement" only affects the oxygen-containing grades when heated in hydrogen or hydrogen-containing gas mixtures and has been known for a long time as the "hydrogen embrittlement" [52–57] and [58] quoted in [51].

The diffusion processes of hydrogen and the reduction of the copper oxide may also occur at lower temperatures, however the generated water vapor pressure is not sufficient to cause material separations. If such parts are subsequently heated to higher temperatures in an inert atmosphere or in vacuum, this may lead to the described damage [59] quoted in [51]. Metallographic examinations regarding this damage pattern are described in [56, 57].

Damage caused by these types of hydrogen embrittlement occur in parts made from oxygen-containing copper grades, e.g. during bright annealing or other heat treatments in a hydrogen-containing atmosphere. For bright annealing a slightly reducing atmosphere needs to be generated and, hence, knowledge of the critical temperature for the occurrence of this damage is important for applications in practice. In the literature operations at temperatures below 400 °C are considered to be non-critical. However, in strongly coarse-grained microstructural regions susceptibility to hydrogen embrittlement must be also expected at temperatures below this temperature limit which is considered safe [57]. Also when processing oxygen-containing copper grades by soldering and welding with the oxyacetylene flame or even during warming up for bending hydrogen damage can occur since, depending on the oxygen/acetylene ratio in the welding flame carbon monoxide and hydrogen are primarily formed. A case of damage caused by hydrogen embrittlement on brazed joints of a rotor is described in [60]. Parts of oxygen-containing copper grades should be soldered with a hydrogen-free flame or inductively.

Examinations of various factors influencing the susceptibility to hydrogen embrittlement when soldering are described in [61]. The content of chemically combined water in the examined fluxing material on the basis of KBF_4 and $K_2B_4O_7$ has turned out to be a major factor.

Also other oxidic components of the copper matrix, e.g. aluminum oxide, which may be formed when heating aluminum-containing copper in air or in oxygen may react with diffusing hydrogen to form water vapor and also lead to embrittlement [53, 62].

The oxygen-free copper grades mentioned in Table 25 were developed to minimize such hydrogen-induced damage. The standards regarding the chemical composition of these grades require the oxygen content be adjusted by the manufacturer such that the material meets the requirements for hydrogen resistance according to DIN EN 1976 [49]. To prove the hydrogen resistance required in this standard, a hydrogen resistance test according to [18] needs to be performed. During this test the prepared specimens are annealed in a furnace in an atmosphere containing not less than 10% hydrogen at a temperature between 825 °C and 875 °C for 30 minutes and thereafter cooled down in the furnace atmosphere or quenched in water. Hydrogen embrittlement, if present, is subsequently assessed by 180° bending or by alternate bending with a subsequent microscopic check for gas pores or cracky microstructures.

Also the ASTM specifications for the oxygen-free copper grades [63, 64] require proof of the sufficiently low content of oxygen and the resistance to embrittlement. The test procedures required to this end are described in [65]. Also their resistance

DIN EN 1976		DIN 1708	
Code	Number	Code	Number
Cu-OF1	CR007A	–	–
Cu-OF	CR008A	OF-Cu	2.0040
Cu-OFE	CR009A	–	–
Cu-PHC	CR020A	SE-Cu	2.0070
Cu-HCP	CR021A	SE-Cu	2.0070
Cu-PHCE	CR022A	–	–
Cu-DLP	CR023A	SW-Cu	2.0076
Cu-DHP	CR024A	SF-Cu	2.0090
Cu-DXP	CR025A	–	–

Table 25: Designation of the oxygen-free copper grades

to cracking during the bending test following an annealing treatment at 850 °C in hydrogen-containing atmosphere is required.

In connection with examinations regarding the influence of tritium on OFHC copper (oxygen-free, high-conductivity) specimens from this material were exposed to hydrogen with a pressure of 100 MPa at 200 °C and 300 °C for a period of 2,900 h [66]. The metallographic examination by means of a raster electron microscope of the specimens charged at 200 °C revealed small voids in the micrometer range at the grain boundaries, being indicative of the fact that also this material is not completely immune to hydrogen attack if exposed to hydrogen for a long time. However, specimens exposed to hydrogen at 300 °C under the same conditions did not show any indication of hydrogen damage. Also the examinations described in [72] following aging in a hydrogen atmosphere at 300 °C did not reveal any negative influence on the mechanical properties of pure copper (refer to A13).

[67] reports comparison of tensile tests of notched and unnotched specimens from OFHC copper in air and in hydrogen at 10,000 psi (700 bar). During the tensile tests of smooth specimens with an extension rate of 0.5 mm/min the stress-elongation curves depicted in Figure 30 were obtained.

Obviously, hydrogen exerts an influence on the curve path and the value of the tensile strength obtained in hydrogen is about 16% lower than the value obtained in air. Also the fracture pattern of the specimens in hydrogen changed more to a shear fracture occurring below 45° although the fractographic image did not show any indication of embrittlement. The notched specimens did not show any influence of hydrogen.

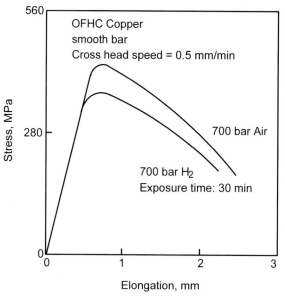

Figure 30: Stress-elongation curves of smooth round test bars from OFHC copper in air and in hydrogen at 700 bar [67]

Due to its marked resistance to hydrogen damage OFHC copper is also selected for the internal lining of high-pressure vessels of A 286 (1.4943) stainless steel used for nuclear fusion, which have to withstand hydrogen pressures of 1,000 bar or 10,000 bar [68].

A 9 Copper-nickel alloys

Since oxygen-free copper is not susceptible to hydrogen embrittlement but nickel can be damaged through absorbed hydrogen, the behavior of the copper-nickel or nickel-copper alloys strongly depends on the nickel content as shown by the examinations described in [69].The examined alloys and their composition are indicated in Table 26. They were obtained by melting copper and nickel pellets in vacuum, rolled to produce sheets and plates and recrystallized at 760 °C for 2 h.

% Cu	52.3	43.7	38.3	34.4	31.2	26.5
% Ni	47.7	56.3	61.7	65.6	68.8	73.5

Table 26: Composition of the examined Cu-Ni test melts [69]

Susceptibility to hydrogen embrittlement was assessed by means of notched round test bars by comparing the test results in compressed hydrogen at 34.5 MPa and in an air/helium mixture. Table 27 contains the values obtained for tensile strength. As can be seen from the strength values and Figure 31, the notch tensile strength in hydrogen continuously decreases with increasing nickel content of the alloy and reaches only about 50% of the values obtained in air at nickel contents of about 70%.

Nickel content %	Tensile strength, R_m in H_2 MPa	Tensile strength, R_m in air/He MPa	Ratio $R_m\,H_2/R_m$ air
47.7	512.3	533.3	0.96
56.3	426.5	576.7	0.74
61.7	402.4	583.6	0.69
65.6	369.3	602.2	0.61
68.8	351.7	608.7	0.58
73.5	313.2	601.5	0.52

Table 27: Notch tensile strengths of the specimens tested in hydrogen and in air/helium [69]

The unfavorable influence of higher nickel contents in binary alloys of the Ni–Cu system is also confirmed by the examinations in [70]. Here, the susceptibility to embrittlement of the alloys indicated in Table 28 was not tested in compressed hydrogen, but after or during cathodic charging.

Ni-Cu	Ni	NiCu10	NiCu20	NiCu30	NiCu40	NiCu50	NiCu60	NiCu70	NCu90	Cu
at.% Cu	0.0	9.1	20	27	38	49	58	68	89	100

Table 28: Composition of binary nickel-copper alloys [70]

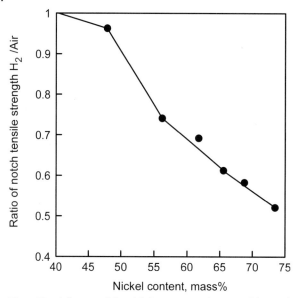

Figure 31: Influence of the nickel content on the ratio of the notch tensile strengths in hydrogen and in air [69]

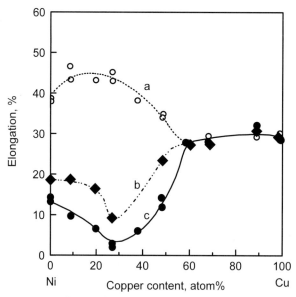

Figure 32: Influence of the copper content on the embrittlement of binary nickel-copper alloys [70]
(a): hydrogen-free, tested in air
(b): precharged with hydrogen, tested in air
(c): precharged with hydrogen, tested while simultaneously charged with hydrogen

Figure 32 shows the values obtained during slow strain rate tensile tests at a strain rate of $\dot{\varepsilon} = 6.67 \times 10^{-4}$ 1/s for the elongation of the specimens as a measure for embrittlement depending on the copper content of the specimens.

For a copper content of 60% identical elongation values are obtained for uncharged and charged specimens, whereas at nickel contents above 50% charged specimens exhibit a clear decrease of elongation as an indication of hydrogen-induced embrittlement. Also the dependence shown in Figure 33 for several examined alloys of the elongation values on current density during electrolytic charging with hydrogen shows that even at higher hydrogen charges embrittlement is not expected for alloys with a copper content of 60%. If the copper contents are lower, susceptibility increases with decreasing copper contents. The decrease of the elongation values occurs earlier and is more distinct at identical charge.

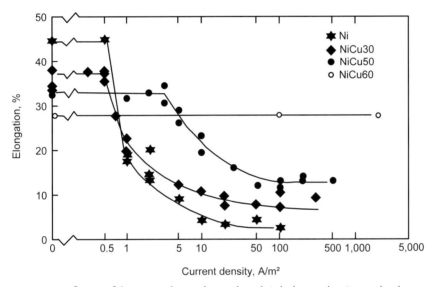

Figure 33: Influence of the current density during electrolytic hydrogen charging on the elongation of the tensile test specimens for various nickel copper alloys [70]

A 10   Copper-tin alloys (bronze)

A 11   Copper-tin-zinc alloys (red brass)

A 12   Copper-zinc alloys (brass)

Hardly any research is reported in the literature regarding the influence of hydrogen on the behavior of alloys of copper with tin and zinc. A study on hydrogen embrittlement of face-centered cubic materials based on a test during which materials were not examined in compressed hydrogen but were electrolytically charged with hydrogen, did not reveal any influence of hydrogen on the mechanical properties of the material CuZn30 [71].

A 13 Other copper alloys

The microstructure of copper-niobium alloys shows strong niobium fiber alignment in the direction of deformation within the copper matrix. Such composite materials exhibit high strength as well as good electrical and thermal conductivity in combination with high thermal stability. Due to this combination these materials are interesting for substituting copper materials with a high resistance to hydrogen when using them in rockets or combustion chambers. On the other hand, the niobium contents with their high susceptibility to hydrogen embrittlement make their potential use in a hydrogen environment questionable. To examine the influence of hydrogen on the deformation behavior of such materials containing phases of different susceptibility to hydrogen embrittlement, slow strain rate tensile tests were performed on round test bars of CuNb20 alloy sheets at a strain rate of 1.7×10^{-4} 1/s prior to and after hydrogen charging, the results being thereafter compared with similarly treated specimens from pure copper and pure niobium [72].

Hydrogen charging was performed by aging the specimens in a hydrogen atmosphere at 300 °C for a period of 72 h. The comparison specimens were tested in the as-received state and following the respective annealing treatment (300 °C, 72 h) in vacuum. Table 29 shows the hydrogen contents of the specimens following annealing in vacuum and in the hydrogen atmosphere.

Material	Hydrogen content 300 °C, 72 h, vacuum at.%	Hydrogen content 300 °C, 72 h, H_2 at.%
Copper	0.13	0.15
Niobium	0.12	1.64
CuNb20	0.42	1.95

Table 29: Hydrogen contents of copper, niobium and CuNb20 specimens prior to and after charging at 300 °C for 72 h [72]

Material/state	R_m at 22 °C MPa	R_m at -78 °C MPa
Cu/rolled	465	
Cu/300 °C, 72 h, vacuum	211	266
Cu/300 °C, 72 h, H_2	213	262
Nb/rolled	745	
Nb/300 °C, 72 h, vacuum	702	802
Nb/300 °C, 72 h, H_2	687	715
CuNb20/ rolled	778	
CuNb20/ 300 °C, 72 h, vacuum	730	697
CuNb20/300 °C, 72 h, H_2	729	695

Table 30: Tensile strengths of specimens consisting of copper, niobium and CuNb20 subjected to different treatment obtained at different test temperatures [72]

Table 30 lists the tensile strength values determined for the various specimens at room temperature and at −78 °C. Strength losses of the specimens charged with hydrogen compared to the uncharged specimens are only found for niobium. Due to its low solubility for hydrogen copper practically did not absorb any hydrogen during aging of the specimens to the hydrogen atmosphere, and hence impairment of the mechanical values was not expected. Although the copper-niobium alloy specimens exhibited higher hydrogen contents after both annealing in vacuum and annealing in the hydrogen atmosphere compared to the niobium specimens, detrimental effects on the strength values cannot be found. As shown by the fracture patterns, the ductile behavior of the composite material is preserved to a major extent by inserting the hydrogen charged niobium in to the ductile copper matrix. A similar effect of the ductile copper matrix was also observed for the material CuCr17.

A 14   Unalloyed and low-alloy steels/cast steel

Two different types of damage depending on the temperature can be observed when compressed hydrogen acts on unalloyed and low-alloy steels. At temperatures above about 200 °C the hydrogen diffused into the material reacts with the carbides according Equation 8 to form methane. At lower temperature this reac-

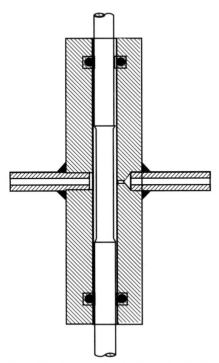

Figure 34:   Pressure chamber for performing tensile tests in a compressed gas atmosphere [74]

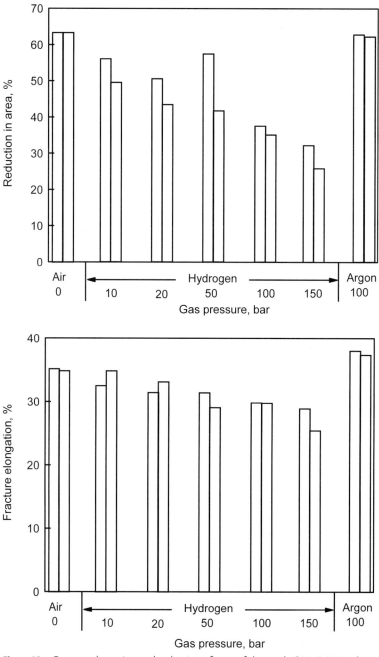

Figure 35: Fracture elongation and reduction of area of the steel Ck22 (1.1151) during tensile tests in air, argon and hydrogen under different pressure conditions (double specimens) [74]

tion cannot take place and the diffused hydrogen causes embrittlement effects as also known for other metals.

In 1896 the VDI magazine reporter about the damage in cold compressed hydrogen for the first time [73]. At the time an explosion of compressed hydrogen cylinders had occurred in a cylinder store of a Royal Aviation division on Tempelhofer Feld in Berlin. However, the potential role of hydrogen in connection with such material damage was not detected.

The damaging effect of cold compressed hydrogen was observed by W. Hoffmann et al. [74–77] as late as in the beginning of the 60ies. They carried out tensile tests on unalloyed and low-alloy steels in a compressed hydrogen atmosphere and, for comparison, in air and in an argon atmosphere using the test apparatus shown in Figure 34. The initial examinations of the unalloyed steel Ck22 (Mat. No. 1.1151, designation C22E according to EN 10083-1, cf. SAE 1020), with about 0.22% carbon content revealed already that the fracture elongation values and, in particular, the reduction of area values obtained in the tensile test are clearly lower when the hydrogen pressure is increased compared to the reference values from the tensile tests in air or in argon (Figure 35).

Apart from numerous secondary cracks in the deformed area of the specimens also fish eyes were found on the fracture surface similar to those in the tensile tests on specimens charged with hydrogen, e.g. hydrogen-containing welded material, or those occurring when hydrogen-charged components burst. In charged specimens the diffused hydrogen accumulates in voids or pores. If the component is plastically deformed, the atomic hydrogen diffusing out of these "voids" leaves a bright diffusion zone called fish eyes and brittle fracture properties around the voids initially filled with molecular hydrogen on the fracture surface are observed, the propagation velocity of the cracks not being excessively high.

During the tensile tests in compressed hydrogen the deformation velocity exerts a clear influence as shown in Figure 36 using the example of Ck22 (1.1151) steel [75]. Fracture elongation and reduction of area increase with increasing deformation velocity, i.e. with decreasing time of exposure to hydrogen.

The influence of the carbon content and, hence, of the hardness was examined with the three grades pure iron (Armco iron), Ck22 (Mat. No. 1.1151, cf. SAE 1020) and C45 (Mat. No. 1.0503, cf. SAE 1043) in the normalized state each. Figure 37 shows the results obtained for fracture elongation and reduction of area of the three grades as a function of hydrogen pressure.

With reference to the values in air, the steel Ck22 (Mat. No.1.1151) exhibits the strongest decrease of values with increasing hydrogen pressure. This is illustrated again in Figure 38 for the results obtained at a hydrogen pressure of 150 bar compared to the values measured in air.

Further examinations of the influence of hydrogen on the processes occurring during the tensile test of low-alloy steels are described in [78]. The tests were performed on normalized and on tempered low-alloy steels in air and in hydrogen at pressures of 50 bar or 100 bar. Whereas, for instance, for steel of type St 34-2 (Mat. No. 1.0032, cf. SAE 1010) the values of the yield strength and the tensile strength decreased only slightly with increasing hydrogen pressure, reduction of area de-

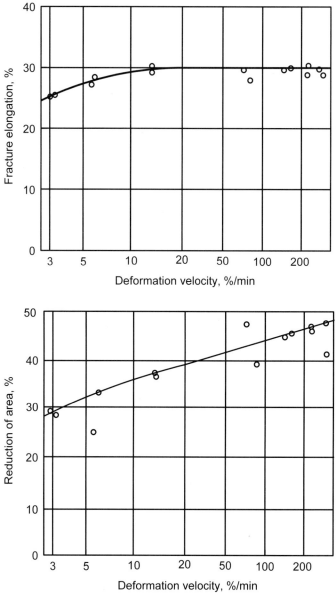

Figure 36: Influence of the deformation velocity on the fracture elongation and reduction of area of normalized steel Ck22 (1.1151) during tensile tests in hydrogen at 160 bar and room temperature [75]

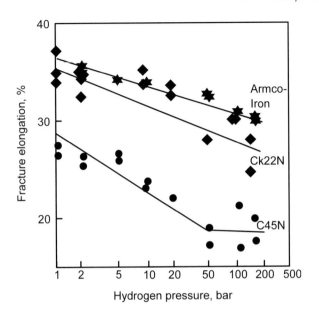
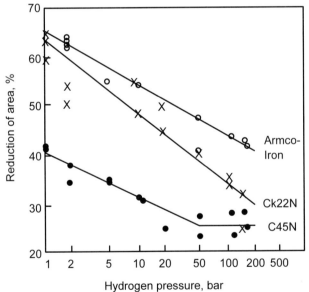

Figure 37: Influence of hydrogen pressure on fracture elongation and reduction of area of the three normalized materials Armco iron, Ck22 (1.1151) and C45 (1.0503) [75]

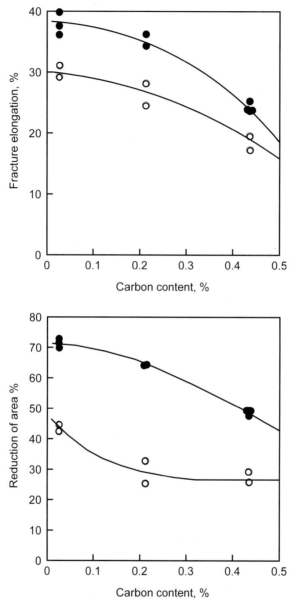

Figure 38: Influence of the carbon content on the fracture elongation and reduction of area of unalloyed steels during tensile tests in air (●) at one bar and in hydrogen (o) at 150 bar [75]

creased from 68% at failure in air to 36% below 100 bar hydrogen. The specimens ruptured in hydrogen exhibited numerous secondary cracks in the necking region, all following the grinding grooves on the surface. The number and depth of the secondary cracks as well as the decrease of the reduction of area values could be clearly reduced by mechanical and electrolytical polishing of the specimen surface. Reduction of area, for instance, at a hydrogen pressure of 100 bar amounted to 58% of the electrolytically polished specimen compared to 36% of the ground specimen. During these tests it could be also established already that the influence of hydrogen depends on the duration of the tensile test, i.e. on the strain rate of the specimens.

Numerous internal cracks were detected in specimens charged with hydrogen in air prior to the breaking test. It is to be assumed that microcracks are initiated in the material already during deformation because dislocations accumulate at grain boundaries. Such microcracks do not propagate in tough materials. Hydrogen dissolved in the specimen accumulates in the triaxial stress areas in front of the crack tip and inhibits the slip processes there, hence reducing the deformation work required for crack propagation and thus enhancing brittle fracture.

In the technical practice damage in high-purity compressed hydrogen at ambient temperature was detected for the first time in high-strength steels in connection with liquid rocket stages of the NASA's space program. The damage when handling hydrogen occurred on transport tanks, pipelines and fittings, all having two features in common: firstly, they were exposed to a higher number of low-frequency alternating load cycles as, for instance, by filling and emptying of compressed gas cylinders or the constantly changing internal pressure of hydrogen buffer tanks, and secondly they had growable notches and grooves on the surface charged with hydrogen.

Although the adsorption heat of hydrogen on pure iron amounting to -100 kJ/mol H_2 is very high such that the surface could be completely covered with hydrogen even at a very low pressure ($\approx 10^{-8}$ bar), the absorption of hydrogen from the gas phase is practically fully prevented in technical material surfaces by deactivation of the surface by oxide layers or the adsorption of other atmospheric gases [79].

In addition, the adsorption process on a normal, uncleaned metal surface under the influence of exclusive elastic elongation in dry molecular hydrogen at common ambient temperatures does not result in any noteworthy concentration of diffusible atomic hydrogen since, practically, the equilibrium of the dissociation according to Equation 6 is fully on the side of the hydrogen molecule. A noticeable dissociation does not occur.

Therefore, embrittlement effects on steels in the practice are found only if plastic deformations occur in areas of the surfaces as a result of mechanical loading in the presence of pure gaseous hydrogen. Such plastic flows may occur both under exclusively static loads as well as loads changing over time.

Thus, the charging effects during the slow strain rate tensile test are considerably increased if active, blank sites are generated on the surface during the plastic deformation of steels and the resulting and mobile dislocations can interact with pure hydrogen. In this way it is possible to transport a relatively high amount of hydrogen to the layers close to the edges of a steel even at low hydrogen pressures. This becomes evident by the reduction of necking of the steels and the formation of numerous incipient cracks. If plastic deformation occurs, the dislocation density is strongly increased, in particular in areas with stronger slip phenomena [80].

Since, obviously, the oxide-free surface areas are necessary for plastic flow to ensure the formation of a sufficient amount of diffusible hydrogen, the purity level of hydrogen plays a decisive role. This is shown by the results summarized in Figure 39 from tensile tests of Ck22 steel (Mat. No. 1.1151) in compressed hydrogen (100 bar) with different contents of contaminant gases.

Whereas the contents of nitrogen and argon in hydrogen practically show no influence on the embrittling effect of hydrogen, the elongation and the reduction of area values obtained during the tensile test in air are reached again at low contents of air or oxygen. The influence of low oxygen contents in the compressed hydrogen on the value in the tensile test of the steel Ck22 (1.1151) is shown in Figure 40.

Consequently, only at oxygen contents below $10^{-4}\%$ do fracture elongation and reduction of area drop to the low values obtained in pure compressed hydrogen, whereas oxygen contents of about 0.02% are already sufficient to prevent the negative effect of the hydrogen. The inhibiting effect of oxygen can be explained by the preferred absorption on the freshly formed iron surface compared to hydrogen.

In his first investigations W. Hoffmann detected already that the damaging effect of hydrogen increases in the pressure range from 1 bar to 150 bar according to a power function. Other examinations revealed that the damage in the range between 7 bar and 700 bar increases with the square root of the hydrogen pressure. Also the inhibiting effect of certain gases detected by Hoffmann in 1960, in particular that of oxygen, has been proven many times. Apart from oxygen, also oxide layers exert an inhibiting effect and, obviously, need to be broken up by straining them or pierced by a slip step before the adsorption and chemisorption of hydrogen can take place. Also other gases, such as oxygen, sulfur dioxide (SO_2), carbon monoxide (CO) and water, also exert an inhibiting effect, whereas carbon dioxide, nitrous dioxide (N_2O) and methane (CH_4) do not inhibit the absorption of hydrogen [80].

Diffusing hydrogen exerts the strongest influence on necking in the tensile test and the notch tensile strength. The elastic properties, the yield strength and frequently also the fracture strength of smooth specimens practically remain unaffected.

Since plastic deformation of the material and concurrent exposure to compressed hydrogen are necessary to initiate damage by molecular hydrogen, mainly two methods are mainly applied for examination purposes. One method is the slow strain rate tensile test (CERT test = constant extension rate tensile test) in a hydrogen atmosphere, providing information on the critical strain rate, the critical temperature ranges, the influence of the surface condition, the gas pressure and the gas purity. The other method comprises fracture-mechanical tests for investigating the hydrogen-induced crack growth. In particular, fracture-mechanical parameters, such as the stress intensity limit value K_{IH} at, or above, which incipient

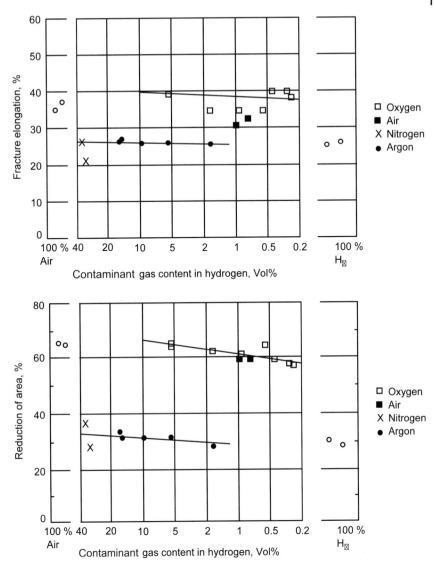

Figure 39: Influence of different contents of contaminant gases on the fracture elongation and the reduction of area of the steel Ck22 (1.1151) during tensile tests in hydrogen at a pressure of 100 bar at room temperature [75]

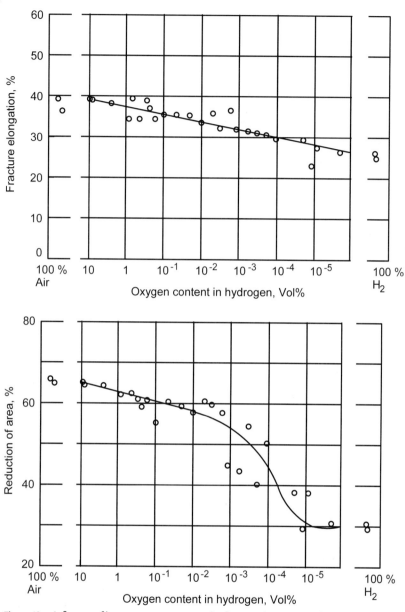

Figure 40: Influence of low oxygen contents on the fracture elongation and the reduction of area of the steel Ck22 (1.1151) during tensile tests in hydrogen at a pressure of 100 bar at room temperature [75]

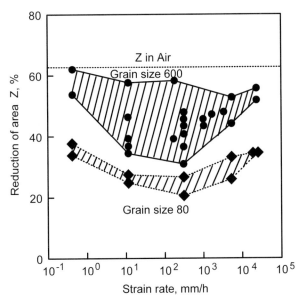

Figure 41: Influence of the surface and the strain rate on the reduction of area of a fine grain structural steel with a yield strength of 500 in compressed hydrogen at 100 bar [80]

cracks grow subcritically, or the propagation velocity of fatigue cracks da/dN depending on the cyclic stress intensity factor Δ K effective at the crack front and the frequency, facilitate predictions regarding the behavior of parts in hydrogen.

Sites at which preferably plastic deformation processes take place as a result of stress concentrations and a relating stress increase under mechanical load, e.g. penetration notches in the weld region or surface roughnesses, are often starting points for hydrogen attacks. Figure 41 shows the results of slow strain rate tensile tests of tensile test specimens from normalized fine grain structural steel with a yield strength of 500, former designation StE 500 (Mat. No. 1.8907, S500N, cf. UNS K02001). (Steels with this yield strength are no longer supplied in the normalized state today, but are produced either as tempered steels with the designation S500Q (Mat. No. 1.8924) or as thermomechanically rolled steels with the designation S500MC (Mat. No. 1.0984).) The figure depicts the influence of the selected strain rate on the necking loss Z (value in air: dashed line) if specimens with different surface roughnesses are exposed to hydrogen gas at a pressure of 100 bar [80–83].

In general, the behavior of the smoother material surface (finished with 600-grit sandpaper) is less critical and leads to quite a large scatter. The rougher surface (80 grit) causes a more marked decrease of necking and a much smaller scatter. It is remarkable that among the applied strain rates of very different magnitudes maximum embrittlement occurs between 10^2 mm/h and 10^3 mm/h. It can be concluded that there is a relationship between the rates of formation and movement of dislocations and of the dissociation of hydrogen and its transport into the metal, leading to an embrittlement maximum under optimal conditions.

The formation and movement of dislocations under the influence of hydrogen and elongation processes can be directly observed under a high-voltage electron microscope as shown by the examinations described in [84] performed on thin sheets of pressure vessel steel of type A 533 (B) (UNS K12539, 0.19% C, 1.28% Mn, 0.61% Ni, 0.55% Mo; cf.: Mat. No. 1.5403, 1.6310). These tests revealed that hydrogen does not only reduce the mobility of dislocations but also the stress required for deformations at the crack tip and for crack propagation.

Apart from the surface roughness notches and similar surface defects exert a clear influence on hydrogen damage. The test performed with round test bars shows a clear influence of the turning grooves where the necking loss (Z) increases with increasing roughness and grooves on the surface [81].

Test atmosphere	Exposure limit	Reduction of area, %	
		StE 51	StE 36
H_2, 100 bar	up to yield strength (point 1)	70	67
	up to 10% above yield strength (point 2)	60	65
	up to 15% above yield strength (point 3)	57	63
	up to maximum load (point 4)	51	51
	up to fracture (point 5)	25	25
Air	up to fracture (point 5)	69	68

Figure 42 shows the influence of hydrogen for two fine grain structural steels in the elastic and plastic deformation ranges. First, slight embrittling effects only oc-

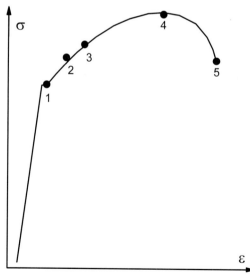

Figure 42: Influence of hydrogen on the reduction of area of tensile test specimens in the elastic and plastic ranges [81].

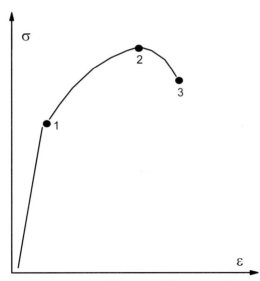

Figure 43: Influence of hydrogen sulfide on the reduction of area of tensile test specimens in the elastic and plastic ranges [82]

cur in the range between yield strength and maximum load. A clear influence reflected by the strong decrease of the reduction of area can be found between the maximum load and fracture. Similar results are also found under loading in dry hydrogen sulfide (H_2S) as shown in Figure 43.

Test atmosphere	Exposure limit	Reduction of area, %
H_2, 16 bar	up to yield strength (point 1)	65
	up to maximum load (point 2)	63
	up to fracture (point 2 to 3)	40
	up to fracture (point 3)	35
Air	up to fracture (point 3)	69

Examinations with the raster electron microscope show that cleavage-type and intercrystalline fractures occur depending on the steel grade; frequently also steps are found, reflecting the stepwise advance of the crack front. The intercrystalline type is preferably detected in tempered steels.

In the course of the hydrogen-induced damage process the absorption of hydrogen first reduces the deformability of a steel reflected by the clear reduction of necking and elongation in the tensile test and becoming more and more marked with increasing strength and hardness of the steels as shown in Figure 44.

The tests described in [85] aimed at examining low-alloy steels used for the manufacture of transport and storage tanks for compressed hydrogen to establish the influence of the surface conditions (ground and electropolished), the alloy compo-

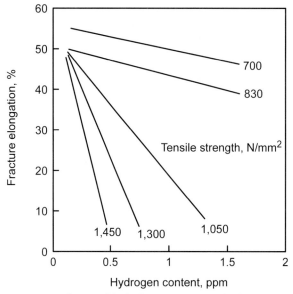

Figure 44: Influence of the hydrogen content on the fracture elongation of low-alloy steels with different strengths [81]

sition, the strength as well as the magnitude and type of load (dynamic, cyclic) on the susceptibility of the examined steels to damage by compressed hydrogen.

The examined steels and their chemical compositions are summarized in Table 31.

By varying heat treatment the three steels were provided with different strength values. Relevant details are contained in Table 32. Each material was provided in three overlapping tempering grades such that, apart from the strength, also the influence of the alloy composition could be examined.

The CERT tests were performed on round test bars both unidirectionally until fracture with strain rates of 1×10^{-7} 1/s to 221×10^{-7} 1/s and cyclically with saw-tooth characteristics. The following load types were applied in the cyclic tests at the constant strain rate of 221×10^{-7} 1/s:

1. $\sigma_{max} = 0.5\,(R_{eL} + R_m)$ $\sigma_{min.} = 0.8$ to $0.95\,\sigma_{max.}$
2. $\sigma_{max} = \sigma_{(F=0)}$ $\sigma_{min.} = 0.8$ to $0.95\,\sigma_{max.}$
3. $\sigma_{max} = \sigma_{(\%\varepsilon ReH)}$ $\sigma_{min.} = \sigma_{(F=0)}$

Steel	C	Si	Mn	P	S	Al	Cr	Ni	Mo	UNS
38Mn6 (Mat. No. 1.1127)	0.36	0.21	1.30	0.017	0.013	0.0074	0.07	0.05	0.01	
34CrMo4 (Mat. No. 1.7220)	0.33	0.21	0.62	0.029	0.013	0.085	0.96	0.024	0.19	cf. G41350
30CrNiMo8 (Mat. No. 1.6580)	0.30	0.25	0.41	0.011	0.0015	0.017	1.90	1.97	0.34	cf. G43400

Table 31: Chemical compositions of the examined steels [85]

A 14 Unalloyed and low-alloy steels/cast steel

Steel	Specimen designation	Heat treatment	R_e N/mm²	R_m N/mm²	Z %
38Mn6 (Mat. No. 1.1127)	A	30 min 860 °C/H₂O 60 min 690 °C/air	490	684	59
	B	30 min 860 °C/H₂O 60 min 650 °C/air	578	765	54
	C	30 min 850 °C/H₂O 60 min 600 °C/air	795	906	48
34CrMo4 (Mat. No. 1.7220)	A	30 min 850 °C/H₂O 60 min 620 °C/air	889	983	57
	B	30 min 850 °C/H₂O 60 min 660 °C/air	776	884	64
	C	30 min 850 °C/H₂O 60 min 725 °C/air	631	753	64
30CrNiMo8 (Mat. No. 1.6580)	A	30 min 840 °C/H₂O 60 min 570 °C/air	1,128	1,188	62
	B	30 min 840 °C/H₂O 60 min 610 °C/air	905	1,024	69
	C	30 min 840 °C/H₂O 60 min 660 °C/air	777	904	72

Table 32: Heat treatment and strength values of the examined steels [85]

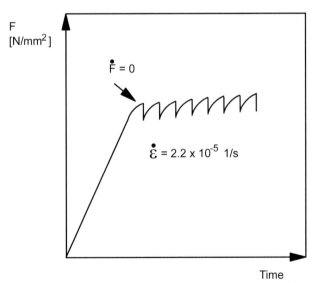

Figure 45: Typical load-time curve for alternating load tests with $\dot{F} = 0$ [85]

In Figure 45, $\dot{F}$ is the rate of load change and $\dot{F} = 0$ marks the beginning of uniform elongation according to the curve path.

The tests were carried out in ultrapure hydrogen (> 99.9999 vol%) and for comparison in ultrapure nitrogen (> 99.999 vol%) at a pressure of 100 bar and a temperature of 20 °C each. To evaluate the influence of hydrogen on fracture processes, the maximum tensile force, breaking load, reduction of area, relative reduction of area (ratio of reduction of area determined in hydrogen and nitrogen) and the brittle fracture component on the fracture surfaces were used.

For testing the influence of different surface roughnesses, specimens with an electropolished surface, notched (0.1 mm depth of notch) and subsequently electropolished surface as well as sanded surfaces of different grades (600 to 80 grit) were used.

The extent of damage turned out to be directly dependent on the surface roughness. The electropolished surface exhibited very minor damage hardly detectable even under the scanning electron microscope, with reduction of area values hardly impaired. Damage of the material caused by hydrogen increases with increasing roughness of the surface. The largest extent of damage was found in the specimens treated with the coarsest sandpapers (80 grit).

The examined materials did not exhibit any influence of the alloying elements manganese, chromium, nickel and molybdenum on the susceptibility to hydrogen. However, the degree of contamination by nonmetallic inclusions was important. On the one hand, inclusions on the surface act like notches, whereas on the other hand banded inclusions preferably serve as paths for hydrogen diffusion. This holds true, in particular, for manganese sulfides. The influence of material strength becomes detectable only at a value $R_m > 1{,}000$ N/mm².

A very clear influence of the strain rate was established during the dynamic test procedure at a constant strain rate until fracture. The damaging effect of hydrogen increases with increasing strain rate. Within the examined range of strain rates from 10^{-7} 1/s to 221×10^{-7} 1/s the fastest strain rate of 221×10^{-7} 1/s turned out to be the most critical rate for all materials.

The tests with specimens under alternating load were performed at the strain rate of 221×10^{-7} 1/s determined to be critical for all materials. Alternating load tests can be performed either as load-controlled or as strain-controlled tests.

Load-controlled swelling tests, during which the upper load was defined by the relation $\sigma_{max} = 0.5\,(R_{eL} + R_m)$, did not yield any satisfactory differentiation between the tests in hydrogen and in nitrogen since the dynamic-plastic deformation required for gaseous hydrogen to diffuse into the metal lattice only occurs during the first load cycles. Due to strain hardening of the specimens under tensile stress, the constant upper load is shifted into regions below the yield strength and any further load cycles occur in the exclusively elastic range.

In contrast, alternating loading with a variable upper load defined by the relation $\sigma_{max} = \sigma_{(F=0)}$ causes plastic deformation of the material at each load cycle. As a result, the effect of hydrogen becomes more visible. Figure 46 summarizes the results.

As the measure for determining the influence of hydrogen on the specimens the relative service life $t_{H2}/t_{N2} \times 100$, where t_{H2} = time to failure in hydrogen and

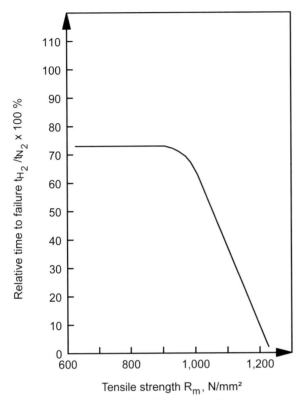

Figure 46: Dependence of the relative service life on strength under alternating load in hydrogen [85]

t_{N2} = service life in nitrogen, was selected. In general, the influence of the loading amplitude turned out to be low and a clear reduction of the relative service life of the specimens in hydrogen can be found only for materials with strength values of $R_m > 1{,}000$ N/mm².

Strain-controlled swelling tests were performed using the steel 34CrMo4 (1.7220) with the three strength values indicated in Table 32 and again at the strain rate of 221×10^{-7} 1/s according to the load characteristics depicted in Figure 47 such that an elongation with a plastic portion (125–280 ε_{ReH}) with reference to the yield strength elongation was used as the upper yield strength.

The number of cycles withstood at loading amplitudes in a range from 115% to 300% of the elongation ($\dot{\varepsilon}_{ReH}$) occurring at the upper yield strength (R_{eH}) is shown in Figure 48.

A clear difference between the results obtained in nitrogen and in hydrogen can be also found among the specimens with the lowest strength. The specimens with the highest strength do not exhibit any difference. Obviously, exclusive mechanical material fatigue masks hydrogen effects, if any, at the selected load characteristics to a major extent.

Hydrogen

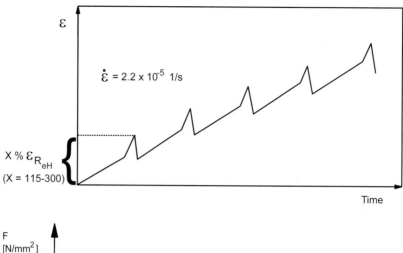
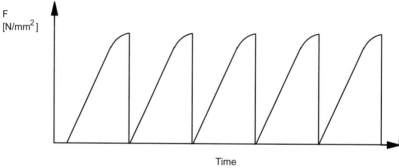

Figure 47: Load characteristics in strain-controlled swelling tests (strain rate with reference to the elongation upon reaching the upper yield strength R_{eH} = 100%) [85]

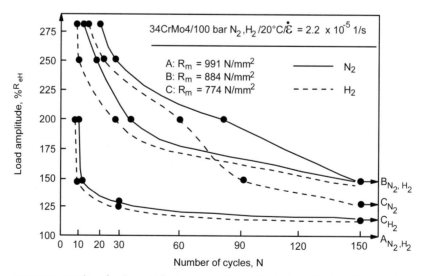

Figure 48: Number of cycles until fracture depending on the load amplitude during a strain-controlled test procedure [85]

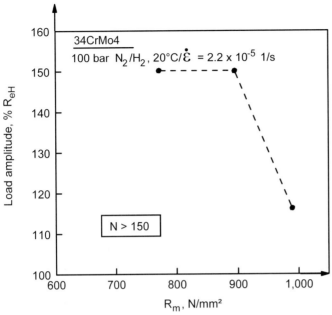

Figure 49: Influence of strength on the amount of the load amplitude withstood after more than 150 load cycles [85]

Evaluating the results shown in Figure 49 to establish at what load amplitude the material, depending on strength, would be able to withstand at least another 150 strain-controlled cycles, it turns out that up to a strength of about 900 N/mm² load amplitudes of about 150% of the elongation (at the upper yield strength) cause fracture of the specimens above 150 load cycles only. At higher strength values the load amplitude permissible for the selected criterion decreases to values around 115%.

In the course of these examinations also the influence of oxygen and other gases, which may occur as potential impurities in technically pure hydrogen, on the susceptibility of higher-strength steels to compressed hydrogen damage was investigated. To this end, CERT tests were performed with the stainless steels listed in Table 31 and in Table 32 in compressed hydrogen at a pressure of 100 bar at 20 °C using a strain rate of 10^{-7} 1/s [86]. The contaminant gas content of hydrogen was determined by means of a highly sensitive quadrupole mass spectrometer facilitating the determination of an oxygen content of less than 1 vpm O_2 in hydrogen at 100 bar.

The susceptibility of the specimens to hydrogen-induced damage was assessed at the end of the tests by analyzing the reduction of area, the occurrence of secondary cracks in the necking regions of specimens and the brittle fracture content on the fracture surface. The results were compared with reference values obtained during tensile tests in pure nitrogen at 100 bar.

The test results confirmed the known effect that low oxygen contents in compressed hydrogen inhibit the absorption of hydrogen atoms on the steel surface and reduce the risk of hydrogen-induced damage to steels. At contents of 10 vpm the inhibiting effect becomes evident, however such low contents are not sufficient to prevent cracking, in particular in steels with a higher strength. In most cases 100 vpm oxygen were enough for the tensile test specimens to reach a reduction of area of at least 90% of the value measured in nitrogen. Only steels with a tensile strength above 1,000 N/mm² exhibited a clear increase of the susceptibility to hydrogen damage. For steels of lower strength an influence of the mechanical values or the alloying elements on the susceptibility to hydrogen damage was not detectable. Also for the high-strength types 100 vpm oxygen were sufficient to prevent secondary cracks in the necking region of the specimens.

The effects of contamination of hydrogen with lower contents of carbon monoxide, carbon dioxide, sulfur dioxide and ammonia were only tested using the steel 34CrMo4 (Mat. No. 1.7220, cf. UNS G41350). The inhibiting effect of these gases strongly varies depending on the strength of the steel, and decreases in the following sequence:

$$NH_3 \ll CO < SO_2 \ll O_2$$

Tests regarding the influence of compressed hydrogen on the fracture behavior of the tempered steel 34CrMo4 (Mat. No. 1.7720) when loaded in the low cycle fatigue range are also reported in [87]. Strain-controlled alternating tests (tension-compression tests) were performed with elongations of about 1.7% and 2.8% at different load frequencies in hydrogen and in nitrogen at a pressure of 150 bar each. As shown in Figure 50 the number of load cycles withstood until fracture clearly decreased in hydrogen.

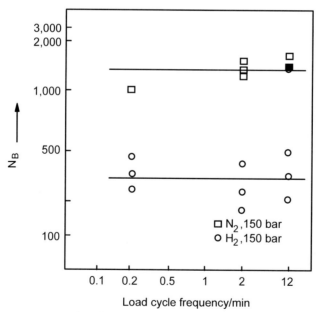

Figure 50: Number of cycles to failure N_B at various load cycle frequencies for the steel 34CrMo4 (1.7720) in nitrogen and in hydrogen (elongation ≈ 2.8%) [87]

Steel	C	Si	Mn	P	S	Cr	Mo	R_{eL} N/mm²	R_m N/mm²	Micro-structure
34CrMo4 (1.7720)	0.32	0.13	0.55	0.025	0.015	1.03	0.20	835	935	tempered
C 75 (1.0605)	0.75	0.18	0.63	0.013	0.021				1,400	pearlite

Table 33: Composition, strength and microstructure of the examined steels [88]

On average the specimens reached a time to failure in nitrogen which was about four times higher than that of the specimens in hydrogen. Tests with lower elongation ($\approx$ 1.7%) yielded comparable results. Although the number of load cycles until fracture almost tripled, there were still differences between nitrogen and hydrogen. The load frequency did not exert any influence on the results in the range under investigation.

Both steels 34CrMo4 (Mat. No. 1.7720, tempered) and C 75 (Mat. No. 1.0605, cold-drawn, cf. SAE 1074) for which compositions, strengths and microstructures are indicated in Table 33 were also used to analyze the influence of precharging with hydrogen gas on the fracture behavior [88]. For hydrogen charging the surfaces of the specimens were first activated with pyrophorous iron powder and thereafter exposed to hydrogen gas at a pressure of 150 bar for 500 h. The specimens were charged until saturation in equilibrium with the external compressed hydrogen of 150 bar. The details of this charging procedure are described in [89].

The reduction of area values obtained for the tensile test specimens at a deformation velocity of 0.4%/min as a measure for hydrogen embrittlement are summarized in Table 34.

During the test in compressed hydrogen both steels exhibited the expected decrease in reduction of area, however their response to hydrogen charging was very different due to their different microstructure. Whereas an effect of charging cannot be detected for 34CrMo4, C 75 shows a strong reduction of the necking values of the charged specimens. This very different behavior may be attributable to the microstructure of both steels. Whereas 34CrMo4 was tested in the tempered and non-deformed state, the pearlitic C 75 steel was a strongly cold-worked steel. Hence, its microstructure is expected to contain a higher number of hydrogen traps and therefore higher contents of dissolved hydrogen. An analysis of the equilibrium contents at a hydrogen pressure of 150 bar revealed a hydrogen content of

Specimen type/test atmosphere	34CrMo4 (1.7720)	C 75 (1.0605)
Uncharged/air	Z = 65%	Z = 46%
Uncharged/hydrogen	Z = 35%	Z = 35%
Charged/air	Z = 65%	Z = 25%
Charged/hydrogen	Z = 35%	Z = 25%

Deformation velocity: 0.4%/min

Table 34: Values of reduction of area Z obtained from tensile tests of charged and uncharged specimens in air and in hydrogen at 150 bar

0.21 Ncm³/100 g (0.19 ppm) for 34CrMo4, however, a content of 1.15 Ncm³/100 g (1.02 ppm) for C 75 of which about 45% are recombined molecular hydrogen and can serve as an internal hydrogen reservoir during deformation.

CERT tests were performed to examine the question of corrosion risks of compressed pipes for hydrogen-containing synthesis gases using higher-strength pipe steels X 65 (Mat. No. 1.8975) and X 70 (Mat. No. 1.8977) as well as the fine grain structural steel StE 500 (Mat. No. 1.8907, S500N). Although natural gas containing 10% hydrogen at a total gas pressure of 70 bar was clearly less aggressive than pure hydrogen at the same pressure, secondary cracks and brittle fracture content were found on all surfaces in all specimens ruptured in the presence of hydrogen after the uniform elongation had been exceeded. The addition of hydrogen at a partial H_2 pressure of about 5 bar to natural gas did not lead to material failure of the examined steels, however it may cause damage under the condition of plastic deformation or alternating loads. However, this should not play a role if a pipeline has been duly designed and is properly operated [80, 90, 91].

The influence of a notch and its role when increasing stress in connection with hydrogen embrittlement was also analyzed in slow strain rate tensile tests using a tempered, low-alloy nickel-chromium-molybdenum steel with the nominal composition indicated in Table 35 [92]. The specimens were austenitized at 840 °C in an argon atmosphere for 1 h, quenched in oil, tempered in argon at 300 °C for 1 h and then cooled in air. The result was a microstructure of predominantly tempered martensite, a tensile strength of 1,720 MPa, a crack resistance of 2,500 MPa, reduction of area of 50% and an elongation of 11%.

C	Mn	Ni	Cr	Mo
0.35–0.45	0.45–0.70	1.3–1.8	0.4–1.4	0.2–0.35

Table 35: Nominal composition of the examined steel [92]

Both notched and unnotched tensile test specimens were tested. The notched specimens were provided with a circumferential 60° V-notch in the middle of the inspection line. Three groups of specimens were tested differing in their outer diameter D, diameter d at the notch root and radius R at the notch root as indicated in Table 36.

The slow strain rate tensile tests were carried out at a strain rate of $\dot{\varepsilon} = 1.2 \times 10^{-5}$ 1/s in purified hydrogen gas at a pressure of 1 bar.

Compared to the specimens tested in air, the tensile strength of the smooth specimens was reduced only slightly (about 2%) during the tensile test in hydrogen,

Group of specimens	D mm	d mm	R mm
A	3.4 to 4.0	3.17	0.21
B	3.95	3.2 to 4.4	0.21
C	3.95	3.17	0.032 to 0.32

D = outer diameter, d = diameter at notch ground, R = radius at notch ground

Table 36: Details of notch geometry of notched tensile test specimens [92]

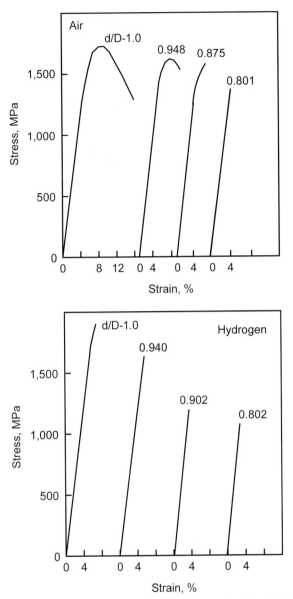

Figure 51: Typical stress-strain curves for smooth and notched tensile test specimens with different d/D values during the tensile test in air and in hydrogen [92]

however the values of both reduction of area and fracture elongation dropped to zero and the reduction of the fracture stress was 34%.

The stress-strain curves in Figure 51 show a clear influence of the notch on the strain behavior also during the tests in air. Plastic deformation decreases with increasing notch factor. During the tests in hydrogen the notched specimens no longer show any plastic deformation and a linear curve path with exclusively elastic deformation of the specimens is obtained. Independent of further details of the notch geometry the highest hydrogen embrittlement was detected in specimens with a ratio $d/D < 0.87$.

These research results confirm the unfavorable influence of surface defects on the susceptibility of a component to damage by gaseous hydrogen. The fresh metal surfaces generated during the flow processes at the notch root facilitate, and do not only enhance, the dissociation of the hydrogen molecule and the absorption of the hydrogen atoms, and the increased stresses and the multiaxial stress conditions in front of the notch tip also lead to the intensified diffusion of the absorbed hydrogen in this area (Gorsky effect: transport into the direction of increasing stress).

Although the damaging influence of compressed hydrogen can be detected in the tensile test, reduction of area and notch tensile strength, i.e. the most affected parameters, do not permit any conclusion regarding the behavior of a component exposed to hydrogen. Due to the complex processes during the absorption of hydrogen, material damage to components at room temperature and atmospheric temperatures is possible only if the surface areas are plastically deformed. In components subjected to normal stress, this precondition is only fulfilled in the area of defects of any kind. In particular, periodical load changes have an impact if fatigue processes can occur in surface defects of a critical size. Fracture-mechanical examinations help to analyze and assess the connection between the size and the type of surface defects and the loading conditions resulting from design and operation.

The typical crack growth curve depicted in Figure 52 for alternating loads shows the connection between the crack growth rate da/dN and the amplitude of the

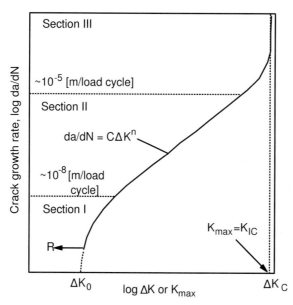

Figure 52: Schematic crack growth curve for a fatigue crack without specific influence of the environment [80]

stress intensity factor ΔK for the case of a fatigue crack propagating in an inert, non-aggressive environment.

The curve path can be divided into three sections.

In Section I crack growth can be proven experimentally if a specific threshold value ΔK_0 is exceeded. In steels this value depends less on the material and more on the cyclic stress ratio R (K_{min}/K_{max}). For Section II the familiar Paris' law (da/dN = C ΔK^n) describes the relationship between da/dN and ΔK.

In Section II crack propagation under cyclic stress can accelerate under the influence of a hydrogen-specific environment if the maximum K value at the crack tip exceeds a certain threshold value $K_{IH\ cycl.}$ which is typical for the relevant material. $K_{IH\ cycl.}$ is a specific parameter depending less on the mechanical strength of the relevant material.

Accelerated crack growth occurs in Section III. The crack grows supercritically if K_{max} has reached the value of the fracture toughness of the material.

Also under static load subcritical crack growth may occur under the influence of hydrogen if a threshold value (K_{IH}) is exceeded. This K_{IH} value rapidly decreases with increasing strength such that crack propagation in high-tensile steels becomes possible even in the presence of relatively small defects. As can be seen from Figure 53, the cyclic threshold value $K_{IH\ cykl.}$ is clearly lower than the threshold value K_{IH} found under static load.

Apart from the low dependence on strength a change in the crack morphology from transcrystalline to intercrystalline was observed in CrMo alloyed steels under static load [81].

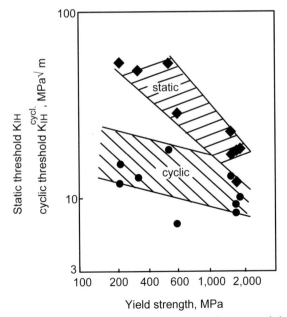

Figure 53: Threshold values K_{IH} and $K_{IH\ cykl.}$ under static and alternating loading [81]

82 | Hydrogen

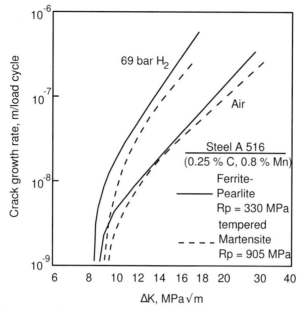

Figure 54: Results of fatigue tests of a type A 516 steel in air and in compressed hydrogen [81]

The influence of hydrogen on the crack growth of a low-alloy C-Mn steel (type A 516, cf.: Mat. No. 1.0346, 1.0356) is shown in Figure 54. The amplitude of the stress intensity factor ΔK in MPa $\sqrt{m}$ is plotted against the crack growth rate da/dN. Under the influence of hydrogen the velocity of crack propagation shifts to smaller ΔK values. It can be also seen that heat treatment involving a clear increase of the yield strength has only a minor effect.

Figure 55 shows the results of measurements performed with different pressure vessel steels in compressed hydrogen at 150 bar and at a load cycle frequency of 1 Hz.

For the range of mean ΔK values the following empirical relation can be developed for the fatigue crack rate da/dN (m/load cycle) in compressed hydrogen:

$$da/dN = 7.5 \times 10^{-10} \Delta K^{2.25}$$

with $\Delta K = \Delta\sigma \times Y \sqrt{\pi a}$

the number of load cycles N required for a specific crack propagation ($a_1 \rightarrow a_f$) can be estimated.

The following holds true:

$\Delta\sigma$ = amplitude of alternating load
y = geometry factor
a = flaw depth

The influence on the crack growth of precracked specimens from a C steel in hydrogen is reflected in Figure 56. The influence of different test atmospheres on

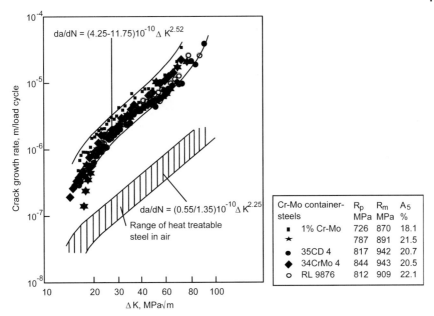

Figure 55: Results of fatigue tests of different pressure vessel steels in compressed hydrogen at 150 bar [81]

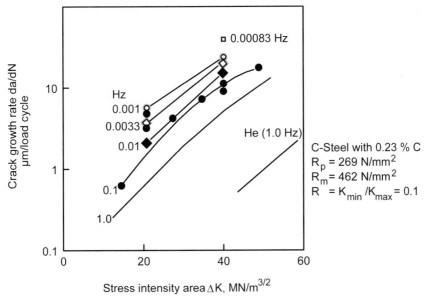

Figure 56: Influence of the frequency on the crack growth rate of a C steel under compressed hydrogen at 1,000 bar and at room temperature [81]

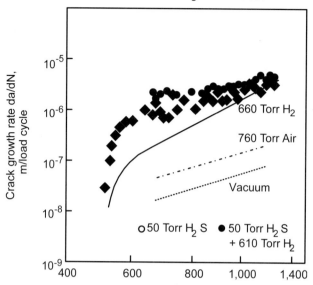

Figure 57: Influence of different test atmospheres on the crack growth of the steel 19Mn3 [81]

the crack growth of the steel 19Mn3 (similar to 17Mn5) is summarized in Figure 57.

The major finding is that both in a hydrogen and in a hydrogen sulfide atmosphere fine incipient cracks can rapidly grow under alternating mechanical load of a low frequency, whereas in air or in an inert gas only very small growth rates are observed.

The technical importance of that examination is that steel surfaces with notches resulting from production can show a critical behavior in a hydrogen atmosphere regarding the occurrence of hydrogen-induced brittle fractures if they are subjected to low-frequency alternating loads. Such loads may be, for instance, varying pressure or repeated slow filling and emptying process.

Damage to carbon steels by hot compressed hydrogen at high pressures and temperatures above about 200 °C is attributable to decarburization, on the one hand, and to the formation, growth and fusion of methane gas bubbles, on the other hand. With increasing temperatures the equilibrium of the Equation 6 shifts more and more into the direction of dissociation to form hydrogen atoms. The atomic hydrogen generated is absorbed on the surface in an adsorption process and by interacting with the electrons of the metal lattice. A typical phenomenon of compressed hydrogen damage to unalloyed and low-alloy steels at high temperatures is the chemical reaction of the diffused hydrogen with dissolved carbon ($C + 4H \rightarrow CH_4$) and with carbon compounds (Fe_3C, cementite) of the material in the reaction according to Equation 8.

These reactions do not only lead to decarburization and hence to a loss of tensile strength of the steel, but also to the accumulation of the insoluble methane gas in the steel in internal voids, building up high pressures such that pores with large volumes are formed and internal material separations occur even without the influence of external forces. This leads to fractures with little deformation as an external damage pattern.

Consequently, compressed hydrogen can damage steels in two ways. At high temperatures and low partial hydrogen pressures surface decarburization occur preferentially, not causing crack formation. Lower temperatures, however above 200 °C, and high hydrogen partial pressures enhance the internal decarburization and the formation of internal cracks, which may also lead to fractures. At high temperatures and high pressures, both damage types occur.

In general, migration of carbon to the surface at high temperatures and its reaction there with hydrogen to form methane is considered to be the mechanism facilitating surface decarburization. Carbon depletion reduces the strength properties of the steel in the surface area.

During internal decarburization and the formation of internal cracks the atomic hydrogen present on the surface diffuses into the material and forms methane, which is not able to diffuse out of the material and accumulates at the grain boundaries and in internal voids. Internal cracks decisively impair the mechanical properties of the material.

The diagrams in Figure 58 and in Figure 59 illustrate the effects of damage by hot compressed hydrogen across the wall cross section of a defective pressure vessel from the beginning of high-pressure hydrogenation [93].

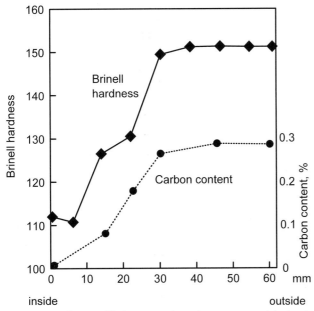

Figure 58: Influence of hydrogen on the carbon content and the hardness values of a damaged pressure vessel on unalloyed carbon steel [93]

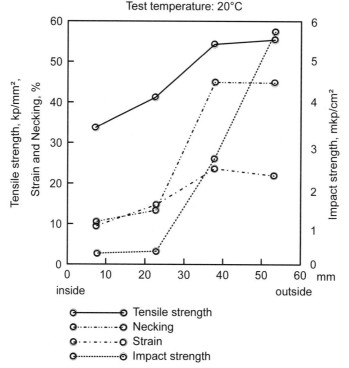

Figure 59: Influence of hydrogen on the mechanical properties of a damaged pressure vessel from unalloyed carbon steel [93]

As can be seen from the development of the carbon content and the hardness values, the hydrogen attack has already reached half of the 60 mm thick vessel wall. Also the strength and toughness values reduce in the decarburized area accordingly, impairing the toughness values, in particular notch toughness, more than the strength.

From 1910 to 1913 damage of carbon steels by hot compressed hydrogen was observed and assessed for the first time during the commercial high-pressure synthesis of ammonia using the Haber-Bosch process. At that time this problem was solved by design measures in that the required pressurized reactors were manufactured as multilayer reactors. For the inner core tube of the reactor unalloyed, soft iron with a very low carbon content was selected, through which hydrogen could diffuse without causing internal damage. The pressure-retaining external steel sheath was provided with numerous holes (Bosch holes), through which the hydrogen leaving the internal tube and recombining in the gap between the internal and the external tube could escape without pressure [94].

After the damage mechanism had been clarified, a way could be found to encounter this problem by using alloying measures during the production of steel. Basic examinations of the influence on alloying elements of the resistance of steels to compressed hydrogen were described already in 1938 [95]. A difference was

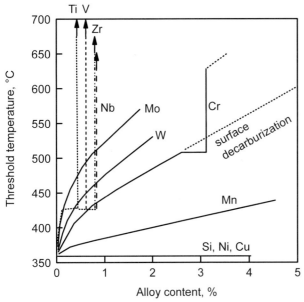

Figure 60: Influence of various alloying elements on the threshold temperature of the resistance of steels with 0.1% carbon in compressed hydrogen at 295 bar [95]

made between the non-carbide-forming alloying elements silicon, nickel, copper, manganese, the carbide-forming elements manganese, chromium, tungsten exhibiting considerable solubility in iron carbide and the special carbide-forming elements vanadium, niobium and tantalum, titanium and zirconium. Figure 60 summarizes the major results of theses studies. Resistance was assessed by means of the reduction of area determined at the tensile test specimens following aging in hydrogen over a period of 100 h at 295 bar and at different temperatures. The temperature during which a drop of the necking values was noticed was selected as the threshold temperature.

The non-carbide-forming elements silicon, nickel and copper do not exert any influence on the hydrogen resistance of the steels. The elements chromium, molybdenum and tungsten cause a strong increase of the hydrogen resistance constantly increasing with increasing content. Since these elements dissolve well in iron carbide, their effect is not only attributable to the stability of their own carbides but also to the stabilization of the iron carbide. Chromium steels exhibit a rapid increase in their resistance from a certain chromium content depending on the carbon content. There is a relationship between this phenomenon and the occurrence of the special carbide $(Cr,Fe)_7C_3$ and the related disappearance of the carbide $(Fe,Cr)_3C$. Manganese does not form any special carbide but also dissolves well in iron carbide leading to a minor increase in the resistance only. The alloying elements titanium, vanadium and zirconium with low solubility in iron carbide also exhibit a rapid increase in their resistance from a certain content, however steels with such high contents of these elements are otherwise disadvantageous in terms of processing and use.

Therefore, the compressed hydrogen resistant steels containing the carbide-forming alloying elements chromium, molybdenum or tungsten which, in contrast to cementite, tie up carbon in the form of much more stable carbides. These compressed hydrogen resistant steels suffer decarburization at temperatures below 450 °C only in the subsurface region. Short-term loading at high temperatures (1,000 h, 600 °C) also leads to a decarburization depth of about 0.5 mm only, which is technically acceptable in most cases. However, grain boundary cracks may occur within the decarburized zone which can widen to form cracks as a result of the internal pressure of the components. This may reduce the creep rupture elongation. For further information on the properties, heat treatment and processing as well as their special applications reference is made to [94, 96].

The damage process due to compressed hydrogen at elevated temperatures and the resistance of the steels can be described by several typical properties [97].

Usually, this process involves a longer incubation period, during which the mechanical properties of steel at room temperature do not change or change only slightly. Then, these properties noticeably worsen, in particular toughness and ductility. The strength and yield strength are impaired only after the damage process has progressed further, this impairment being only of a minor extent. The incubation period becomes rapidly shorter if the temperature is increased. Therefore, any steel exhibits a threshold temperature at a given hydrogen pressure and a limiting pressure at a specific temperature, below which the incubation period is longer than the operating life of the part, i.e. a hydrogen-induced damage is not expected during the operating life. This threshold temperature permits the evaluation of the material's resistance and is the basis of the **Nelson diagram**. A common method for testing the susceptibility of a steel to an attack by compressed hydrogen is therefore aging specimens under constant hydrogen pressures at increasing temperatures and subsequently testing their mechanical properties. Usually, a complete loss of ductility is found within a temperature interval of 50 °C.

Even in unalloyed carbon steels an attack is not found, unless the partial hydrogen pressure exceeds 1 MPa. At higher hydrogen pressures the application temperature drops very steeply, this decrease being slow as the pressure increases further (Figure 64).

Basically, chromium and molybdenum are used as alloying elements forming stable carbides and, hence, improve the resistance to decarburization and methane formation. Reduction of the carbon content is not an effective measure since also steels with only a few hundredths of a percent carbon can be damaged.

The best behavior can be expected from tempered CrMo steels, however also in these steels the behavior in the weld region and after cold forming can be impaired. An example are the examinations described in [98, 99] regarding the behavior of welded joints of two tempered steels of type 2.25% Cr and 1% Mo. The chemical compositions of the base material and of the welding material are indicated in Table 37. This material type corresponds, to a great extent, to the steel 10CrMo9-10 (Mat. No. 1.7380) according to DIN EN 10028-2 [100].

The welding material 1/S was used for gas shielded welding and the welding material S-2 was used for submerged arc welding.

	C	Mn	Si	P	S	Al	Cr	Mo	Cu	As	Sb	Sn
G-1	0.13	0.51	0.25	0.006	0.008	0.019	2.32	0.98	0.13	0.031	0.002	0.012
G-2	0.15	0.56	0.38	0.005	0.005	0.003	2.40	0.90	0.12	–	0.002	0.011
1/S	0.055	0.69	0.30	0.010	0.012	0.014	2.42	1.01	0.14	0.022	0.002	0.012
S-2	0.084	0.074	0.30	0.009	0.005	0.007	2.25	0.97	0.12	–	0.002	0.008

G = base material S = welding material

Table 37: Chemical compositions of the base material and the welding material of the examined 2.25%Cr-1%Mo steels [98]

The test was performed in hydrogen at pressures of 10 bar to 31 bar and at temperatures from 460 °C to 590 °C. The influence on the test specimens taken separately taken from the base material, the weld metal and in case of the G-1/1/S combination also from the heat affected zone, were assessed with a highly sensitive dilatometer on the basis of the volume increase resulting from the formation of methane gas bubbles and indicated as the strain rate. The strain rate of control specimens tested at 575 °C and low hydrogen pressure (2 MPa) was lower than the

Temperature	Hydrogen pressure	Strain rate			
°C	MPa	10^{-8} 1/h			
		G-1	G-2	1/S	S-2
460	20.3			6.9	
480	20.3			3.9	
500	20.3	1.4; 1.9	< 1; < 1; 1.3	15.6; 15.6	1.2
520	20.3		2.5	35.7	
540	20.3			47.9	
550	20.3	3.8; 10.9	6.9; 6.2; 11.4		10.3; 21.0
565	20.3	10.2; 20.8		22.7	
570	20.3			20.9	28.0
580	20.3	19.8	19.4; 22.6		
590	20.3	37.8			
500	16.9			4.8	
500	32.7			20.6	
550	13.5		3.3; 3.4		3.7
550	16.9		4.2		7.6; 8.3
550	23.7				23.4
580	16.9	12.9	10.0		
580	13.5	7.9	3.6		

Table 38: Strain rates for evaluating the methane formation in the base material and in the weld metal of both examined 2.25%Cr-1%Mo steels [98]

Test temperature °C	Hydrogen pressure MPa	Strain rate 10^{-8} 1/h
530	20.5	41
550	20.5	11; 71
550	24.5	20
550	28.0	33
550	31.0	55
565	20.5	50
580	17.5	86
580	20.5	87; 184
580	24.5	164
580	27.5	301

Table 39: Strain rates for evaluating the methane formation in the heat affected zone of one of the two examined 2.25%Cr-1%Mo steels [99]

resolution of the dilatometer of 10^{-8} 1/h. Hence, it was ensured that the values measured at higher hydrogen pressures were attributable to the reaction of hydrogen. Following a test duration of about 100 h the strain value became constant and was used for evaluation. The results summarized in Table 38 and Table 39 show that at identical temperatures and pressures the hydrogen attack in the weld material is clearly stronger than in the base material. In these tests the heat affected zone turned out to be the area with the most sensitive response to the hydrogen attack. The attack was much more influenced by the temperature compared to the hydrogen pressure.

The examinations described in [101] show that, by increasing the contents of chromium to about 3% and of molybdenum to about 1.5%, the response to hot compressed hydrogen could still be clearly improved. For the tests sheets welded using a submerged arc welding process with the composition indicated in Table 40 were selected. The results were compared with the values obtained with a 2.25% Cr-1%Mo steel. The composition of this steel is also indicated in the table.

The 100 mm thick sheets of the 3Cr-1.5Mo steel were austenitized at 955 °C, quenched in water and tempered at 705 °C for 20 h. From the hardness values a tensile strength of 590 MPa was calculated. The welded specimens were stress-relieved at 635 °C for 4 h, resulting in a strength of 740 MPa. The 2.25Cr-1Mo steel

Steel	C	Mn	Si	P	S	Cr	Mo	Ni	Cu	V
3Cr-1.5Mo	0.12	0.84	0.27	0.011	0.002	2.86	1.48	0.14	0.06	0.09
3Cr-1.5Mo*	0.054	1.10	0.34	0.005	0.002	2.93	1.40	0.18	0.006	0.08
2.25Cr-1Mo	0.13	0.54	0.23	0.01	0.01	2.23	0.95	0.18		

* welding material

Table 40: Chemical composition of the examined steels [101]

was austenitized at 910 °C, quenched in water and tempered at 690 °C for 12 h to a strength of 590 MPa. Specimens from the base material, the weld metal and the heat affected zone were aged in hydrogen at a temperature of 600 °C and a pressure of 13.8 MPa up to 400 days.

Some of the specimens were also aged in hydrogen under tensile stress. The specimens from the welded sheets were taken such that the area of the weld metal and the area of the heat affected zone could be examined. Table 41 lists the various test conditions. Also specimens from the base material following cold forming by 33% were tested.

Specimen	Test temperature °C	Hydrogen test pressure MPa	Test stress MPa	Aging time h
B	600	13.8	0	96,000
B	600	13.8	0	4,800
B	600	13.8	110	304
B	600	13.8	96	599
B	600	13.8	83	1,186
B*	600	13.8	0	2,400
W	600	13.8	0	4,800
HAZ	600	13.8	0	960

B = base material, W = weld material, HAZ = heat affected zone, * = 33% cold forming

Table 41: Test conditions for aging the specimens from 3Cr-1.5Mo steel [101]

Following their aging, 15 sites on the specimens were examined in the scanning electron microscope and the number of the bubbles formed as a result of methane formation at the grain boundaries was determined. Bubbles were not found in any of the 3Cr-1.5Mo steel specimens, whereas bubbles occurred in the 2.25Cr-1Mo steel already after an aging period of a few days under the same test conditions. Prestressing of the specimens from the 2.25Cr-1Mo steel led to significanctly increased bubble formation and the reduction of the creep strength as well as fracture toughness. Figure 61 shows the results obtained from the base material specimens from both steels.

Whereas cold forming of the 2.25Cr-1Mo steel increased the susceptibility to compressed hydrogen, bubbles were not found in the cold formed specimens from 3Cr-1.5Mo steel (Figure 62).

Also the specimens from the weld metal and the heat affected zone from the 3Cr-1.5Mo steel did not exhibit any bubbles after a test duration of 200 days (weld metal) and 40 days (heat affected zone), whereas the response of the 2.25Cr-1Mo steel specimens both from the weld metal and the heat affected zone was clearly more sensitive to the hydrogen attack.

Based on the phase diagram for the iron-carbon-chromium system the better behavior of the higher alloyed steel is attributed to the fact that in the 2.25Cr-1Mo

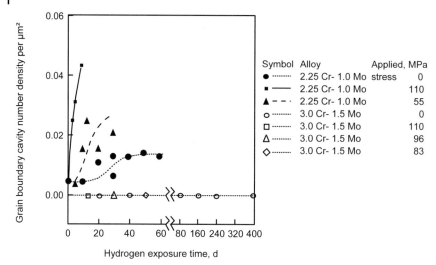

Figure 61: Total area of methane bubbles at the grain boundaries of the 2.25Cr-1Mo steel and the 3Cr-1.5Mo steel after aging in compressed hydrogen (13.8 bar) at 600 °C [101]

steel the reaction of hydrogen occurs preferentially on the carbides of type M_3C. In the 3Cr-1.5Mo steel the formation of this carbide type is suppressed in favor of the more stable carbides. Although these more stable and more coarse carbides reduce the creep strength to a certain extent, they ensure clearly better resistance to hot compressed hydrogen due to their lower carbon activity. As early as in 1938 the

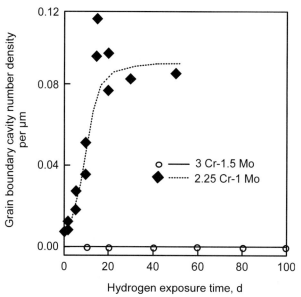

Figure 62: Influence of 33% cold forming on the formation of methane bubbles in 2.25Cr-1Mo steel and 3Cr-1.5Mo steel [101]

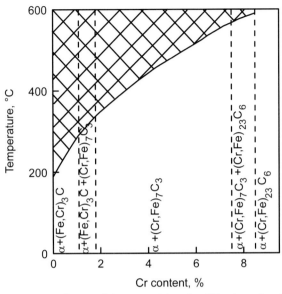

Figure 63: Influence of the composition of carbide phases in a 0.20% Cr steel on the resistance to hydrogen at 300 bar at temperatures up to 600 °C [102]

better resistance of the steels with a higher chromium content observed in the tests described in [95] was attributed to the avoidance of the M_3C carbide type.

Also the examinations described in [102] deal with the formation of different carbide phases depending on the alloying element and its content as well as the influence of these phases on the resistance in hot compressed hydrogen. As a consequence again, the decisive process for increasing hydrogen resistance is the suppression of the M_3C phase by the elements chromium, molybdenum and tungsten by forming corresponding mixed carbides with Fe_3C. This is illustrated in Figure 63 using the example of increasing chromium contents in a steel with 0.20% carbon and its behavior in hydrogen with a pressure of 300 bar at different temperatures.

The hydrogen resistant steels most often used in Germany were formerly listed in the Stahl–Eisen–Werkstoffblatt (SEW) 590-61 "Druckwasserstoffbeständige Stähle" (Material Sheet "Compressed Hydrogen Resistant Steels") [103] as well as in DIN 17176 "Nahtlose kreisförmige Rohre aus druckwasserstoffbeständigen Stählen" (Seamless circular Tubes of compressed hydrogen Resistant Steels) [104] (Table 42 and Table 43).

SEW 590-61 was abandoned on 11 August 2003. The Material Sheet SEW 595 "Stahlguß für Erdöl- und Erdgasanlagen" (Steel Castings for Crude Oil and Natural Gas Installations) [105] only mentions the following three materials as compressed hydrogen resistant steels among the cast steel grades listed in the tables:

G12CrMo9-10	Mat. No. 1.7380	UNS K21390
GX12CrMo5	Mat. No. 1.7363	UNS J42045
GX12CrMo10-1	Mat. No. 1.7389	UNS J82090

Steel	Mat. No.	C	Si	Mn	Cr	Mo	Others
25CrMo4	1.7218	0.22	0.15	0.50	0.90	0.15	
		0.29	0.35	0.80	1.2	0.25	
16CrMo9-3	1.7281	0.12	0.15	0.30	2.0	0.30	
		0.20	0.35	0.50	2.5	0.40	
26CrMo7	1.7259	0.22	0.15	0.50	1.5	0.20	
		0.30	0.35	0.70	1.8	0.25	
24CrMo10	1.7273	0.20	0.15	0.50	2.3	0.20	Ni < 0.80
		0.28	0.35	0.80	2.6	030	
20CrMo 9		0.16	0.15	0.30	2.1	0.25	Ni < 0.80
		0.24	0.35	0.50	2.4	0.35	
10CrMo11	1.7276	0.08	0.15	0.30	2.7	0.20	
		0.12	0.35	0.50	3.0	0.30	
17CrMoV10	1.7766	0.15	0.15	0.30	2.7	0.20	V 0.10–0.20
		0.20	0.35	0.50	3.0	0.30	
20CrMoV13-5-5	1.7779	0.17	0.15	0.30	3.0	0.50	V 0.45–0.55
		0.23	0.35	0.50	3.3	0.60	
X20CrMoV11-1	1.4922	0.17	0.10	0.30	11.0	0.80	Ni 0.30–0.80
		0.23	0.50	0.80	12.5	1.2	V0.25–0.35
X8CrNiMoVNb16-13	1.4988	< 0.10	0.30	1.0	15.5	1.1	Ni 12.5–14.5
			0.60	1.5	17.5	1.5	V 0.60–0.85
							N 0.07–0.13
							Nb > 10%C

Table 42: Compressed hydrogen resistant steels according to SEW 590-61 [103]

Moreover, this Material Sheet provides the following information under points 1 and 2 in connection with compressed hydrogen resistance:

Apart from the cast steel grades dealt with in this Material Sheet, also the cast steel grades GP240GH (1.0619), G17CrMo5-5 (1.7357) and GX23CrMoV12-1 (1.4931) can be used according to EN 10213-2 and DIN 17245 if resistance to compressed hydrogen is required.

Steel castings for crude oil and natural gas installations are considered to be cast steel grades being resistant to compressed hydrogen or having a good corrosion resistance under the operating conditions in crude oil and natural gas installations. The resistance of the cast steel grades to compressed hydrogen depends on the ratio of the weight content of carbon and the weight content of carbide-forming elements. The good corrosion resistance to carburization or attacks of oleic acids, sulfur compounds, caustic solutions and other chemicals depends on the chemical composition of the relevant cast steel grades.

In August 2002 DIN 17176 was substituted by DIN EN 10216-2 [106] which, thereafter, was revised in July 2004. Both versions contain the following information in their comments on changes to previous standards: "Compressed hydrogen resistant steels are not specifically marked". Obviously, it is assumed that the user

Steel	Mat. No.	C	Si	Mn	P	S	Cr	Mo	Others	cf. UNS
25CrMo4	1.7218	0.22–0.29	≤ 0.40	0.60–0.90	max. 0.035	max. 0.030	0.90–1.20	0.15–0.30		G41300
13CrMo4-5	1.7335	0.10–0.18	0.10–0.35	0.40–1.00	max. 0.025	max. 0.020	0.70–1.10	0.40–0.65		K11547
10CrMo9-10	1.7380	0.08–0.15	≤ 0.50	0.40–0.70	max. 0.025	max. 0.020	2.0–2.5	0.90–1.20		K21390
12CrMo9-10	1.7375	0.0.10–0.15	≤ 0.30	0.30–0.80	max. 0.015	max. 0.015	2.0–2.5	0.90–1.10	Ni < 0.30 Al > 0.010 Cu < 0.20	K21390
12CrMo12-10	1.7381	0.06–0.15	≤ 0.50	0.30–0.60	max. 0.025	max. 0.020	0.65–3.35	0.80–1.06		K31545
12CrMo 19 5	1.7362	0.08–0.15	≤ 0.50	0.30–0.60	max. 0.025	max. 0.020	4.0–6.0	0.45–0.65		S50100
X11CrMo9-1	1.7386	0.07–0.15	0.25–1.00	0.30–0.60	max. 0.025	max. 0.020	8.0–10.0	0.90–1.10		S50400
20CrMoV13-5-5	1.7779	0.17–0.23	0.15–0.35	0.30–0.50	max. 0.025	max. 0.020	3.0–3.3	0.50–0.60	V 0.45–0.55	–
X20CrMoV11-1	1.4922	0.17–0.23	≤ 0.50	≤ 1.00	max. 0.025	max. 0.020	10.0–12.5	0.80–1.20	Ni 0.30–0.80 V 0.25–0.35	–

Table 43: Compressed hydrogen resistant steels according to DIN 17176 [104]

has the necessary knowledge to identify whether it is resistant to hydrogen based on the designation of the steel or its chemical composition.

For compressed hydrogen resistant steels also the following VdTÜV material sheets [107] are available:

VdTÜV WB 007/1
07/2001
Compressed hydrogen resistant and heat resisting steel 12 CrMo 19 5 G and 12 CrMo 19 5 V – Mat. No. 1.7362 (cf. UNS S50100)
VdTÜV WB 007/2
04/1983
Compressed hydrogen resistant and heat resisting steel 12 CrMo 19 5 G and 12 CrMo 19 5 V I or II (1.7362)
VdTÜV WB 007/2 supplementary sheet
09/2001
Compressed hydrogen resistant and heat resisting steel 12 CrMo 19 5 G and 12 CrMo 19 5 V I or II – Mat. No. 1.7362 (applies only in conjunction with VdTÜV WB 007/2 (1983-04)
VdTÜV WB 007/3
12/2001

Compressed hydrogen resistant and heat resisting steel 12 CrMo 19 5 G and 12 CrMo 19 5 V – Mat. No. 1.7362
VdTÜV WB 007/3 supplementary sheet
06/2002
Compressed hydrogen resistant and heat resisting steel 12 CrMo 19 5 G and 12 CrMo 19 5 V – Mat. No. 1.7362 (applies only in conjunction with VdTÜV WB 00713 (2001-06)
VdTÜV WB 109
06/2001
heat resisting and compressed hydrogen resistant steel X 12 CrMo 9 1 G and X 12 CrMo 9 1 V – Mat. No. 1.7386 (cf. UNS S50400)
VdTÜV WB 109 supplementary sheet
06/2001
heat resisting and compressed hydrogen resistant steel X 12 CrMo 9 1 G and X 12 CrMo 9 1 V – Mat. No. 1.7386 (applies only in conjunction with VdTÜV WB 109 (2001-06))
VdTÜV WB 166
09/2001
Compressed hydrogen resistant steel, heat resisting rolled steel 17 CrMoV 10 – tempering condition l and II – Mat. No. 1.7766

Today the resistance limits of the various compressed hydrogen resistant steels regarding temperature and partial hydrogen pressure are well known and specified in the aforementioned Nelson diagram. This diagram, the first version of which was published in 1949 [108], is based on the very comprehensive experience regarding the suitability of low-alloy steels when exposed to compressed hydrogen at higher temperatures, this experience having been accumulated since the implementation of catalytic high-pressure hydrogenation in practice in the various fields of the chemical and petrochemical industry in connection with failure analyses, monitoring of operations and examinations in labs.

This worldwide experience has been collected by the American Petroleum Institute (API), evaluated and published in the form of the Nelson diagram mentioned above, explained and adapted to the newest findings (see Figure 64 [109]).

For the specification of operating limits damage by both surface decarburization and internal decarburization and cracking are considered. Therefore, the operating limits indicate the critical process conditions regarding operating temperature and hydrogen partial pressure, above which the types of damage mentioned can occur. The precondition to be considered for the specification of curves is a damage-free behavior of specimens or installations over a period of not less than one year. Damage, on the other hand, is considered independent from the time after which it occurred.

The dashed lines in the upper part of Figure 64 indicate the limit of the operating conditions, above which surface decarburization of the steel must be expected. Under operating conditions marked by the area above the solid lines damage to the respective steels due to internal decarburization and internal cracks has been

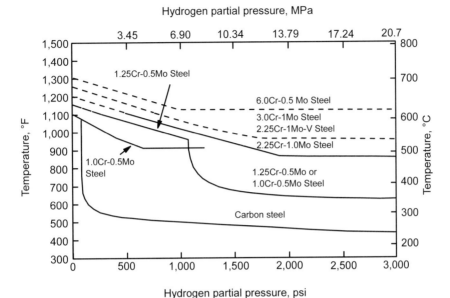

Figure 64: Operating limits for the use of low-alloy steels in compressed hydrogen depending on partial hydrogen pressure and temperature [109]

experienced. There is only a dashed line for the highest alloyed steel with 6% Cr and 0.5% Mo since for this material only damage as a result of surface decarburization has become known to date.

For the operating conditions below the solid lines and on their right hand side satisfactory operating experience over a period of up to fifty years has been made with the relevant steel.

When selecting a material on the basis of the information from the Nelson diagram care must be exercised since only the load caused by compressed hydrogen was used for evaluating the behavior of the material. Other factors, which may influence the material behavior at high temperatures, such as other corrosive components in the medium, the creep behavior or any temper embrittlement of the material have not been considered.

This regular revision of the Nelson curve may also lead to downward corrections, i.e. corrections down to lower temperatures, of the resistance lines of certain steels due to new damage or new test results. The result may be that, according to the new findings, older installation components are operated at excessively high temperatures. To ensure the safe continuation of operation stricter test methods or shorter inspection intervals may be necessary to evaluate potential damage at regular intervals by performing nondestructive tests. Suitable tests to this end are, for instance, ultrasound test methods [109, 110]. The values for steel with 0.5% molybdenum, for instance, have been altered several times following the last revisions. In part, separate curves are indicated for such steels [109, 111].

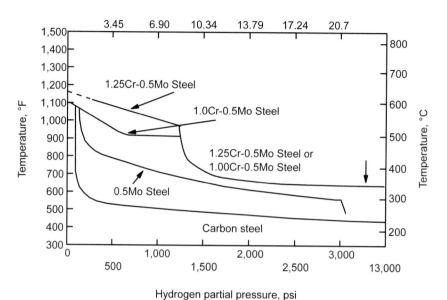

Figure 65: Operating conditions for the use of C-0.5%Mo steels and Mn-0.5%Mo steels in compressed hydrogen depending on partial hydrogen pressure and temperature [109]

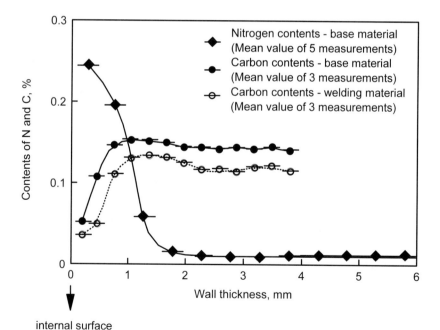

Figure 66: Carbon and nitrogen contents in the area of the internal surface of a 2.25%Cr-1%Mo steel ammonia converter [112]

In APPI-941 the 0.5%-Mo steels are not contained in the general survey (Figure 64) but are shown in a separate diagram (Figure 65).

Certain conditions in synthesis gases during the production of ammonia lead to the formation of thin, nitrided surface layers in compressed hydrogen resistant steels. Due to the thermodynamic stability of carbides and nitrides, carbides are modified and release carbon which can form methane with the diffused hydrogen and lead to a decarburized sub-surface zone as shown in Figure 66 by illustrating the carbon and nitrogen contents across the wall cross section of an ammonia converter.

Depending on the amount and the direction of the total stresses in the components, small microstructural separations parallel to the surface may occur or cracks may slowly grow through the wall. Hence, the Nelson diagram is suited for the selection of materials for ammonia-containing synthesis gases only to a limited extent [112].

A 17 Ferritic chromium steels with < 13% Cr

On the basis of hydrogen-induced crack formation models the resistance of the stainless ferritic steel 12Cr-1Mo to hydrogen embrittlement was calculated under the operating conditions of fusion reactors [113]. Materials in fusion reactors can be exposed to different hydrogen sources:

– exposure to plasma (ionized hydrogen)
– hydrogen formed inside the material as a result of nuclear reactions (n, p)
– cathodic hydrogen separation in water-cooled systems.

The influence of the temperature on the crack growth rate in steel 12Cr-1Mo exposed to a plasma is illustrated in Figure 67, to cathodic hydrogen separation in Figure 68 and to hydrogen formed inside the material as a result of nuclear reactions in Figure 69. These results permit conclusions as to the temperatures for the maximum susceptibility to hydrogen embrittlement and temperature limits for this resistance.

The following conclusions can be derived from these results:

– Under the operating conditions of a fusion reactor only the hydrogen formed inside the material as a result of nuclear reactions can lead to crack propagation.
– Under the influence of a plasma cracks do not propagate further at temperatures $\geq 127\,°C$.
– For cathodically generated hydrogen the highest susceptibility is found at room temperature.
– Plasma and cathodic hydrogen can play a role in fusion reactors only if the plant is decommissioned and the temperature is reduced.

Since hydrogen is formed inside the material during nuclear reactions, the hydrogen content needs to remain below the critical concentration c_H^* for crack initia-

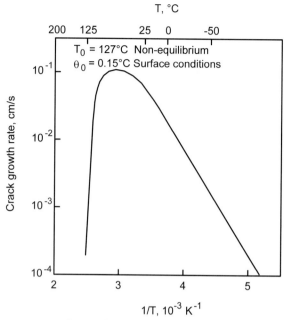

Figure 67: Influence of the temperature on the calculated crack growth rate in 12Cr-1Mo steel exposed to plasma (ionized hydrogen) [113].
$\Theta_0 = 15$ = degree of coverage with H at T_0; if $T > T_0$, crack propagation approaches zero.

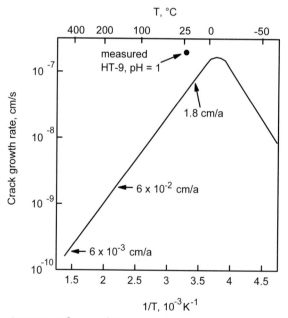

Figure 68: Influence of the temperature on the calculated crack growth rate in 12Cr-1Mo steel exposed to cathodically formed hydrogen [113]

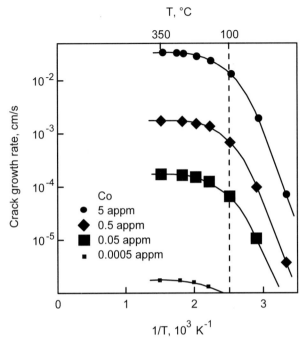

Figure 69: Influence of the temperature and the hydrogen contents resulting from nuclear reactions in 12Cr-1Mo steel on the calculated crack growth rate [113]

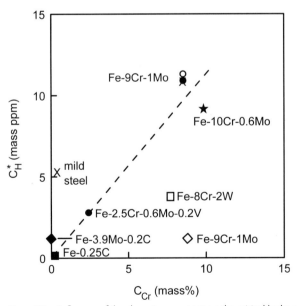

Figure 70: Influence of the chromium content on the critical hydrogen concentration c_H^* [114]

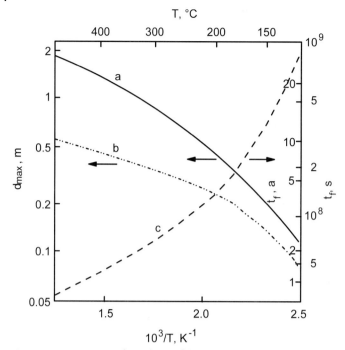

Figure 71: Temperature dependence of the maximum allowable diffusion distance d_{max} to keep the hydrogen generated during nuclear reactions below the critical concentration of 10 ppm (curves a and b) and the effusion time t_f to lower the hydrogen content in a 0.5 m thick plate to 10% (curve c) [114]

tion [114]. Also the rate at which hydrogen effuses again plays a role. To answer these questions the influence of the chromium content on the critical hydrogen concentration $c_H{}^*$ was literarily determined for a number of low-alloy and high-alloy steels (Figure 70). For the chromium steels with 9–12% chromium, which are of interest for fusion reactors, the critical hydrogen concentration $c_H{}^*$ amounts to about 10 ppm. With the help of diffusion equations the temperature dependence of the maximum permissible thickness of the structure (maximum diffusion path) d_{max} was determined at which the hydrogen content caused by the nuclear reactions remains below the critical concentration $c_H{}^* = 10$ ppm. The calculations refer to two different assumptions regarding the hydrogen formation rate: exponential behavior (Figure 71, curve a) or constant formation rate (Figure 71, curve b). After the reactor has been shut down, the hydrogen concentration decreases exponentially. The effusion time t_f required to reduce the hydrogen content of a 0.5 m thick plate to 10% is plotted in Figure 71, curve c. Hence, two years are necessary, for instance at 300 °C, to lower the hydrogen concentration to 10% in case of unidirectional effusion and at 150 °C the effusion time is 10 years.

A 18 Ferritic chromium steels with ≥ 13% Cr

The solubility of hydrogen in steels depends on the physical state (solid, liquid) of the crystal lattice, the alloy content, the temperature and the partial pressure of the atmosphere. According to Table 44 the type of the lattice, for instance, exerts a much greater influence on the hydrogen solubility than the alloy content [115]. The face-centered cubic (fcc) lattice can absorb much more hydrogen than the body-centered cubic (bcc) lattice at identical temperature and pressure. In bcc iron materials chromium reduces hydrogen solubility, whereas nickel increases it slightly (up to 8%). Therefore, hydrogen solubility in bcc ferritic chromium steels with 13% Cr is lower than in pure iron. Also the diffusion rates of hydrogen in steel decisively depends on the crystal lattice. Table 45 shows several characteristic values of the diffusion coefficient for bcc and fcc high-alloy stainless steels compared to pure iron. As can be seen, the diffusion rate in the fcc lattice is several orders of magnitude lower than in the bcc lattice.

Due to their bcc lattice structure stainless ferritic steels are susceptible to hydrogen embrittlement. A super ferrite X1CrMoNi29-4-2 with the chemical composition indicated in Table 46 was used to perform closer investigations of the hydro-

Material	Crystal lattice	Solubility cm³ H/100 g metal
Pure iron (alpha)	bcc*	0.7
13% Cr, balance Fe	bcc	0.4
13% Cr, balance Fe	fcc**	4.8
18% Cr-10% Ni, balance Fe	fcc	5.8

* bcc: body-centered cubic, ** fcc: face-centered cubic

Table 44: Solubility of hydrogen in iron materials at 400 °C and a pressure of 1 mm Hg [115]

Material	Crystal lattice	D at 25 °C cm²/s
Pure iron (alpha)	bcc	1.6×10^{-5}
Pure iron (gamma)	fcc	5.4×10^{-10}
Steel (α-Fe + Fe₃C)	bcc	3.0×10^{-7}
Ferrite: 27% Cr balance Fe	bcc	6.7×10^{-8}
Austenite: 18% Cr-9% Ni balance Fe	fcc	3.5×10^{-12}

Table 45: Comparison of the diffusion coefficients D of hydrogen in iron and stainless steels under consideration of the lattice structure [115]

C	Mn	P	S	Si	Cr	Ni	Mo	N
0.0029	0.10	0.01	0.009	0.10	29.5	2.23	3.93	0.12

Table 46: Chemical composition of the super ferrite X1CrMoNi29-4-2 in wt% [116]

gen-induced crack propagation velocity [116]. The crack growth was studied using sheet specimens (1.5 × 25 × 150 mm) notched on one side and provided with an incipient fatigue crack under constant load over time in air or 108 kPa hydrogen gas at 25 °C. A portion of the specimens was precharged with hydrogen in a salt melt (KOH + NaOH of a ratio of 59:41 with steam purge) at 230 °C for 4 h. The hydrogen content determined by hot extraction amounts to 12 ppm following precharging and only 2 ppm after the crack propagation tests. Due to the high diffusion rate of hydrogen of 10^{-11} m^2/s at room temperature in the steel X1CrMoNi29-4-2 it is assumed that the absorbed hydrogen quickly effuses and only the amount in the traps (2 ppm) remains in the material.

Figure 72 shows the crack growth rate in specimens obtained in air and in compressed hydrogen, which were precharged with hydrogen (2 ppm) in a salt melt before the test or were not precharged. Hydrogen precharging of the specimens tested in air leads to the reduction of the threshold value K_{IH} – below which crack propagation is not possible – by about 10% (20 MPa $\sqrt{m}$). In addition a plateau

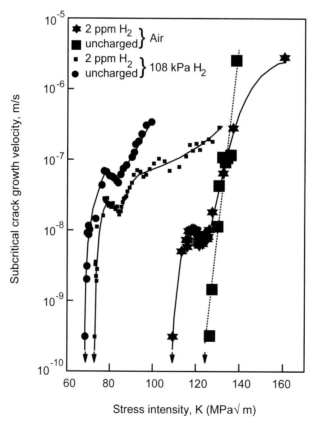

Figure 72: Comparison of the crack growth rates in X1CrMoNi29-4-2 steel in air and in 108 kPa compressed hydrogen without and with hydrogen precharging (at 25 °C) (2 ppm) [116]

area II with a crack propagation velocity of 7.5 × 10⁻⁹ m/s occurs. If the crack growth is measured in 108 kPa compressed hydrogen, the threshold value K_{IH} is considerably reduced in both non-precharged and precharged specimens compared to air and a marked area II can be observed. This means that the compressed hydrogen offered from outside is more harmful than hydrogen dissolved in the material.

To simulate the condition at the tip of a hydrogen-induced crack, the diffusion capacity of the plastically deformed steel X1CrMoNi29-4-2 was examined [117]. The steel specimens with the chemical compositions indicated in Table 46 were cold-rolled or rolled with an interannealing step up to a deformation degree of 25, 50 and 75%. The specimen thicknesses were 396, 185 and 137 µm. For the permeation tests a palladium film was sputter cooled on the surfaces after cleaning with Ar. The diffusivity was examined by means of the gas phase permeation technology from 110 to 260 °C. To this end, two ultra high vacuum chambers were separated by a membrane of the test material. Following the presetting of hydrogen in one chamber from 1 to 26 kPa, the permeated hydrogen in the other chamber was determined by means of a calibrated ion pump. The diffusion rate of the hydrogen in the X1CrMoNi29-4-2 steel in the soft-annealed and plastically deformed state can be derived from Figure 73. Compared to the soft-annealed state, the diffusion rate is considerably reduced by the deformation. Below 200 °C it is reduced by a factor of 10–30. Hydrogen solubility, on the other hand, is slightly increased [117]. The hydrogen pressure hardly influences the diffusion rate in the examined range from 1–26 kPa.

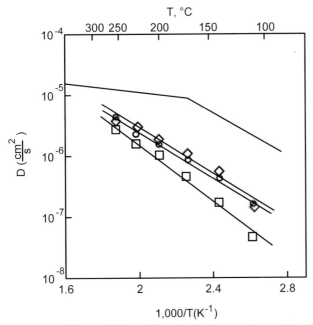

Figure 73: Influence of plastic deformation and temperature on the effective hydrogen diffusivity D in X1CrMoNi29-4-2 steel [117]

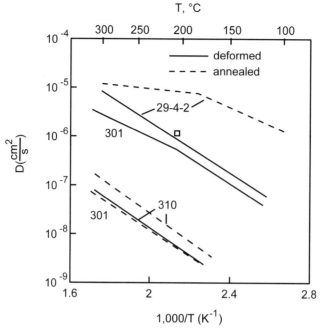

Figure 74: Comparison of the hydrogen diffusivity D in deformed (solid lines) and soft-annealed (dashed lines) stainless steels SAE 301 (1.4310), SAE 310 (1.4841) and X1CrMoNi29-4-2 [117]

Figure 74 summarizes the experimental findings of X1CrMoNi29-4-2 steel compared to the unstable austenitic steel SAE 301 (1.4310, X10CrNi18-8) and the stable austenitic steel SAE 310 (1.4841, X15CrNiSi25-21). In the soft-annealed state the diffusivity in the steels SAE 301 and SAE 310 is almost identical but is more than three orders of magnitude lower than in the ferritic steel X1CrMoNi29-4-2. By cold rolling of the material SAE 301 with a thickness reduction of 67% the diffusion rate is increased by two orders of magnitude, attributable to the conversion of 86% austenite to α'-martensite. Cold rolling of the stable austenite SAE 310 by 80% leads to an increase of the diffusion rate by a factor of two only, whereas a deformation degree of 75% of the ferritic steel X1CrMoNi29-4-2 causes a considerable reduction (factor 20) of the hydrogen diffusion. This different behavior of the ferritic and austenitic stainless steels regarding hydrogen transport and hydrogen solubility can be also used to explain the differences found in the propagation behavior of hydrogen-induced cracks [117].

Since ferritic stainless steels are susceptible to hydrogen embrittlement, care must be exercised when welding this material group as to ensure that the shielding gas does not contain any hydrogen [118]. Also the hydrogen generated during welding from water vapor or oils needs to be kept away from the weld region since it may cause rapid embrittlement. In ferritic steels hydrogen can quickly effuse again; to this end, a heat treatment from 90 to 200 °C is suited. At room temperature the effusion can take several days or weeks depending on the presence of inhibiting surface films. Susceptibility to hydrogen embrittlement also depends on the material composition. In general, steels with a high chromium and molybdenum content, e.g. the super ferrite AL 29-4C (1.4592, X2CrMoTi29-4: 29.5 Cr, 3.8 Mo, 0.48 Ni, 0.018 C), are more susceptible than high-purity CrMo steel E-BRITE 26-1 (UNS S44627) with a lower Cr and Mo content (25.9 Cr, 1.01 Mo, 0.14 Ni, 0.0057 C).

A 19 Ferritic/pearlitic-martensitic steels

A 19.1 Martensitic steels

The high-temperature steel 1.4914 (Table 47) is considered a suitable material for wall and structural components of the DEMO (demonstration power) reactor. It exhibits low susceptibility to helium embrittlement and more favorable thermophysical properties compared to the austenitic steel SAE 316 L (cf: 1.4404, 1.4435) [119]. To ensure a purely martensitic microstructure, the steel was heat-treated as follows: 2 h, 970 °C + 0.5 h, 1,075 °C/H_2O + 2 h, 750 °C/slow cooling. The hydrogen transport in the material was examined in permeation tests and by means of isovolumetric desorption.

C	Si	Mn	P	S	Cr	Mo	N	Zr	Nb	Ni	V
0.11	0.18	0.85	0.005	0.004	10.3	0.58	0.030	0.014	0.14	0.65	0.19

Table 47: Chemical composition of the steel 1.4914 in wt% [119]

During the isovolumetric desorption the specimen, including the tank wall, is charged with 10^5 Pa hydrogen at a specific temperature, then the pressure is reduced shortly to a value below 10^{-3} Pa, and then the path of hydrogen degassing over time is determined by registering the pressure increase. By performing a parallel test without the specimen, the share of the wall could be eliminated. These findings were used to determine the diffusion coefficient D.

During the permeation method two chambers are separated by a film of the test material. In one chamber a constant hydrogen pressure is maintained and the pressure increase by the permeated hydrogen is measured in the other chamber following its evacuation to determine the diffusion coefficient. The results of the diffusion capacity of hydrogen (H_2) and deuterium (D_2) in the material 1.4914 are summarized in Figure 75. The curves a (H_2), b (D_2) and f (H_2) were determined by means of isovolumetric desorption and the curves c (D_2), d (H_2) and e (H_2) were determined by applying the permeation method.

Irrespective of the test method and of the fact whether the gas is hydrogen or deuterium, the diffusion behavior of the specimens a to e is similar. Only specimen f exhibited different behavior. It could be proven that specimen f exhibited an oxide film with a thickness of 300 angstroms which, of course, inhibited the absorption of hydrogen.

108 | Hydrogen

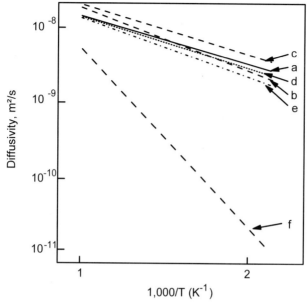

Figure 75: Influence of the inverse temperature on the diffusivity of hydrogen (H_2) and deuterium (D_2) in martensitic steel 1.4914 [119]. The curves a, d, e and f refer to H_2 and the curves b and c to D_2

A 19.2 Ferritic-austenitic steels/ duplex steels

Due to their two microstructural components, the ferritic-austenitic steels are also called duplex steels. The content of alloying elements is chosen such that the ferritic and austenitic phases in the microstructure are balanced (usually at a ratio of 1:1). Most frequently, these steels consist of 50% austenite in a ferritic matrix. Due to their excellent corrosion resistance they are often used in the chemical industry, for oil and gas production and in seawater. When using ferritic-austenitic steels in hydrogen-containing fluids, e.g. for oil and gas production, the question arises whether hydrogen is absorbed and such absorption causes damage [120]. This problem is not easy to handle because the ferritic phase and the austenitic phase exhibit different solubilities and diffusion rates for hydrogen.

Duplex steel Ferralium® 255 (26% Cr, 3% Mo, 5.55% Ni, 1.6% Cu, 0.16% N; cf.: 1.4507) with about 50% austenite in a ferritic matrix was used to perform a closer investigation of the hydrogen-induced crack growth in hydrogen gas from 0 to 100 °C [120]. Austenitic steel SAE 301 (1.4310) was used for comparison, which was partially predeformed yielding different contents of martensite (bcc phase). To this end it was rolled at room temperature (specimen 301C with 57% α′) and 110 °C (specimen 301H with ≤ 0.5% α′). Several 301C specimens were then heat-treated in vacuum at 450 °C for 24 h with the α′-content being increased again (specimen 301A with 65% α′).

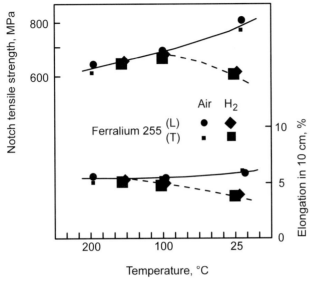

Figure 76: Influence of the temperature on the mechanical properties of Ferralium® 255 in air (full symbols) and 108 kPa hydrogen gas (open symbols). Results of tensile tests of notched specimens in L and T direction [120]

The mechanical properties of the Ferralium® 255 material in air and 108 kPa hydrogen gas for 25 to 200 °C are shown in Figure 76. They were determined by slow strain rate tensile tests (8.5×10^{-5} cm/s) using notched specimens (10 cm gauge length) in the L and T direction (L: longitudinal and T: transverse to the rolling direction). Compared to air both tensile strength and elongation are clearly reduced in hydrogen gas at temperatures T < 100 °C. The specimen orientation (L or T) has no major influence. Whereas the specimens tested in air exhibited ductile fracture surfaces, the specimens in hydrogen exhibited cleavage fractures with secondary cracks at 25 °C. Above 100 °C they also exhibit ductile fractures.

Table 48 shows a comparison of the mechanical properties of the duplex steel Ferralium® 255 with the solution-annealed austenitic and predeformed martensite-containing steel SAE 301 (specimens 301 H, 301 C, 301 A) in air and 108 kPa hydrogen at 25 °C. In all tested specimens both the tensile strength and the fracture elongation are clearly reduced in hydrogen gas. The specimens exhibited cleavage fractures with transcrystalline and intercrystalline secondary cracks.

Hydrogen-induced crack growth of Ferralium® 255 was examined under constant load over time using notched specimens [120]. Subcritical crack growth did not occur in air. In 108 kPa hydrogen the ln v-K curves shown in Figure 77 for the Ferralium 255 (T) material were measured for temperatures from 25 to 100 °C. The K_{IH} value above which the hydrogen-induced crack growth starts is lower the lower the temperature is. A comparison of the results obtained with the austenite SAE 301 (1.4310) and the super ferrite Al 29-4-2 revealed higher resistance of the

Hydrogen

Material	Tensile strength			Fracture elongation		
	Air MPa	H_2 MPa	Loss in H_2 %	Air %	H_2 %	Loss in H_2 %
SAE 301[2])	379	289	24	7.4	2.3	69
301 H[3])	757	624	18	3.4	2.3	33
301 C[3])	979	732	25	2.9	2.2	23
301 A[3])	1,016	845	17	2.8	2.1	26
Ferralium® 255 L	795	606	24	6.0	4.0	34
Ferralium® 255 T	769	614	20	6.3	4.1	35

1) Tensile tests of notched specimens
2) 1 h, 110 °C /H2O
3) 301 H, 301 C, 301 A: predeformed

Table 48: Mechanical properties[1]) of various stainless steels in air and 108 kPa hydrogen at 25 °C [120]

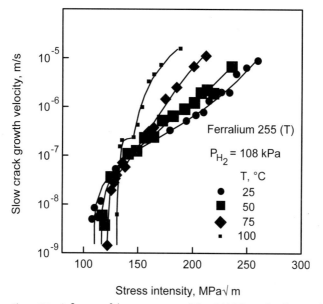

Figure 77: Influence of the temperature (25 to 100 °C) on the slow crack growth rate in the duplex steel Ferralium® 255 (T) in 108 kPa hydrogen [120]

duplex steel [120]. The K_{IH} value at 25 °C amounts to 45 MPa $\sqrt{m}$ for the austenite SAE 301, 70 for the super ferrite AL 29-4-2 and 110 MPa $\sqrt{m}$ for the duplex steel Ferralium® 255. The results shown in Figure 78 were obtained in 108 kPa hydrogen for Ferralium 255 loaded in the longitudinal direction L. Here again, an analog influence of the temperature on the K_{IH} and K_{Ic} values was found, however with a wider range of scattering of the values.

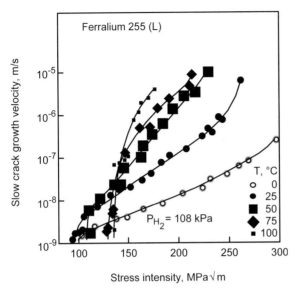

Figure 78: Influence of the temperature (0 to 100 °C) on the slow crack growth rate in the duplex steel Ferralium® 255 (L) in 108 kPa hydrogen [120]

Examinations of the influence of hydrogen pressure (54, 108 and 216 kPa) on the crack growth in the duplex steel Ferralium® 255 (L) at 25 °C revealed a clear increase of the slow crack growth rate as a function of hydrogen pressure (Figure 79). In area II (K = 110 to 130 MPa $\sqrt{m}$) variation of the slow crack growth rate was almost linear with respect to the hydrogen pressure [120].

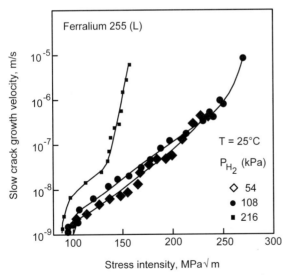

Figure 79: Influence of the hydrogen pressure (54 to 216 kPa) on the slow crack growth rate in the duplex steel Ferralium® 255 (L) at 25 °C

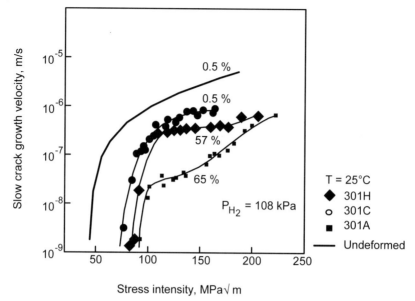

Figure 80: Slow crack growth rate in the solution-annealed and predeformed steel SAE 301 (1.4310) in 108 kPa hydrogen at 25 °C. The martensite content is indicated as the parameter

For comparison also examinations of the steel SAE 301 (1.4310) in the solution-annealed and predeformed condition were carried out. Figure 80 shows the influence of the material condition on the slow crack growth rate in 108 kPa hydrogen at 25 °C. The solution-annealed austenite exhibited the highest slow crack growth rate with the lowest K_{IH} value. For the predeformed steels the following sequence applies to increasing resistance:

301 H (rolling at 110 °C: 0.5% α'-martensite)
301 C (rolling at RT: 57% α'-martensite)
301 A (rolling at RT + 24 h 450 °C : 65% α'-martensite)

These results show that for steel SAE 301 (1.4310) the threshold value K_{IH} for the initiating crack growth increases and the slow crack growth rate in compressed hydrogen decreases as the α'-martensite content increases.

Finally, the behavior of the duplex steel Ferralium® 255 was examined in 108 kPa hydrogen at 25 °C compared to the super ferrite AL 29-4-2 and the austenite SAE 301. The results of the slow crack growth rates da/dt as a function of the stress intensity factor K are summarized in Figure 81 for the various materials. The following sequence of increasing susceptibility to hydrogen-induced crack propagation in compressed hydrogen can be found:

duplex steel – super ferrite– austenite.

This behavior is attributed to the following cause [120]:
The microstructure in front of the crack tip is more important to the slow crack growth rate of stainless steels in compressed hydrogen than the microstructure in-

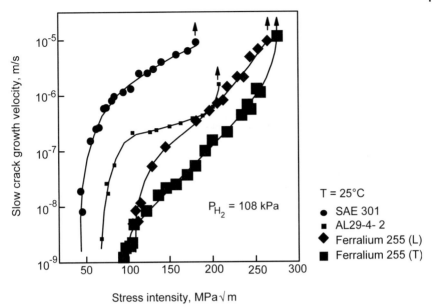

Figure 81: Comparison of the slow crack growth behavior of four stainless steels in 108 kPa hydrogen at 25 °C [120]

side the material. Since strain-induced martensite is exclusively formed in front of the crack tip in notched austenite specimens under mechanical tensile stress, hydrogen quickly accumulates in front of the crack tip. The austenite outside the crack tip, i.e. inside the material, forms a sort of a diffusion barrier for hydrogen. This means that the rate of the hydrogen transport into the crack tip area is the rate determining step for the slow crack growth rate. In a ferritic steel hydrogen is also quickly transported into the crack tip area but, due to the high diffusion rate (refer to Table 45), the hydrogen also quickly distributes inside the material such that finally the accumulation in front of the crack tip is lower than in the austenite. This also explains why the hydrogen-induced slow crack growth rate in mechanically deformed austenites is inversely proportional to the martensite content inside the material (cf. Figure 80).

SIMS (secondary ion mass spectrometer) examinations were performed to clarify the spatial hydrogen distribution during the crack growth in a duplex steel [121, 122]. To this end, specimens from X2CrNiMo25-6-4 steel (chemical composition in Table 49) were deformed in a slow strain rate tensile test (10^{-6} 1/s) while concurrently being continuously exposed to 30 bar hydrogen or deuterium. The tensile tests were interrupted prior to fracture to be better able to identify the crack path. As soon as the addition of hydrogen in the autoclave was interrupted, a crack arrest occurred. This means that the crack growth by hydrogen is initiated in the immediate surrounding of the crack tip. The crack path was analysed by a scanning electron microscope. A cleavage fracture occurs in the ferrite phase. As soon as the crack reaches the α/γ-phase boundary, a microcrack is initiated in the adjacent aus-

Element	C	Ni	Cr	Mn	Cu	P	N	Si	Mo	S	Fe
Wt%	0.02	6	25	1.2	1.5	0.01	0.25	0.34	3.7	< 5 ppm	balance

Table 49: Chemical composition of the duplex steel X2CrNiMo25-6-4 [122]

tenite grain in the surrounding of which the SIMS analysis revealed the highest hydrogen concentration. This explains why crack arrest occurs when the hydrogen supply from outside is interrupted.

The question arises as to the extent of hydrogen embrittlement caused by internal hydrogen, i.e. hydrogen dissolved inside the material, compared to hydrogen acting from outside. To this end, the duplex steel Ferralium® 255 was subjected to hydrogen absorption by the cathodic polarization in a salt melt and the findings regarding hydrogen embrittlement were then compared with the results obtained in external compressed hydrogen [123].

The duplex steel with the chemical composition indicated in Table 50 was first precharged with hydrogen by cathodic polarization in a salt melt (KOH + NaOH

Material	C	Mn	P	S	Si	Cr	Ni	Mo	N	Cu
Ferralium® 255	0.03	0.61	0.021	0.003	0.55	25.56	5.59	3.01	0.16	1.63

Table 50: Chemical composition (wt%) of the duplex steel Ferralium® 255 [123]

of a ratio of 59:41 with steam purge) at 265 °C for 30 h. Subsequently, the specimens were electropolished, notched on one side and provided with an incipient fatigue crack and the crack growth was immediately measured in air under constant load over time [123]. The tests performed under constant load over time in air took place at 0, 25 and 50 °C with hydrogen contents of 3 to 15 ppm. Figure 82 exemplarily shows the results regarding the influence of the hydrogen content and Figure 83 indicates the results regarding the influence of the temperature on the slow crack growth rate as a function of the stress intensity factor. As can be seen, the threshold value K_{tH} for the subcritical growth rate is lowest at 25 °C, corresponding to the experience made with hydrogen embrittlement. In addition, the K_{tH} value decreases with increasing hydrogen content. For further results reference is made to the original literature [123].

Figure 84 shows the influence of the hydrogen content in the duplex steel Ferralium® 255 on the threshold value K_{tH} for various temperatures. Whereas the temperature in the examined area from 0 to 50 °C exhibited a minor influence only, the K_{tH} value clearly decreases when the hydrogen content increases.

Comparing the embrittlement effect in the duplex steel Ferralium® 255 caused by hydrogen dissolved inside and the effect caused by compressed hydrogen from outside, the external compressed hydrogen is more harmful [120, 123]. The threshold value K_{tH} for subcritical crack growth amounts to 110 MPa $\sqrt{m}$ [120] in 1 bar compressed hydrogen at 25 °C and, hence, is lower than the K_{tH} value for 15 ppm dissolved hydrogen (cf. Figure 84) [123]. When extrapolating the K_{tH} values at 25 °C to 110 MPa $\sqrt{m}$ in Figure 84, a content of about 25 ppm hydrogen would be

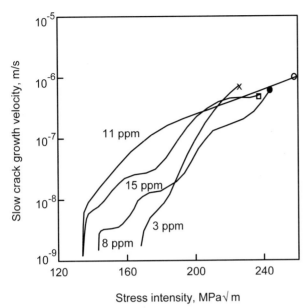

Figure 82: Influence of the hydrogen content on the slow crack growth rate in the duplex steel Ferralium® 255 as a function of the stress intensity factor at 0 °C [123]

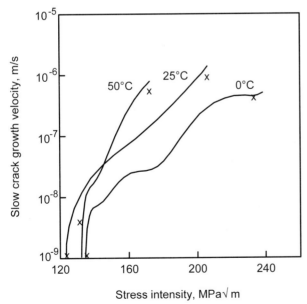

Figure 83: Influence of the temperature on the slow crack growth rate in the duplex steel Ferralium® 255 as a function of the stress intensity factor at a hydrogen content of 15 ppm [123]

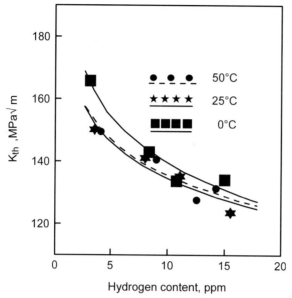

Figure 84: Influence of the hydrogen content and the temperature on the threshold value K_{tH} for the subcritical crack growth in the duplex steel Ferralium® 255 [123]

necessary to obtain the same K_{tH} value as that in compressed hydrogen. However, with the high-temperature charging technology such high hydrogen concentrations could not be reached.

Two commercially available pipe steels from duplex steel were used to study the influence of the microstructure on embrittlement in hydrogen gas [124]. The chemical composition of the test materials is shown in Table 51 and Table 52 depicts the mechanical properties. Whereas there is hardly any difference between the mechanical properties, the microstructures differ completely. In steel A the austenite

Steel	Cr	Ni	Mo	Mn	Si	V	Cu	Co	C	P	S
A	22.3	5.7	2.90	1.62	0.35	0.06	0.06	0.05	0.027	0.021	< 0.002
B	22.9	5.2	3.12	0.99	0.50	0.04	0.03	0.07	0.016	0.019	0.002

Table 51: Chemical composition of the stainless pipe steels from duplex steel in wt% [124]

Steel	Orientation	$R_{p0.2}$ N/mm²	R_m N/mm²	A %	Z %
A	longitudinal	651	795	42	84
	transverse	634	785	41	74
B	longitudinal	620	740	36	85
	transverse	600	710	39	83

Table 52: Mechanical properties of the examined duplex steels (strain rate: 10^{-4} 1/s) [124]

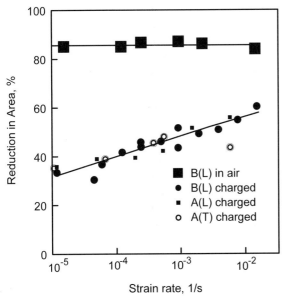

Figure 85: Influence of the strain rate on reduction of area of duplex pipe steels A and B in longitudinal and transverse direction following hydrogen charging at a pressure of 170 bar and at 350 °C [124]

grains are aligned longitudinally to the rolling direction and steel B shows an irregular arrangement. The ferrite content was about 63% for steel A and about 65% for steel B.

Following the evacuation of the autoclave charging with hydrogen was carried out at 350 °C for a period of 48 h at a hydrogen pressure of 170 bar. Thereafter the specimens were cooled under pressure and stored for immediate examination in liquid nitrogen. The hydrogen content was 15 ppm. Subsequently, some of the specimens were subjected to tensile tests at strain rates of 1.7×10^{-2} to 1.2×10^{-5}/s in air. The results are shown in Figure 85. Hydrogen charging resulted in a clear decrease of reduction of area, which became even stronger with decreasing strain rate. The microstructure exerted only an insignificant influence in this connection.

If the duplex steels are not precharged with hydrogen, but are strained in 1 or 2 bar hydrogen gas, material A and material B exhibit a different behavior. The embrittlement index EI served as the measure for embrittlement, the index being calculated from the reduction of area Z in air and in hydrogen as follows:

Equation 16 $\qquad EI = 100(Z_{air} - Z_H)/Z_{air}$

In this connection Figure 86 shows the influence of the strain rate on the embrittlement index EI of material A and B (cf. Table 51) for longitudinal and transverse specimens each in 2 bar compressed hydrogen. As can be seen, the duplex steel A is less susceptible and its susceptibility depends on the specimen orientation, which is not the case for steel B. Therefore, crack propagation is determined by the

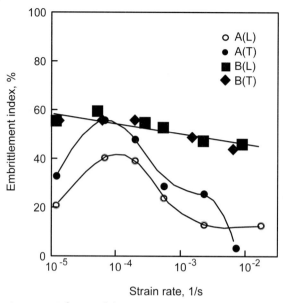

Figure 86: Influence of the strain rate on the embrittlement index EI for longitudinal and transverse specimens from commercially available duplex steels A and B in 2 bar compressed hydrogen [124]

orientation of the less susceptible austenite phase. If the austenite grains are aligned in the longitudinal direction to the steel pipe (specimen AL), the lowest hydrogen embrittlement is observed. In the case of a statistical distribution of the austenite phase in specimen B a difference in the longitudinal and transverse direction is not expected, which was also proven by experiments [124].

Hydrogen-induced crack initiation and crack growth were subjected to more detailed examinations by using the duplex steel (1.4462) [125]. The chemical composition of the test material (C 0.016; Mn 0.99; Si 0.50; S 0.002; P 0.019; Cr 22.9; Ni 5.2; Mo 3.12; N 0.13; Co 0.07; V 0.04; Cu 0.03 wt%) roughly corresponds to the material B in Table 51, however, the mechanical properties ($R_{p0.2}$: 577 MN/m²; R_m: 766 MN/m²; A: 36%; Z: 87%) differed slightly. The solution-annealed material (1,050 °C/water) exhibits a statistical distribution of the longitudinally formed austenite phase (35%) in the ferrite matrix (65%). Smooth, unnotched round test bars were used for performing tensile tests at strain rates of 1.7×10^{-2} 1/s to 1.2×10^{-5} 1/s in 0.5 to 2 bar compressed hydrogen. The microstructure was radiographically examined prior to and after this test.

The results of the CERT tests of the duplex steel (1.4462) are shown in Figure 87 for the hydrogen pressures 0.5, 1.0 and 2 bar compared to air. Whereas the reduction of area Z in air was about 85%, it was considerably reduced in 2 bar hydrogen and was further reduced with decreasing strain rate. At 0.5 and 1 bar hydrogen the ductility assumes a minimum value between about 3×10^{-4} 1/s and 6×10^{-4} 1/s. Examinations using a scanning electron microscope revealed that cracks are always initiated in the ferrite phase, starting from small inclusions or micronotches. Also

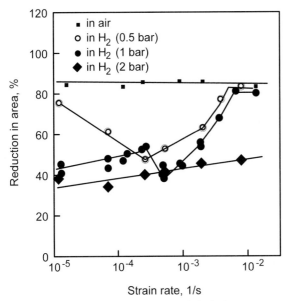

Figure 87: Influence of the strain rate and the hydrogen pressure on the reduction of area of the duplex steel 1.4462 [125]

the crack propagation first starts in the ferrite phase in the form of a transcrystalline cleavage fracture. After the crack has reached the phase boundary of the austenite, there are basically several options: either the crack stops or, if the stress is high enough, the austenite breaks more or less ductily due to overloading. Another option is that, as a result of the excessive stress in front of the crack tip and under the concurrent influence of hydrogen, the metastable austenite is converted in ε-martensite and, hence, also an H-induced cleavage fracture becomes possible in the converted austenite. It was possible by means of X-ray tests to clearly prove the existence of the ε-martensite peak following loading of the duplex steel in hydrogen, which had not been there before the test [125]. This proves that the crack propagation in the austenite phase of a duplex steel is facilitated by the conversion of austenite into ε-martensite in front of the crack tip.

In a subsequent study the hydrogen embrittlement of a pipe steel from duplex steel (23 Cr, 5 Ni, 3 Mo, 1 Mn, 0.13 Ni in wt%) was tested in various ambient fluids [126]. The material contained 35% austenite predominantly aligned in the longitudinal direction of the pipe. Slow strain rate tensile tests (6.5×10^{-6} to 6.4×10^{-3} 1/s) were performed in various atmospheres:

– hydrogen gas at pressures up to 0.2 MPa
– air following thermal hydrogen charging at 350 °C and 32 MPa H_2 for 48 h.

In addition, tests were performed in aqueous fluids, which are reported elsewhere. Reduction of area Z or the embrittlement index EI served as the measure for the embrittlement.

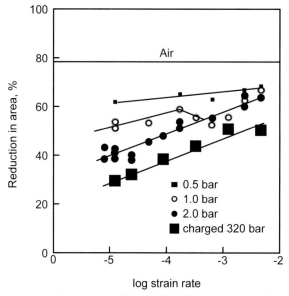

Figure 88: Dependence of the ductility of a duplex steel on the external hydrogen pressure (0.5, 1 and 2 bar) and following thermal hydrogen charging (350 °C and 320 bar) as a function of the strain rate in the slow strain rate tensile test of longitudinal specimens [126]

Figure 88 summarizes the results of the slow strain rate tensile tests of duplex steel in hydrogen gas at pressures up to 0.2 MPa and in air following thermal hydrogen charging at 350 °C and 32 MPa hydrogen for 48 h compared to air. The ductility clearly decreases with increasing hydrogen pressure and is lowest under the conditions of the thermal loading specified here.

Comparing the embrittlement index EI (Equation 16) of longitudinal and transverse duplex steel specimens, a difference can be found only at a lower hydrogen pressure (0.1 MPa) and lower strain rates ($< 10^{-3}$ 1/s) (Figure 89a). A longitudinal alignment of the austenite grains in the longitudinal direction of the specimens makes a crack propagation more difficult since the elongated austenite grains form an effective barrier for cracks to propagate. The values come closer to each other as the hydrogen pressure increases (Figure 89b).

Another study examined the influence of the austenite content in the duplex steel on the hydrogen embrittlement [127]. The specimens were taken from a duplex steel pipe (wall thickness 12.7 mm) and had the chemical composition indicated in Table 53 as well as an austenite content of 35% in a ferritic matrix.

Different austenite contents could be obtained by performing specific heat treatments (1 h, 1,250 °C/H_2O; 1 h, 1,300 °C/H_2O; 1 h, 1,200 °C/slow cooling to 1,000 °C/H_2O). Quenching was always performed in ice water. The obtained austenite contents and their influence on the mechanical properties are listed in Table 54. The ferrite content was determined by means of a calibrated ferrite meter. With increasing ferrite content the strength increases and the ductility decreases.

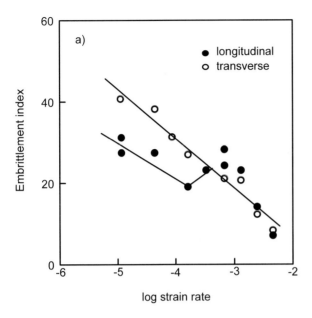
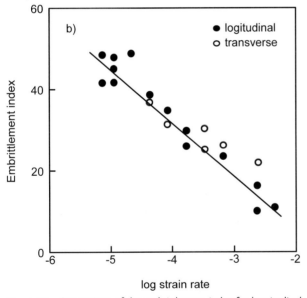

Figure 89: Comparison of the embrittlement index for longitudinal and transverse specimens from duplex steel in hydrogen gas at 0.1 MPa (a) and 0.2 MPa (b) [126]

Cr	Ni	Mo	Mn	N
23.0	5.0	3.0	1.0	0.13

Table 53: Chemical composition of the duplex steel pipe in wt% [127]

Heat treatment	Austenite content %	$R_{p0.2}$ MPa	R_m MPa	Z %	A %
1 h, 1,200 °C → 1,000 °C/H$_2$O	50	592	758	80.5	39.1
1 h, 1,250 °C/H$_2$O	15	704	807	64.7	30.6
1 h, 1,300 °C/H$_2$O	0	743	844	51.7	19.9

Strain rate 3.7 × 10-6 1/s ; →: slow cooling to 1,000 °C: 5 K/min

Table 54: Influence of heat treatment on the austenite content and the mechanical properties of the examined duplex steel [127]

The influence of heat treatment of the duplex steel on its susceptibility to hydrogen embrittlement was determined in slow strain rate tensile tests (6.5 × 10^{-6} to 6.4 × 10^{-3} 1/s) in 0.2 MPa compressed hydrogen and tensile tests in air following hydrogen charging (in 25 MPa H$_2$ at 350 °C for 48 h). The embrittlement index EI = 100 (A$_{air}$ − A$_H$)/A$_{air}$) resulting from the fracture elongation A in air and in hydrogen served as the measure for the susceptibility. The results are summarized in Figure 90. As can be seen, embrittlement increases with increasing heat treatment temperature (1,000–1,300 °C), which however is attributable to the increasing fer-

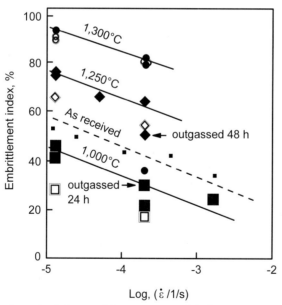

Figure 90: Influence of the strain rate and the heat treatment of the examined duplex steel on the embrittlement index EI in 0.2 MPa hydrogen (open symbols) and in air following hydrogen precharging in 25 MPa hydrogen (full symbols) [127]

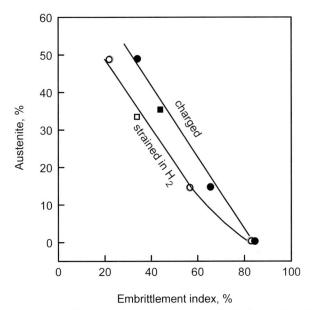

Figure 91: Influence of the austenite content of a duplex steel on the embrittlement index EI in 0.2 MPa hydrogen (open symbols) and in air following hydrogen precharging in 25 MPa hydrogen (full symbols). Result in the delivery state for comparison (square symbol). Strain rate: 10^{-4} 1/s [127]

rite content. Moreover, it can be seen that the embrittlement of the specimens precharged in 25 MPa compressed hydrogen and broken in air is slightly higher than that under the direct influence of 0.2 MPa hydrogen. This is confirmed by the findings in [126].

In this respect Figure 91 illustrates the influence of the austenite content (0 to 50%) on the embrittlement index EI for specimens either deformed until fracture in 0.2 MPa hydrogen or in air following hydrogen precharging in 25 MPa compressed hydrogen. For comparison the result obtained for steel in its delivery state (35% austenite).

Also these results confirm that the austenite/ferrite content of a duplex steel exerts a major influence on the susceptibility to hydrogen embrittlement. The most favorable behavior of the duplex steel is found if austenite and ferrite are present at a ratio of 50/50, such steel being used most frequently.

Duplex steel Zeron® 100 (Mat. No. 1.4501) was used to examine the hydrogen-based fatigue behavior in gaseous hydrogen [128]. The chemical composition of the duplex steel is indicated in Table 55. The material was rolled at 1,150 °C and quenched in water and, thereafter, exhibited a microstructure containing 50 vol%

Cr	Ni	Mo	Mn	Cu	W	Si	P	S	C	N	Cr
24.04	6.827	3.770	0.770	0.626	0.635	0.175	0.025	0.002	0.024	0.215	24.04

Table 55: Chemical composition of the duplex steel Zeron® 100 in wt% [128]

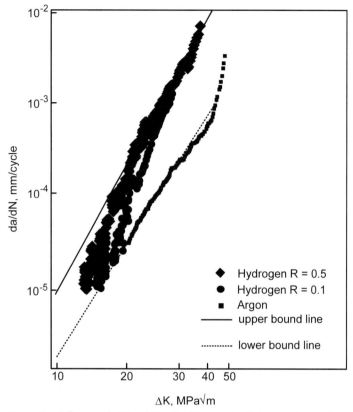

Figure 92: Influence of ΔK and R = K_{min}/K_{max} on the fatigue crack growth rates in the duplex steel Zeron® 100 in hydrogen gas [128]

austenite in a ferritic matrix. The fatigue behavior was examined using CT (compact tension) specimens with incipient fatigue crack in hydrogen gas at RT, frequencies of 0.1 and 5 Hz and R = K_{min}/K_{max} values of 0.1 and 0.5. During the tests at 5 Hz the hydrogen pressure of the pre-evacuated specimen chamber was 950 mbar and the test duration was less than 4 h. At 0.1 Hz hydrogen was absorbed in a flow meter apparatus with a continuous hydrogen flow slightly above atmospheric pressure. Here, the test duration was 1 week.

Figure 92 shows the crack growth as a function of the amplitude of the stress intensity factor ΔK for R = 0.5 and R = 0.1 in hydrogen. Argon served for comparison. At higher ΔK values the crack propagation velocity in hydrogen is considerably higher than in argon and the influence of the R values gradually disappears. Although the crack propagation per load cycle da/dN at R = 0.5 is higher than that at R = 0.1 at lower ΔK values (< 25 MPa √m), the da/dN values gradually approach the results in argon with ΔK further decreasing. Fractographic examinations proved that the cleavage fracture content in the ferritic phase increases with increasing ΔK values and is higher at R = 0.5 than at R = 0.1. When plotting the

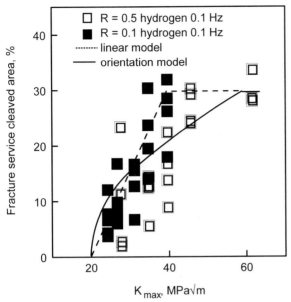

Figure 93: Influence of the maximum stress intensity K_{max} on the cleavage fracture surface area [128]

cleavage fracture content as a function of the maximum stress intensity K_{max}, all values are in the same scatter band and are independent from R (Figure 93). This means that the cleavage fracture content directly depends on K_{max}. There is no cleavage fracture at $K_{max} < \sim 20$ MPa $\sqrt{m}$ and at $K_{max} > \sim 40$ MPa $\sqrt{m}$ some sort of a "saturation" is reached since then a cleavage fracture occurs in all ferrite grains.

The super duplex steel Zeron® 100 (Table 55) was also used to examine the influence of aging on the embrittlement behavior in hydrogen gas [129]. In its as-received state the material had a yield strength R_p of ~ 645 MPa and a tensile strength R_m of ~ 840 MPa and the austenite content was 50 vol%. The specimens were aged at 400 °C for 100, 500, 1,000 and 5,000 hours. The specimens were subjected to alternating mechanical sinusoidal loads and the crack depth was determined by using the potential probe technique. The tests were performed in hydrogen flowing at atmospheric pressure, 30 °C, 0.1 Hz and R = K_{min}/K_{max} = 0.5. In addition, tests were performed in air at 5 to 20 Hz at 25 °C each. The results of the Vickers hardness tests (100 kg load) and microhardness tests (0.1 to 0.25 N) are indicated in Table 56. The microhardness increases with the aging period. However, hardening occurred only in the ferrite phase since only in this phase the microhardness increases with aging.

The crack propagation velocity in air increases with aging (Figure 94). The cleavage fracture surface content also increases with increasing ΔK values and increasing macrohardness. In hydrogen gas the crack propagation velocity is up to one order of magnitude higher than in air (Figure 94). There is no difference between the H-induced transcrystalline cleavage fracture in the ferritic matrix and that in air, however the fracture content is clearly higher.

Material	Vickers hardness	Microhardness, HV	
	HV	Austenite	Ferrite
as-received state	255	298	405
100 h at 400 °C	320	–	–
500 h at 400 °C	340	–	–
1,000 h at 400 °C	352	–	–
5,000 h at 400 °C	354	404	604

Table 56: Vickers hardness and microhardness of the super ferrite Zeron® 100 [129]

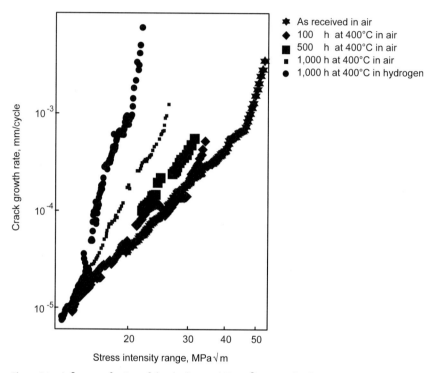

Figure 94: Influence of aging of the duplex steel Zeron® 100 on the fatigue behavior in air and hydrogen gas [129]

To clarify the influence of the austenite/ferrite ratio on hydrogen embrittlement, examinations of FeMnAl alloys were performed [130]. In these steels chromium was replaced by manganese and nickel was replaced by aluminum. Whereas all these duplex steels exhibit similar manganese and aluminum contents, the austenite/ferrite ratio can be varied by the carbon content (Table 57). Increasing the carbon content from 0.33 to 1.047% changes the microstructure from almost purely ferritic (98 vol% ferrite) to almost purely austenitic (0.5% ferrite). The ferrite content was determined by means of a ferrite meter and by performing EPMA (electron probe microanalysis). The sheet specimens (150 × 20 × 1.5 mm) were solution-annealed in argon at 1,050 °C for 1 h and quenched in water. The mechan-

A 19 Ferritic/pearlitic-martensitic steels

Material	Mn	Al	C	Cr	Ferrite content (%)
A	28.52	9.97	1.047	–	0.5
B (medium)	29.60	10.19	0.832	–	10
	24.39	11.26	–	–	–
	30.04	9.62	–	–	–
C (medium)	28.63	10.45	0.498	–	35
	26.66	9.79	–	–	–
	31.36	8.22	–	–	–
D (medium)	29.99	10.19	0.305	–	65
	28.13	10.33	–	–	–
	32.25	8.81	–	–	–
E	21.50	9.86	0.330	6.23	98

Table 57: Chemical composition (wt%) and ferrite content of the FeMnAl alloys [130]

ical properties are indicated in Table 58. Whereas the tensile strength R_m increases with the austenite content and also with the C content, the yield strength first decreases with the ferrite content and increases again above 35% ferrite. The microhardness is always higher in the ferrite phase compared to the austenite phase.

Material	R_p MPa	R_m MPa	A %	Microhardness	
				Austenite	Ferrite
A	570	910	45	–	302
B	485	790	53	333	300
C	400	725	36	327	292
D	425	705	33	306	270
E	438	455	3	320	–

Table 58: Mechanical properties of the FeMnAl alloys [130]

To determine the susceptibility of the FeMnAl alloys to hydrogen embrittlement CERT tests (constant extension rate tensile tests) were performed using longitudinal tensile test specimens notched on one side at a strain rate of 5.8×10^{-4} mm/s and 5.8×10^{-5} mm/s in air and 1 atmosphere hydrogen gas at 25, 50 and 75 °C. The results of the CERT tests for the specimens A (99.5% austenite, 0.5% ferrite) and E (98% ferrite, 2% austenite) are shown in Figure 95. Whereas the quasi austenite (specimen A with 0.5% ferrite) is mainly resistant to hydrogen embrittlement, hydrogen causes a clear reduction of the tensile strength of the quasi ferrite (specimen E with 98% ferrite) at all test temperatures (25 to 75 °C). The results of the test materials with the duplex microstructure – specimen B (10% ferrite), C (35% ferrite) and D (65% ferrite) – are shown in Figure 96. The strongest hydrogen influence is found at 25 °C. The reduction of the strength of specimen C at 25, 50

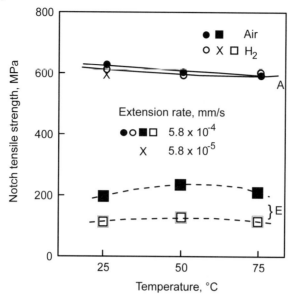

Figure 95: Comparison of the mechanical properties for the test materials A (99.5% austenite) and E (98% ferrite) in air and 108 kPa hydrogen gas at different temperatures [130]

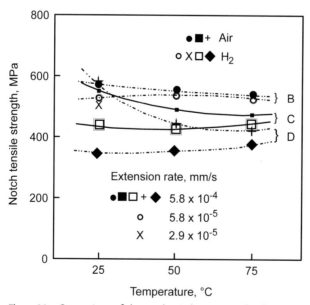

Figure 96: Comparison of the mechanical properties for the test materials B (10% ferrite), C (35% ferrite) and D (65% ferrite) in air and 108 kPa hydrogen gas at different temperatures [130]

and 75 °C is 20, 11 and 2%, respectively. A comparison of the results in Figure 95 and Figure 96 makes clear that the susceptibility of duplex steels to hydrogen embrittlement clearly increases with increasing ferrite content.

In addition, tests were performed under constant load over time in 108 kPa hydrogen on notched specimens from FeMnAl alloys provided with an incipient fatigue crack [130]. The maximum load was 70% of the yield strength. If no crack growth occurred within 24 hours, the load was increased by 1 to 5%. The crack growth was observed using a special microscope with a resolution of 6.2 × 10^{-2} mm. The tests were limited to the materials B to D. Specimen B (10% ferrite) exhibited a crack arrest immediately after crack initiation, attributable to a strong plastification in front of the crack tip. The conclusion was made that there is no hydrogen-induced crack growth under constant load over time in specimen B. The results in Figure 97 and Figure 98 show the influence of the crack intensity and the temperature on the slow crack growth rate in the specimens C (35% ferrite) and D (65% ferrite). It should be noted that the slow crack growth rate increases with increasing temperature in material C and reduces in material D. The explanation in [130] is as follows: The rate-determining step in specimen C is the diffusion of hydrogen through the ferrite phase to the crack tip. Specimen D rather behaves like a ferritic steel, the negative temperature influence being attributed to the distribution of hydrogen on the various slip planes. Comparing the material influence (specimen C and D) at 27 °C, it becomes clear that an increasing ferrite content reduces the threshold value K_{IH} for the starting crack growth and considerably increases the slow crack growth rate (Figure 99).

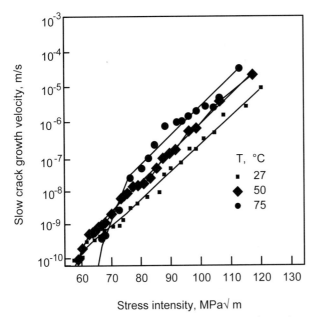

Figure 97: Influence of the temperature on the slow crack growth rate in the alloy C (35% ferrite) in 108 kPa hydrogen gas [130]

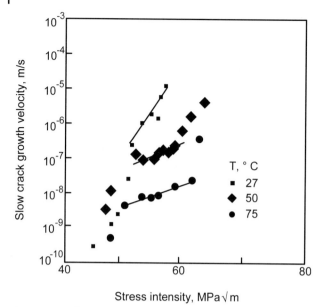

Figure 98: Influence of the temperature on the slow crack growth rate in the alloy D (65% ferrite) in 108 kPa hydrogen gas [130]

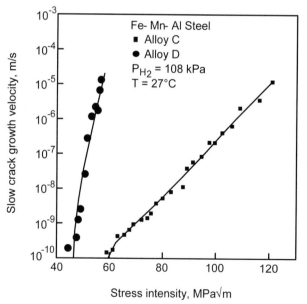

Figure 99: Comparison of slow crack growth rates of alloys C (35% ferrite) and D (65% ferrite) in 108 kPa hydrogen gas at 27 °C [130]

A 20   Austenitic CrNi steels

A 21   Austenitic CrNiMo steels

A 22   Austenitic CrNiMo(N) and CrNiMoCu(N) steels

In contrast to the ferritic steels, austenitic steels are resistant to hydrogen embrittlement to a major extent. This justifies the widespread use of austenites in hydrogen-containing atmospheres. However, hydrogen resistance only applies to stable austenitic steels. If a partial conversion into martensite occurs in unstable austenites, e.g. by forming processes, these steels also become susceptible to hydrogen.

In [131] the influence of hydrogen gas on the strength and ductility of various austenitic steels was examined. In addition, tests were performed regarding hydrogen diffusion and permeation. The test materials and their chemical compositions are listed in Table 59 and Table 60 contains the strengths. To stabilize the austenitic microstructure, part of the nickel was replaced by manganese and nitrogen in the material NITRONIC® 40 (cf.: 1.3965, 1.4454) and by copper and nitrogen in 311DQ. The examined FeMnAl alloy exhibited a stable austenite microstructure. The CrNi steels were solution-annealed (1,065 °C/H$_2$O) and the FeMnAl alloy was vacuum-annealed at 950 °C/ H$_2$O for 2 h with subsequent aging (1 h, 550 °C).

Material	C	Mn	P	S	Si	Cr	Ni	N	Cu	Al	Mo
311DQ	0.05	2.72	0.025	0.012	0.53	16.55	4.77	0.13	2.39		0.21
311DQ	0.04	2.50			0.50	17.25	4.50	0.15	2.40		
NITRONIC® 40 (cf.: 1.3965, 1.4454)	0.04	9.00			1.00	20.50	6.00	0.30			
FeMnAl	0.94	30.7								8.9	
SAE 301 (cf.: 1.4310)	0.052	1.28		0.009	0.48	17.1	7.25	0.038			0.24
SAE 304 (cf.: 1.4301)	0.062	1.31	0.023	0.020	0.57	18.35	9.19			0.27	0.14

Table 59:   Chemical compositions of the test materials in wt% [131]

Material	$R_{p0.2}$, MPa	R_m, MPa
311DQ	345	793
NITRONIC® 40	469	772
FeMnAl	n.d.	1,242

n.d.: not determined

Table 60:   Mechanical properties of several test materials [131]

To evaluate the austenite stability, the specimens were cold-rolled with different deformation degrees and subsequently the content of α′-martensite was determined by means of a ferrite meter (Figure 100). As can be seen, the austenite of the steel 311DQ is slightly more stable than that of SAE 301, however considerably more unstable than that of the steel SAE 304. The materials NITRONIC® 40 and

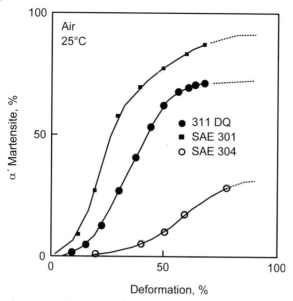

Figure 100: Comparison of the amount of martensite produced by cold rolling of unstable austenitic steels [131]

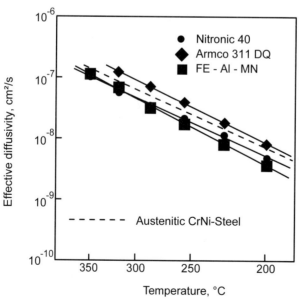

Figure 101: Influence of hydrogen diffusivity in the austenitic steels NITRONIC® 40, 311 DQ and FeMnAl. The dashed curve applies to the materials SAE 301 (1.4310), 304 (1.4301) and 310 S (1.4845), [131]

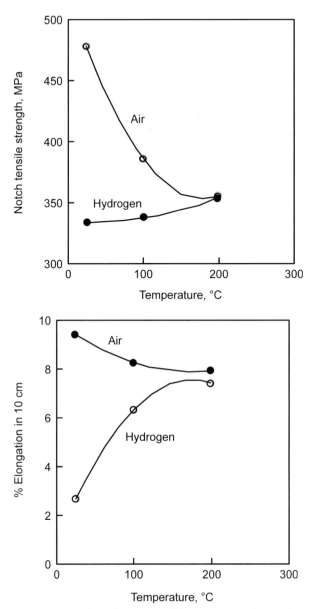

Figure 102: Mechanical properties of the austenitic steel 311 DQ in air and in 108 kPa hydrogen gas [131]

FeMnAl did not even exhibit any magnetizing phase following cold forming up to 60% [131].

The hydrogen transport in the austenites was examined by means of the gas phase permeation technique at 200 to 350 °C. To this end, two ultra high vacuum chambers were separated by a membrane of the test material. Both sides of the

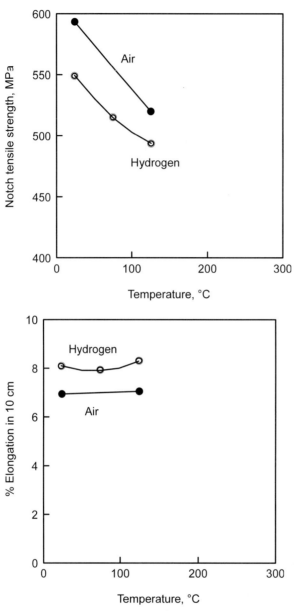

Figure 103: Mechanical properties of the austenitic steel NITRONIC® 40 in air and in 108 kPa hydrogen gas [131]

membrane were vapor-coated with palladium. Following the presetting of hydrogen in one chamber from 0.1 to 30 kPa, the permeated hydrogen in the other chamber was determined by means of a calibrated ion pump. Figure 101 shows the influence of the temperature on the diffusion coefficient D of the three auste-

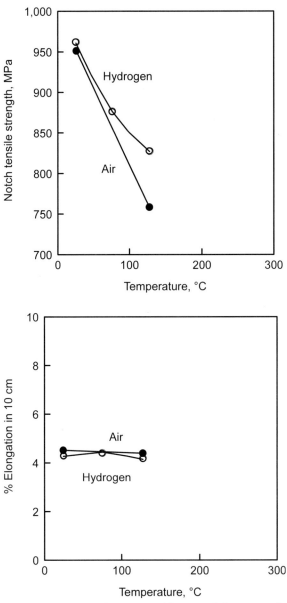

Figure 104: Mechanical properties of the FeMnAl alloy in air and in 108 kPa hydrogen gas [131]

nitic steels NITRONIC® 40, 311 DQ and FeMnAl (Fe-Al-Mn) in comparison to the materials SAE 301 (1.4310), 304 (1.4301) and 310 S (1.4845), identified by dashed curves. Since the measured values differ from the standard austenites by a factor 2 only, the hydrogen diffusion rate is considered to be almost identical.

The susceptibility of the austenitic steels to hydrogen embrittlement was examined by performing slow strain rate tensile tests (8.5×10^{-7} 1/s) using specimens notched on one side in 108 kPa hydrogen gas [131]. The atmosphere for comparison was air. The temperatures covered a range from 25 to 200 °C. The tensile strength and fracture elongation served as a measure for the susceptibility. The results are shown in Figure 102, Figure 103 and Figure 104. The 311 DQ steel showed a considerable reduction of the tensile strength by 30% and of the ductility by 72% under load in 108 kPa compressed hydrogen at 25 °C. The examination using a scanning electron microscope revealed a quasi cleavage fracture with secondary cracks. At higher temperatures a ductile fracture behavior occurred. The strength of the steel NITRONIC® 40 in hydrogen in a temperature range from 25 to 125 °C was reduced only by 4 to 7% and its ductility was even slightly higher than in air. In contrast, the strength of the FeMnAl alloy in hydrogen slightly increased (0.6 to 9.5%) and fracture elongation slightly decreased (1.2 to 5.2%). All examined NITRONIC® 40 and FeMnAl specimens broke ductily in air and in hydrogen at 25 to 125 °C. The higher resistance of the NITRONIC® 40 and FeMnAl steels is attributed to the higher manganese content (9 and 30.7%, respectively). In contrast, part of the nickel of the steel 311 DQ was replaced by copper (2.4%), obviously resulting in a negative effect on the resistance to hydrogen.

To clarify whether hydrogen-induced subcritical crack growth occurs in the austenitic steels, tests under constant load over time were performed in 108 kPa hydrogen at 25 °C using notched specimens provided with an incipient fatigue crack [131]. Whereas no crack growth was found in NITRONIC® 40 and FeMnAl, a subcritical hydrogen-induced crack growth occurred in the steel 311 DQ (Figure 105). The standard austenite SAE 301 (1.4310) served for comparison. The threshold

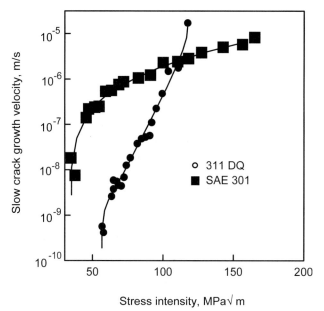

Figure 105: Comparison of the slow crack growth rate in the steels 311 DQ and SAE 301 (1.4310) in 108 kPa hydrogen gas at 25 °C [131]

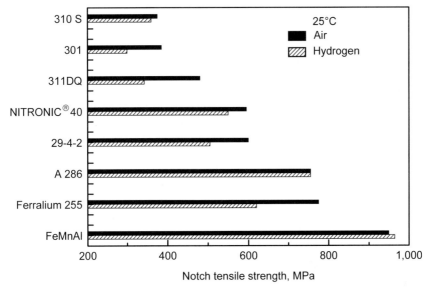

Figure 106: Comparison of the notch tensile strengths of various stainless steels in air and hydrogen gas at 25 °C [131]
Ferrite: AL 29-4-2
Austenite: SAE 301 (1.4310), SAE 310 S (1.4845), 311 DQ, NITRONIC® 40 (1.3965), A 286 (1.4943), FeMnAl
Duplex steel: Ferralium® 255

value K_{IH} for the initial subcritical crack growth is slightly higher in 311 DQ, however the stress intensity factor K_{Ic} for the unstable crack growth is clearly lower than that of SAE 301. The different K_{Ic} values are attributed to the different strengths of the two materials. The slow crack growth rate in the area II of 311 DQ is lower than that of SAE 301 and, hence, the standard austenite is considered to be less resistant. It is assumed that the reason for that is the faster γ/α'-conversion of the unstable SAE 301 during plastic deformation [131].

Figure 106 compares the strength values determined for various stainless steels in air and in hydrogen. The materials include a ferritic (AL 29-4-2), several austenitic (SAE 301, SAE 310 S, 311 DQ, NITRONIC® 40, A 286, FeMnAl) and a duplex steel (Ferralium® 255). All values were determined under the condition of identical specimen geometries and by applying the same testing technique. All steels in brackets "}" had a more or less high ferrite content during the test and a stronger hydrogen influence compared to air was observed for these materials. Summarizing the above, only unstable austenitic steels are susceptible to hydrogen embrittlement.

A type CrNiTi18-9 stainless steel was examined to establish the influence of a directional solidification (DT) compared to cast steel (CC) and forged steel (WA) on the resistance to hydrogen embrittlement [132]. Directional solidification took place at a high solidification rate (1.2 × 10⁻⁴ m/s). Table 61 shows the chemical composition of the directionally solidified steel. Following heat treatment (1 h

C	Cr	Ni	Mn	Ti	Si	P	S	Fe
0.08	18.38	10.27	1.00	0.77	0.51	0.0027	0.006	balance

Table 61: Chemical composition of the directionally solidified steel [132]

1,050 °C/H$_2$O) smooth round test bars were manufactured and charged with hydrogen in 10 MPa compressed hydrogen (99.999% H$_2$) at 300 °C for 140 h. The hydrogen content was 2.2 to 2.3 ppm prior to charging and 28.8 ppm after charging of the cast steel and 42 to 43 ppm after charging of the other steels (refer to Table 62). Immediately after the charging with hydrogen tensile tests (4.2 × 10^{-5} m/s) were performed at room temperature. The strength values obtained for uncharged specimens and specimens charged with hydrogen are summarized in Table 62 for forged steel (WA), cast steel (CC) and steel with directional solidification in the longitudinal and transverse direction {DT and DL}. Whereas the strength slightly increases as a result of the hydrogen absorption, the ductility of all examined materials noticeably decreases. The resistance to hydrogen embrittlement is better in case of directional solidification compared to cast or forged steel of identical chemical composition.

Specimen	$R_{p0.2}$ MPa	$R_{p0.2}$ (H$_2$) MPa	R_m MPa	R_m (H$_2$) MPa	Z %	Z (H$_2$) %	A %	A (H$_2$) %	H$_2$ ppm
WA	261.7	283.2	635.0	636.0	74.3	59.6	61.0	38.4	43.2
CC	255.8	266.6	525.3	530.2	49.5	32.9	47.6	37.7	28.8
DT	231.3	234.2	485.1	502.7	64.4	53.2	61.5	49.6	43.2
DL	242.1	249.9	489.0	543.9	73.3	65.0	43.0	39.5	42.3

WA = forged steel; CC= cast steel; DT = directional solidification in transverse direction; DL = directional solidification in longitudinal direction

Table 62: Mechanical properties of the test materials with and without hydrogen charging [132]

Following hydrogen absorption, the ferrite content of the steels was radiographically determined (Table 63). As can be seen, only the forged steel remains ferrite-free. All other test materials exhibited almost the same ferrite content of 4.2 to 4.3%, however they clearly differed in terms of their reduction of area Z (refer to Table 62). The conclusion was drawn that it is not only the ferrite content, but also the ferrite distribution – isolated (cast steel) or continuous (directional solidification) which plays a role in the hydrogen embrittlement of the examined steel [132]. In case of directional solidification the continuous "ferrite paths" can serve as short diffusion channels for hydrogen diffusion and effusion and, hence, reduce hydrogen embrittlement [132].

Specimen	Forged steel	Cast steel	Directional solidification (T)	Directional solidification (L)
Ferrite content	0.0	4.2	4.2	4.3

Table 63: Ferrite content of the test materials in wt% [132]

Ni	Cr	Mo	Ti	Al	V	C	B
30	15	1.5	2.0	0.25	0.25	0.025	0.002

Table 64: Chemical composition of the steel JBK-75 [133]

The aged austenitic steel JBK-75 (type X3NiCrMoTi30-15) was examined to establish the influence of the microstructure on the susceptibility to hydrogen embrittlement [133]. Table 64 shows the nominal chemical composition of the test material.

The steel, which was first solution-annealed (1 h, 980 °C/H$_2$O), exhibited an austenitic microstructure with < 0.1% TiC and traces of oxidic inclusions. Then, the specimens were aged in a temperature range from 670–800 °C with air cooling for 16 h. The result was the precipitation-hardening fcc γ′-phase Ni$_3$(TiAl) and in the case of overaging also the hcp η-phase Ni$_3$Ti, along the grain boundaries identified by η$_{KG}$ and η$_{intra}$ inside the grain.

The aged round test bars (5 mm diameter) were charged with high-purity 10 MPa compressed hydrogen at 300 °C over a period of 7 days, leading to a hydrogen saturation. Immediately thereafter, the tensile tests were performed with a strain rate of 2.5 mm/min in air. The influence of the aging temperature on the strength values R$_m$ and R$_{p0.2}$ is clearly marked, however the charging with hydrogen remains without almost any effect (Figure 107). In contrast, hydrogen charging impairs the ductility. The ductility loss in percent as a function of the aging temperature is shown in Figure 108. The following conclusions can be drawn from these findings:

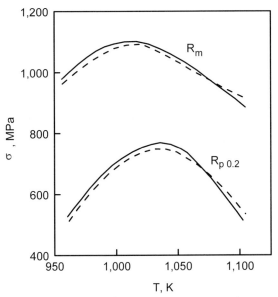

Figure 107: Influence of the aging temperature on the strength of hydrogen-charged (---) and uncharged (——) JBK-75 specimens [133]

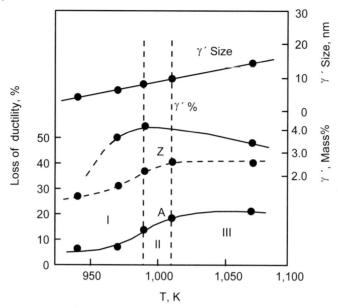

Figure 108: Influence of the aging temperature of the steel JBK-75 on the ductility loss [$Z = (Z_0 - Z_H)/Z_0 \times 100$ and $A = (A_0 - A_H)/A_0 \times 100$] and the formation of the γ'-phase [133].

- If there is no η-phase, the hydrogen-induced damage to the material increases in case of the aged austenitic steel JBK-75 with the magnitude of the γ'-precipitates and the content of the γ'-phase.
- Precipitation of the η-phase at the grain boundaries increases the hydrogen-induced crack formation along the grain boundaries, whereas its precipitation inside the grain remains without almost any influence.

Also the austenitic grain size exerts an effect on the ductility of hydrogen charged specimens. Results obtained from the aged (16 h 720 °C/air) steel JBK-75 are available (Figure 109). Fracture elongation and reduction of area are clearly reduced with increasing grain size.

The austenitic steels X2CrNiMn21-6-9 and X10CrNiTi18-9 were examined to establish the influence of the internal hydrogen content on the H-induced embrittlement [134]. The different hydrogen content was obtained by a different charging duration in compressed hydrogen. Both test materials with the chemical compositions indicated in Table 65 were solution-annealed (1,050 °C/H_2O and 1,100 °C/H_2O) first. Hydrogen charging was performed using notched round test bars in 24 MPa compressed hydrogen at 200 °C for 0.7, 1.6, 2, 6 and 14 days. Immediately after hydrogen charging slow strain rate tensile tests were performed at a strain rate of 6.7×10^{-5} 1/s in air and concurrently the internal hydrogen content was determined.

The hydrogen content increases with charging time (Figure 110). However, the steel X2CrNiMn21-6-9 absorbs hydrogen more rapidly than X10CrNi18-9. After a charging time of 14 days hydrogen contents of 64 ppm and 38 ppm were obtained

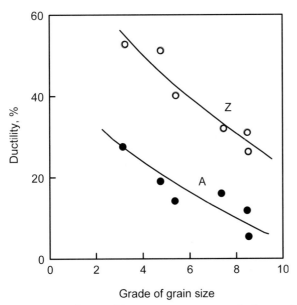

Figure 109: Influence of the austenite grain size on the fracture elongation and the reduction of area of the steel JBK-75 [133]

Steel	Chemical composition, wt%							
	C	Cr	Ni	Mo	Ti	Mn	Si	S	P
X2CrNiMn21-6-9	0.019	20.25	7.42	0.23		9.43	0.48	0.0026	0.031
X10CrNi18-9	0.097	18.17	9.58		0.42	1.48	0.47	0.0073	0.033

Table 65: Chemical composition of the test materials [134]

for both steels. Figure 111 shows the influence of the internal hydrogen concentration on the reduction of the ductility $[Z = (Z_0 - Z_H)/Z_0 \times 100]$ in the slow strain rate tensile test. Here again the results of both steels differ considerably. At 40 ppm hydrogen the ductility loss of X2CrNiMn21-6-9 is about 10% and that of X10CrNi18-9 is about 42%. Therefore, the stable austenitic steel X2CrNiMn21-6-9 is much more resistant to hydrogen embrittlement than the metastable X10CrNi18-9 in which the absorption of hydrogen leads to a reduction of the stacking fault energy and finally to the conversion of austenite into ε-martensite [134].

The influence of hydrogen on the microstructure, the mechanical properties and the phase transformation of austenitic steels was examined more closely using the material NITRONIC® 40 (cf.: 1.3965, X8CrMnNi18-8) [135]. The steel with the chemical composition indicated in Table 66 was solution-annealed (1 h, 1,050 °C/ H_2O) first. The specimens were charged with hydrogen in 10 MPa hydrogen gas (> 99.999%) at 300 °C for 14 days. The hydrogen content of the steel NITRONIC® 40 was 3.9 ppm prior to charging and about 65 ppm after charging.

Immediately after charging with hydrogen tensile tests were performed in air. Figure 112 shows the influence of hydrogen charging on the strength in the tem-

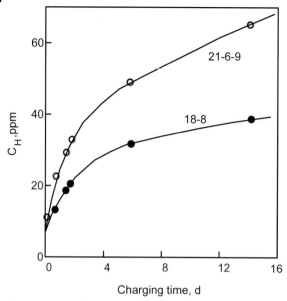

Figure 110: Hydrogen concentration (C_H) in the austenitic steels X2CrNiMn21-6-9 and X10CrNi18-9 as a function of the hydrogen charging time [134]

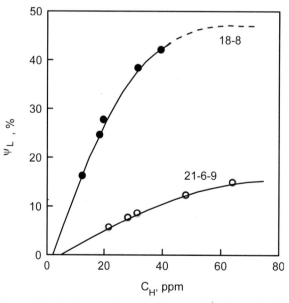

Figure 111: Hydrogen-induced ductility loss (Ψ_L) of the steels X2CrNiMn21-6-9 and X10CrNi18-9 as a function of the hydrogen concentration. $Z = (Z_0 - Z_H)/Z_0 \times 100$ [134]

C	Cr	Ni	Mn	Si	N	S	P
0.034	20.52	7.60	9.24	0.40	0.24	0.005	0.032

Table 66: Chemical composition of the austenitic steel Nitronic® 40 in wt% [135]

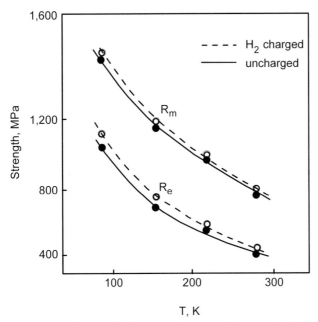

Figure 112: Influence of hydrogen precharging on the strength of the steel NITRONIC® 40 in the temperature range from –196 °C to 27 °C [135]

perature range from –196 °C to 27 °C. Hydrogen charging leads to a slight increase of the strength. Accordingly, a slight decline in the ductility was found. Also the steel SAE 316 L (cf.: 1.4404, 1.4435) only shows a minor influence of hydrogen charging on the mechanical properties [135]. In the toughness test, precharging of smooth and notched NITRONIC® 40 specimens with hydrogen hardly impairs the toughness (Figure 113).

The influence of the δ-ferrite content on the hydrogen embrittlement of austenitic steels was examined using Cr-Ni-Mn-N steels [135]. The Cr, Mn, C and Si contents were changed only to such extent that the composition still corresponded to a NITRONIC® 40. The composition of the test materials and the respective δ-ferrite content are indicated in Table 67. The specimens were charged with hydrogen in 10 MPa hydrogen gas (> 99.999%) at 300 °C for 4 days. The hydrogen content of the specimens was 3 to 4 ppm prior to charging. After charging the hydrogen content of the specimens 1 to 9 was: 46, 46, 48, 46, 53, 49, 40, 43 and 43 ppm. The calculated hydrogen embrittlement index EI characterized as the loss of necking (EI = 100% $(Z_0 - Z_H)/Z_0$) is also indicated in Table 67. As can be seen, the δ-ferrite content has a clear influence on hydrogen embrittlement.

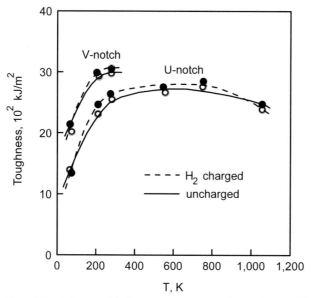

Figure 113: Influence of hydrogen precharging on the toughness of the steel NITRONIC® 40 at various temperatures [135]

To vary the δ-ferrite content of the steel with identical chemical composition, the specimens were subjected to different heat treatments (950 to 1,250 °C) [135]. This treatment was performed with the specimens 2 and 3 since they considerably differed in their δ-ferrite content (6.63 and 0.57%) after solution annealing at 1,050 °C. According to Figure 114 hydrogen embrittlement of both test materials considerably increases with increasing δ-ferrite content.

Also the influence of a sensitization on hydrogen embrittlement was tested using the steel NITRONIC® 40. To this end specimens were aged following solution annealing at 1050 °C and the chromium depletion caused by $M_{23}C_6$ precipitates at the grain boundaries was determined adjacent to the grain boundary (dis-

No.	Cr	Mn	C	Si	S	P	Ni	N	δ	El*
1	20.42	9.32	0.016	0.48	0.015	0.009	4.09	0.21	7.72	38.96
2	20.02	9.29	0.020	0.51	0.011	0.011	6.13	0.15	6.63	47.72
3	20.15	9.35	0.035	0.58	0.010	0.013	6.20	0.22	0.57	30.91
4	20.25	9.25	0.017	0.51	0.013	0.009	8.02	0.20	0.11	18.27
5	20.43	9.00	0.014	0.48	0.013	0.009	10.17	0.20	0.01	21.27
6	20.60	9.21	0.018	0.37	0.013	0.008	10.76	0.23	0	24.72
7	20.36	9.30	0.023	0.39	0.014	0.010	11.67	0.16	0	20.45
8	20.38	9.18	0.028	0.37	0.014	0.011	11.63	0.22	0	18.48
9	20.37	9.22	0.018	0.40	0.015	0.010	11.77	0.28	0	22.63

* El = 100% $(Z_0 - Z_H)/Z_0$, mean values from 3 tests

Table 67: Chemical composition, δ-ferrite content and necking loss after hydrogen charging of the test materials [135]

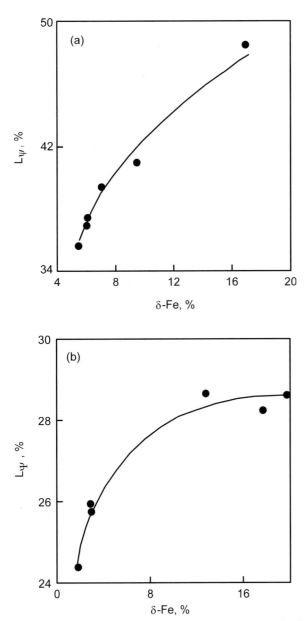

Figure 114: Influence of the δ-ferrite content on hydrogen embrittlement (L_ψ) of the test materials 2 (a) and 3 (b) [135]

146 | Hydrogen

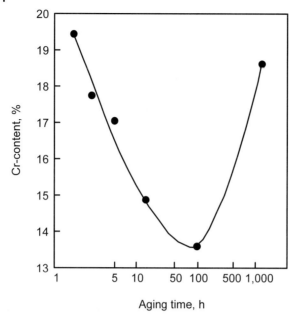

Figure 115: Chromium content at a distance of 0.1 μm from the grain boundary as a function of the aging time of the steel NITRONIC® 40 at 650 °C [135]

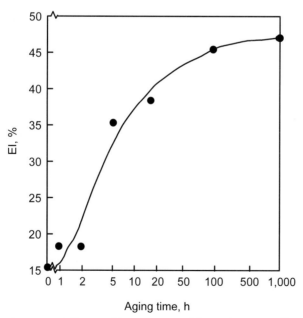

Figure 116: Influence of the aging time of the steel NITRONIC® 40 at 650 °C on the hydrogen embrittlement index $EI = 100\% \, (Z_0 - Z_H)/Z_0$ [135]

tance 0.1 µm) [135]. The influence of the aging time at 650 °C on chromium depletion is shown in Figure 115. As can be seen, chromium depletion increases up to 100 h aging and, thereafter, the chromium concentration gradually returns to the original concentrations. On the other hand, the susceptibility of NITRONIC® 40 to hydrogen embrittlement constantly increases with increasing aging time up to 1,000 h (Figure 116). This proves that the hydrogen embrittlement of the steel does not depend on chromium depletion but on the precipitation of the carbides.

Moreover, three different Cr-Ni-Mn-N steels (Table 68) were examined to establish the influence of deformation and hydrogen charging on the formation of α'- and ε-martensite [135]. The mechanical properties altered by hydrogen charging are shown in Table 69; they were determined in another study [136]. The influence of deformation and dissolved hydrogen on the content of ε- and α'-martensite in Cr-Ni-Mn-N steels is summarized in Table 70 [135]. These results are important since the formation of martensite increases the susceptibility to hydrogen embrittle-

Material	Cr	Ni	Mn	N	C	Si	S	P
21	20.42	6.90	8.92	0.28	0.039	0.36	0.013	0.005
18	18.74	5.54	6.83	0.17	0.017	0.038		
12	12.35	5.59	8.36	0.10	0.028	0.42	0.012	0.009

Table 68: Chemical composition of the Cr-Ni-Mn-N steels in wt% [135]

Material	H_2 uncharged				H_2 charged			
	$R_{p0.2}$ MPa	R_m MPa	A %	Z %	$R_{p0.2}$ MPa	R_m MPa	A %	Z %
21	387	742	72.0	75.1	442	769	71.3	63.6
18	346	739	79.1	70.6	385	732	41.8	35.2
12	292	877	61.3	64.7	342	565	29.3	low

Table 69: Strength values of the examined Cr-Ni-Mn-N steels [135]

Material	State	ε (%)		α' (%)	
		uncharged	H_2-charged	uncharged	H_2-charged
21	undeformed	0	0	0	0
21	deformed	0	0	0	0.03
18	undeformed	little	little	0.2	0.23
18	deformed	7.5	16.9	4.6	1.8
12	undeformed	little	little	0.09	
12	deformed	17.5	59.5	26.7	1.5

Table 70: Influence of the deformation and of dissolved hydrogen on the content of ε- and α'-martensite in Cr-Ni-Mn-N steels [135]

ment. In steel 21 neither deformation nor hydrogen charging caused the formation of martensite since this steel is a stable austenitic.

In contrast, martensite is formed in the metastable austenitic steels 18 and 12 by both deformation and hydrogen charging. If the hydrogen contents are sufficient, the strain-induced ε-martensite formation is enhanced compared to the α'-formation (Table 70).

Summarizing the results contained in [135], the following conclusions can be drawn:

- Hydrogen charging of the steels SAE 316 L (cf.: 1.4404, 1.4435) and NITRONIC® 40 (cf.: 1.3965, 1.4454) in compressed hydrogen leads to a slight increase in the strength and a minor reduction of the ductility.
- The susceptibility of the austenitic Fe-Cr-Ni-Mn-N steels to hydrogen embrittlement clearly grows with increasing δ-ferrite content.
- Sensitized steels, e.g. following aging at 650 °C, exhibited a higher susceptibility to hydrogen embrittlement. The susceptibility of the steel NITRONIC® 40 clearly increases with the aging time. $M_{23}C_6$ precipitates are considered to be the cause.
- Tensile stress initiates the formation of α'- and ε-martensite in metastable austenitic steels.
- The formation of ε-martensite in austenitic steels leads to a noticeable increase of the susceptibility to hydrogen embrittlement.
- If the hydrogen content is sufficiently high, the strain-induced formation of ε-martensite is enhanced in metastable austenitic steels, lowering the formation of α'-martensite.

The embrittlement accelerated by tritium charging compared to hydrogen was proven by using the steel Nitronic® 40 with the chemical composition indicated in Table 71 [137]. To this end, mechanically polished round test bars (Ø 3 mm, measuring section 15 mm) were charged with tritium at 573 K (300 °C) and a total pressure of 82.75 MPa for 110 days (saturation). Due to the tritium decay the partial tritium pressure was 77 MPa and the rest consisted of ^{3}He and deuterium (D). The initial tritium content was 4,800 appm. The tritium decay results in c_T = 4,800 − C (^{3}He) and c_T + 250 appm D (deuterium) for the total hydrogen concentration. Following charging, the specimens were cooled under pressure for 8 h to 300 K (27 °C). Another 72 h were necessary to reach a storage temperature of 233 K (−40 °C) of the specimens taken from the surge tank, ensuring a minimum tritium desorption and the formation of ^{3}He. Subsequently, tensile tests were performed at a strain rate of 8.5 × 10^{-7} m/s at room temperature. The results of the tensile tests as a function of the helium concentration are summarized in Table 72.

Ni	Cr	Mn	C	P	Si	N	S	P	Fe
6.99	19.65	8.57	0.031	0.03	0.32	0.29	0.015	< 0.003	balance

Table 71: Chemical composition of the steel Nitronic® 40 in wt% [137]

³He appm	Rp₀.₂ MPa	Rₘ MPa	A %	Z %
0	673	832	38.8	82
153	742	875	36.3	54.2
465	805	908	20.4	45
774	835	926	14.6	30
950	846	904	9.9	20.5
1,128	882	944	7.3	18.5
1,438	889	939	5.0	15

The tritium concentration for the indicated ³He concentration results from $c_T = 4{,}800 - C(^3He)$ and the total hydrogen concentration results from $c_T + 250$ appm D.

Table 72: Results from tensile tests of the steel Nitronic® 40 following tritium charging [137]

Figure 117 shows the influence of the helium concentration (³He) on the strength and Figure 118 depicts the influence on the ductility properties of the steel NITRONIC® 40. The values obtained from charging with pure hydrogen are included for comparison. As can be seen, the strength clearly increases with the increasing enrichment of ³He in the material and its ductility decreases accordingly. It is stated that the increase in strength by helium enrichment in the steel NITRONIC® 40 is about 8 to 12 times higher than in the case of charging with hydrogen [137]. Helium enrichment preferably occurs at dislocations and grain boundaries and the fracture changes from transcrystalline to intercrystalline under the helium influence of about 774 appm ³He.

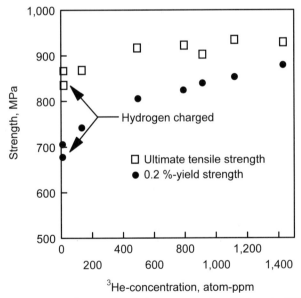

Figure 117: Influence of the helium concentration (³He) on the 0.2% yield strength and the tensile strength of the steel NITRONIC® 40. Also the values for charging with pure hydrogen are shown [137]

150 | Hydrogen

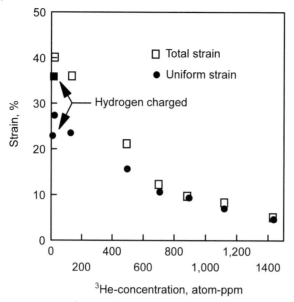

Figure 118: Influence of the helium concentration on the uniform elongation and the total elongation to failure of the steel NITRONIC® 40 [137]

By using the tritium autoradiography it is possible to determine the local hydrogen distribution in metallic materials [138]. To this end, a specimen charged with tritium is provided with a monolayer of a β-sensitive silver bromide film. Results from examinations of the welded austenitic steel SAE 316 (1.4401) are available. The 1 mm thick steel specimen was charged in pure tritium gas (2.5 MPa, 6 h, 450 °C) with up to 700 atomic ppm tritium and subsequently aged at low temperatures to form ^{3}He. Another heat treatment at 350 °C led to desorption and furnished proof of the local tritium distribution. The tritium autoradiography also revealed the highest tritium concentration in the laser-welded steel SAE 316 in the area of the heat affected zone. Hence, this zone is especially susceptible to hydrogen embrittlement.

The austenitic 18–10CrNi steel (Table 73) was examined to determine the influence of dissolved hydrogen on the mechanical properties [139]. Following the solution annealing treatment (1,050 °C/oil) of the 0.3 mm thick foils in argon, the microstructure was purely austenitic. Hydrogen charging of the steel was performed at 1,050 °C for 12 h in an argon/hydrogen mixture (97% Ar + 3% H_2) shortly above atmospheric pressure under the action of an electric arc and subsequent quenching in water. The hydrogen content determined by applying the vacuum extraction method was 340 to 350 ppm. In the solution-annealed state the hydrogen content

C	Si	Mn	S	P	Cr	Ni	Fe
0.10	0.50	1.50	0.05	0.05	18.40	10.20	balance

Table 73: Chemical composition of the austenitic steel [139]

State	H ppm	$R_{p0.2}$ N/mm²	R_m N/mm²	A %
Solution-annealed	5	211	508	43
Hydrogen-charged	350	233	471	13
Aging (3h, 400 °C) after H-charging	7	207	494	41

Table 74: Influence of the hydrogen content on the mechanical properties of the austenitic steel [139]

varied between 3 and 5 ppm. After hydrogen charging, a phase transformation could neither be detected radiographically nor magnetically.

The influence of the hydrogen content on the mechanical properties determined by tensile tests (0.05 cm/min) at room temperature is shown in Table 74. Hydrogen charging causes a slight increase in the yield strength, a slight reduction of the tensile strength and a clear decrease of the fracture elongation. The fracture surfaces exhibit a predominantly ductile character. The ductility loss of the austenitic steel 304 by hydrogen absorption also depends on the strain rate. In this connection Figure 119 shows the influence of the strain rate on the fracture elongation of the hydrogen-containing steel compared to the solution-annealed material. As can be seen, strain rates below 0.5 cm/min have a particularly negative impact on hydrogen-containing specimens.

The load(K)-extension(L) diagram obtained from the slow strain rate tensile test (strain rate = 0.1 cm/min) shows the considerable reduction of the fracture work

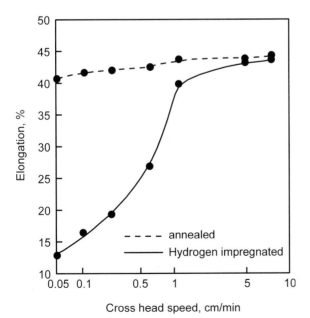

Figure 119: Influence of the cross head speed on the fracture elongation of the solution-annealed and hydrogen-charged steel by comparison [139]

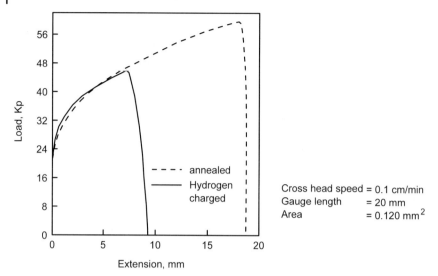

Figure 120: Load-extension diagram of the solution-annealed and hydrogen-charged steel by comparison [139]

(area below the K-L-curve) of the hydrogen steel (Figure 120). In contrast, no difference is found between H-charged and uncharged specimens at higher strain rates (e.g. 10 cm/min) [139].

Figure 121 shows the influence of the temperature on the fracture elongation of H-charged and uncharged specimens. The susceptibility of the steel to hydrogen

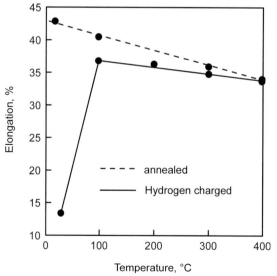

Figure 121: Influence of the temperature on the fracture elongation of solution-annealed and hydrogen-charged steel specimens from [139]

Steel	C	Si	Mn	P	S	Cr	Ni	Mo	Fe
304	0.060	0.59	1.01	0.018	0.009	18.33	8.35	–	balance
316	0.040	0.48	0.66	0.010	0.002	17.10	10.05	2.02	balance
310S	0.060	1.24	1.53	0.002	0.009	24.63	20.21	–	balance

Table 75: Chemical composition of stainless austenitic steels in wt% [140]

embrittlement strongly increases below 100 °C with decreasing temperature and above 100 °C the hydrogen influence gradually vanishes with increasing temperature.

[140] deals with the influence of deformation martensite on the hydrogen embrittlement of sensitized austenitic steels at low temperatures. The commercial austenitic steels 304 (1.4301), 316 (1.4401, 1.4436) und 310 S (1.4845) with the chemical compositions indicated in Table 75 were subjected to several heat treatments (Table 76) and solution-annealed, sensitized (S) and desensitized (SD).

Steel	Solution-annealed	Sensitized	Desensitized
304	1 h, 1,100 °C		
304 (S)	1 h, 1,100 °C	24 h, 700 °C	
304 (SD)	1 h, 1,100 °C	24 h, 700 °C	8 h, 920 °C
316	1 h, 1,080 °C		
316 (S)	1 h, 1,080 °C	24 h, 700 °C	
316 (SD)	1 h, 1,080 °C	24 h, 700 °C	8 h, 920 °C
310S	1 h, 1,060 °C		
310S (S)	1 h, 1,060 °C	48 h, 700 °C	

S = sensitized, (SD) = desensitized

Table 76: Heat treatment of stainless austenitic steels [140]

Smooth round test bars (Ø 4 mm, direction of stress parallel to the rolling direction) were used to perform tensile tests at a strain rate of 4.17×10^{-5} 1/s in high-purity hydrogen (99.9999%) and helium (99.999%) at a pressure of 1 MPa each. The temperatures covered a range from 22 °C to −193 °C. The load-extension diagrams obtained for the solution-annealed state of the steel SAE 304 (1.4301) in helium and in hydrogen at −193 °C, −53 °C and 22 °C are shown in Figure 122. At room temperature (22 °C) hydrogen caused a slight reduction of the tensile strength, however a strong reduction of elongation. At −53 °C both the tensile strength and the fracture elongation were strongly reduced and no influence was found at −193 °C.

The influence of heat treatment of the steel SAE 304 on the behavior in 1 MPa helium and hydrogen is depicted in Figure 123. In contrast to helium, hydrogen exerts a strong influence on the fracture elongation of the steel, in particular if the steel is sensitized.

The relative reduction of area $Z_{hydrogen}/Z_{helium}$ was used for the quantification of hydrogen embrittlement. Figure 124 shows the influence of the temperature on

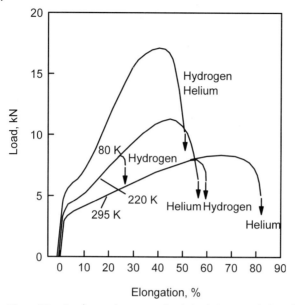

Figure 122: Load-extension curves for the solution-annealed steel SAE 304 (1.4301) in 1 MPa hydrogen and helium at temperatures of 22 °C to −193 °C [140]

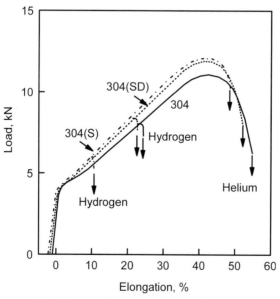

Figure 123: Influence of the heat treatment of the steel SAE 304 (1.4301) on the behavior in 1 MPa hydrogen and helium at −53 °C. 304: solution-annealed; 304(S): sensitized; 304(SD): desensitized [140]

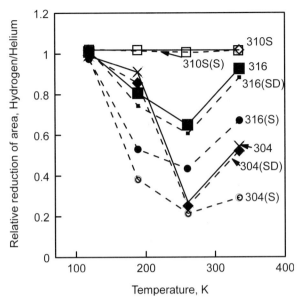

Figure 124: Influence of the temperature on the relative reduction of area [140]

the relative reduction of area of the steels 304, 316 and 310 S in the solution-annealed, sensitized (S) and desensitized (SD) state. The stable austenitic steel 310 S with $Z_H/Z_{He} = 1$ remains resistant to hydrogen embrittlement. In contrast, the steels 316 and 304 are susceptible to hydrogen embrittlement. The most susceptible steel is steel 304 in the sensitized state. As these results show, the highest susceptibility to hydrogen embrittlement occurs at about −53 °C.

The fracture surfaces of the steels ruptured in 1 MPa hydrogen at −73 °C were fractographically examined. The stable austenitic steel SAE 310 S (1.4845) exhibits a ductile fracture behavior. The metastable austenitic steels SAE 304 (1.4301) and SAE 316 (1.4401, 1.4436) rupture in the solution-annealed state in a transcrystalline manner along the martensite formed as a result of deformation. In contrast, the failure of sensitized steels is of an intercrystalline type, attributable to the preferable formation of the deformation martensite along the grain boundaries. Due to the desensitization the fracture becomes transcrystalline again, since then the martensite formation is no longer supported at the grain boundaries. Specific mention is made that the intercrystalline fracture of the sensitized steels is not caused by the formation of the $M_{23}C_6$ carbides but by the martensite formation along the grain boundaries [140].

The austenitic stainless steel SAE 316 is a potential material for the construction of vacuum vessels for fusion research reactors [141]. The vessels are electron beam welded (EBW) or electric arc welded (EAW). Specimens welded by applying these methods were used to examine the influence of hydrogen on the embrittlement of the welded joint. The chemical composition and the mechanical properties of the steel SAE 316 are indicated in Table 77. Three different states of the material were examined:

C	Si	Mn	P	S	Ni	Cr	Mo	$R_{p0.2}$ N/mm²	R_m N/mm²	A %	HB
0.04	0.68	1.30	0.030	0.002	10.85	16.88	2.07	267	579	58	78

Table 77: Chemical composition and mechanical properties of the steel SAE 316 (1.4401, 1.4436) [141]

- welded
- 24 h, 650 °C/air carbide precipitation
- 6 h, 850 °C/air.

Hydrogen charging was performed in an autoclave under the following conditions: 48 h, 450 °C in 220 bar compressed hydrogen with subsequent cooling (200 °C/h) to 150 °C and repeated cooling to room temperature within 2 h. The delta ferrite content in the weld seam amounted to 1.9% following electron beam welding (EBW), 6.6% after electric arc welding (EAW) and 0.1% in the base material (BM). Heat treatment converts the delta ferrite to carbide precipitates or sigma phase. The mean hydrogen content in the weld zone and in the base metal is indicated in Table 78.

The results from tensile tests (0.5 mm/min) of the base metal (BM) and the welded joints with or without additional heat treatment and without (−) and with (+) hydrogen charging are summarized in Table 79.

A clear reduction of the ductility of the steel SAE 316 is found as a result of

- Welding
- inappropriate heat treatment
- hydrogen charging.

The embrittlement index EI served as a measure for embrittlement:

Equation 17 $EI = (Z_0 - Z_x)/Z_0 \times 100$

Z_0: Reduction of area in the state without heat treatment and without hydrogen charging
Z_x: Reduction of area under the condition of inappropriate heat treatment and hydrogen charging.

The results are summarized for different heat treatments of the base metal (BM) and welded joints (EBW: electron beam welded; EAW: electric arc welded) with and without hydrogen charging in Figure 125. The following conclusions can be drawn:

Specimen	Mean hydrogen content, ppm	
	Prior to H-charging	H-charged
EBW	1	41
EAW	3	41
BM	4	41

EBW: electron beam welding; EAW: electric arc welding; BM: base metal

Table 78: Mean hydrogen content in the weld zone and in the base metal [141]

A 22 Austenitic CrNiMo(N) and CrNiMoCu(N) steels

Specimen	Heat treatment	H-charging	R_m N/mm²	A %	Z %	Fracture in
BM	as-received	−	603	82.0	77.6	−
		+	594	73.2	67.1	−
	24 h, 650 °C	−	615	75.5	76.7	−
		+	595	71.0	60.4	−
	6 h, 850 °C	−	618	77.7	70.4	−
		+	589	74.0	58.2	−
EBW	welded	−	628	53.1	81.4	HTZ
		+	671	51.1	48.5	WM
	24 h, 650 °C	−	641	53.2	66.3	HTZ
		+	649	40.4	34.2	HTZ
	6 h, 850 °C	−	635	53.1	74.2	WM
		+	647	42.3	34.4	HTZ
EAW	welded	−	593	42.3	60.1	WM
		+	664	34.3	34.1	WM
	24 h, 650 °C	−	680	27.8	44.2	WM
		+	594	25.1	31.3	WM
	6 h, 850 °C	−	590	36.4	34.1	WM
		+	585	25.9	23.6	WM

BM: base metal; EBW: electron beam welded; EAW: electric arc welded; WM: welding material; HAZ: heat affected zone; strain rate: 0.5 mm/min

Table 79: Results from tensile tests with steel SAE 316 (1.4401, 1.4436) and different welded joints with and without additional heat treatment and with and without hydrogen charging [141]

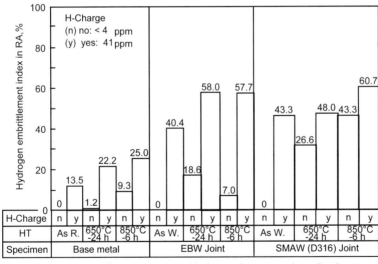

Figure 125: Influence of heat treatment and hydrogen charging on the embrittlement index EI of the base metal SAE 316 (1.4401, 1.4436) and of welded joints (EBW, EAW) in the tensile test [141]
* As R. = as received
** As W. = as welded

Hydrogen embrittlement of the welded specimens is decisively higher than that of the base material SAE 316.

The embrittlement effect clearly increases both in the specimens with hydrogen charging and without hydrogen charging as a result of heat treatment leading to carbide or sigma phase precipitation.

A 23   Special iron-based alloys

Although the austenitic steels have a very low susceptibility to hydrogen-induced damage due to their face-centered cubic lattice and relating higher solubility for hydrogen and a lower diffusion rate, high-strength iron-based super alloys may exhibit hydrogen embrittlement. Also these materials have an austenitic matrix with good resistance to hydrogen, but hardening precipitates, such as the γ-precipitates, may lead to susceptibility to hydrogen embrittlement.

In [142] examinations are described regarding the influence of hardening precipitates in the Fe-Ni-Co super alloy Incoloy 907 on the behavior towards hydrogen. Table 80 shows the chemical composition of the alloy.

Ni	Co	Nb	Ti	Al	Mn	Si	C	S	P	B	Fe
37.4	12.9	4.77	1.47	0.08	0.36	0.21	0.02	0.004	0.002	0.002	balance

Table 80: Chemical composition (wt%) of the alloy Incoloy 907 [142]

Firstly the material was solution-annealed at 980 °C for 1 h and cooled in air, resulting in an austenite grain size of 50 µm. To vary the microstructure, tensile test specimens were annealed at 620, 720 and 770 °C for 1, 2, 4, 8, 12 and 16 h. The properties obtained for the specimens subjected to different annealing processes are summarized in Table 81.

At an aging temperature of 620 °C the strength values increase with increasing aging time and the toughness values decrease accordingly. If a critical value of the γ′-phase is reached at higher temperatures, also the strength values decrease again.

Temperature	620 °C				720 °C				770 °C			
Annealing time h	R_e MPa	R_m MPa	A %	Z %	R_e MPa	R_m MPa	A %	Z %	R_e MPa	R_m MPa	A %	Z %
1	715	1,037	38.2	57.5	767	1,143	28.2	41.9	599	936	33.4	46.6
2	758	1,076	36.9	56.3	787	1,158	26.5	36.3	664	1,009	29.6	40.2
4	786	1,119	34.3	54.3	802	1,148	23.8	32.9	623	966	30.3	43.1
8	855	1,186	28.3	47.4	799	1,135	22.3	31.8	577	949	30.7	39.0
12	882	1,227	27.4	43.5	833	1,115	20.7	25.5	562	921	30.6	38.8
16	890	1,230	26.3	41.1	799	1,110	20.8	26.9	572	924	26.7	31.6

Table 81: Mechanical properties of Incoloy 907 specimens subjected to different annealing processes [142]

To check the susceptibility to hydrogen, tensile test specimens were aged in compressed hydrogen (99.99%) of 10 MPa at 300 °C for 14 days. Then the mean hydrogen content of the specimens was 16.5 ppm. These specimens charged with hydrogen were used to perform slow strain rate tensile tests at room temperature at a strain rate of 6.7×10^{-4} 1/s. The reduction of these values for the necking of the charged specimens compared to the values of the uncharged specimens was used as the measure for hydrogen damage. As shown by the values in Table 82, the hydrogen-induced impairment of the toughness clearly increases with increasing aging time, i.e. with an increasing share of hardening precipitates.

	1 h	2 h	4 h	8 h	12 h	16 h
620 °C	29.6	38.1	38.6	49.8	64.9	66.6
720 °C	61.1	59.5	62.2	69.4	61.6	70.9
770 °C	57.8	61.8	62.2	65.5	71.5	68.5

Table 82: Influence of the annealing temperature and the annealing time on the reduction of reduction of area (%) of the specimens charged with hydrogen [142]

Investigations regarding the influence of hydrogen on the mechanical properties of the high-strength Fe-Ni-Co super alloy Incoloy alloy 903 (UNS N19903) with similar composition are described in [143]. The chemical composition of the alloy with the face-centered cubic structure, which is also γ'-precipitation-hardening, is indicated in Table 83.

Ni	Co	Nb	Ti	Al	Mn	Si	C	Mo	Fe
37.8	15.25	3.07	1.33	0.07	0.15	0.7	0.04	0.1	balance

Table 83: Chemical composition (wt%) of the alloy Incoloy alloy 903 [143]

To obtain a microstructure with different grain size, the examination material was solution-annealed at different temperatures between 940 °C and 1,180 °C, quenched in water and subjected to a double tempering treatment 720 °C/8 h plus 620 °C/8 h. The resulting grain sizes are indicated in Table 84. The size of the γ'-precipitates was independent from the grain size and had a diameter of 20 nm.

Temperature, °C	940	1,000	1,060	1,060	1,100	1,120	1,180
Duration, h	1	1	1	4	1	1	1
Grain size, μm	28	49	114	172	200	226	350

Table 84: Solution annealing temperatures and resulting grain sizes of the examination material from Incoloy alloy 903 [143]

For charging with hydrogen the tensile test specimens were aged in hydrogen gas at 300 °C for 60 days at a pressure of 136 MPa. The result is a hydrogen concentration of 0.5 at.%. The specimens with the grain size of 28 μm were also charged with hydrogen at lower pressures, yielding the following hydrogen contents:

at 68.0 MPa: 0.2900 at.%
at 34.0 MPa: 0.1876 at.%
at 6.9 MPa: 0.0767 at.%.

The slow strain rate tensile tests were performed at a strain rate of 3.3×10^{-4} 1/s. The results obtained from charged and uncharged specimens are summarized in Table 85.

Grain size μm	Yield strength MPa		Tensile strength MPa		Reduction of area %	
	uncharged	charged	uncharged	charged	uncharged	charged
28	1,081	1,102	1,315	1,314	37	13
49	1,088	1,091	1,327	1,324	36	13
114	1,080	1,082	1,295	1,257	34	11
172	1,065	1,073	1,289	1,239	32	11
200	1,054	1,072	1,276	1,201	15	6
226	1,007	1,052	1,248	1,166	13	4
350	1,000	982	1,119	1,001	4	3

Table 85: Results of the slow strain rate tensile tests of charged and uncharged Incoloy alloy 903 specimens [143]

In both the charged and uncharged specimens the grain size has little influence on the yield strength and only a slightly stronger influence on the tensile strength. However, larger grain sizes cause a decrease in the elongation values even in uncharged specimens. As usual, the negative influence of hydrogen charging is noticeable to a minor extent only in terms of the yield strength and the tensile strength, whereas it turns out clearly in the elongation values.

Fracture toughness was tested in CT specimens with a loading rate of 42 MPa $\sqrt{m}$/min. The results are shown in Figure 126.

Charging with 5,000 ppm hydrogen reduces the fracture toughness of Incoloy alloy 903 from 95 MPa $\sqrt{m}$/min to 48 MPa $\sqrt{m}$/min. Fracture toughness is independent of the grain size both in charged and uncharged specimens. The specimens with the grain size 28 μm charged with lower hydrogen contents exhibited a clear reduction of the fracture toughness even at the lowest hydrogen content of 770 ppm (Figure 127).

For fracture-mechanical examinations regarding the crack propagation WOL specimens were aged at 25 °C in hydrogen gas at a pressure of 207 MPa. The specimens had been prestressed with a stress intensity factor of 70 MPa $\sqrt{m}$. During aging the load was continuously registered to identify crack propagation and crack arrest. After four month a crack propagation was no longer detectable, the specimens were removed, stored for several weeks to allow the absorbed hydrogen to effuse and then broken up. The critical stress intensity values were determined from the measured crack length and the load at crack arrest. The critical stress

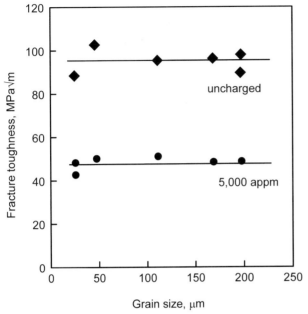

Figure 126: Fracture toughness of charged and uncharged specimens [143]

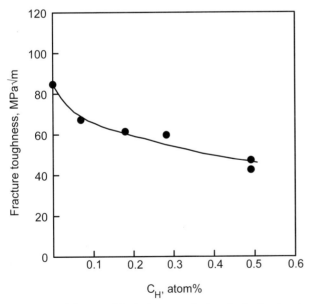

Figure 127: Influence of the hydrogen content on the fracture toughness of the specimens with a grain size of 28 μm [143]

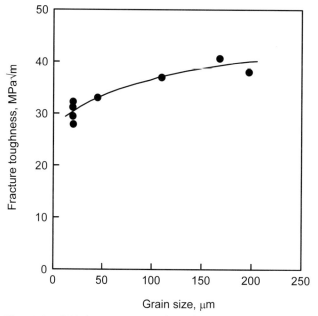

Figure 128: Critical stress intensity values of the WOL specimens from Incoloy alloy 903 depending on the grain size [143]

intensity values were lower than the fracture toughness values and exhibited only a low dependence on the grain size of the microstructure of the specimens. They increased from 30 MPa $\sqrt{m}$ for the specimens with the grain size of 28 µm and to 40 MPa $\sqrt{m}$ for the specimens with the grain size of 200 µm (Figure 128).

The conclusion can be drawn that the determination of the critical stress intensity factor for the crack arrest are the lowest and safest limiting values for the susceptibility to hydrogen embrittlement.

Other examinations using the same material (composition in Table 83) show that the fracture toughness, the limiting values for slow crack growth and the related fracture types vary with the loading conditions and the source of hydrogen [144]. The specimen material with the heat treatment 940 °C/water/720 °C/620 °C was selected, resulting in a yield strength of 1,080 MPa and a grain size of 28 µm. Again, the specimens were aged in hydrogen gas at 300 °C and pressures of 6.9 to 136 MPa for 60 days, leading to the hydrogen contents indicated above.

The material subjected to such a treatment was used to determine the fracture toughness of notched three-point bending specimens with a load of 42 MPa $\sqrt{m}$. To determine the stress limit for a slow crack growth, 22.2 mm thick WOL specimens were selected. Prior to the test the specimens were provided with an incipient fatigue crack, again charged in hydrogen gas at 300 °C and pressures of 23.1 to 136 MPa and then prestressed with 95% of the fracture toughness measured in similarly charged specimens. In addition, the stress limit for slow crack growth was measured using WOL specimens not having been charged before, which had

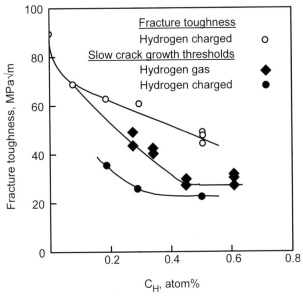

Figure 129: Fracture toughness and limiting values for slow crack growth as a function of the hydrogen content [144]

been aged in hydrogen gas at pressures of 20.7 MPa (0.2735 at.% H), 44.8 MPa (0.3355 at.% H), 103.5 MPa (0.4505 at.% H) and 207 MPa (0.6135 at.% H). The hydrogen contents indicated in parentheses are calculated gas phase equilibrium concentrations based on the effective hydrogen fugacity and hydrogen solubility in the lattice. These specimens were prestressed at different stress intensity values such that the range between 90 MPa $\sqrt{m}$, i.e. the fracture toughness value for uncharged Incoloy alloy 903 specimens, and 30 MPa $\sqrt{m}$, i.e. the value determined in hydrogen of 207 MPa before, was covered. All tests were performed at 22 °C. Figure 129 illustrates the measured values of the fracture toughness and the slow crack growth as a function of the hydrogen content of the specimens.

Both the values of the fracture toughness and the limiting values of the slow crack growth decrease with increasing hydrogen content. The limiting values of the slow crack growth of the specimens precharged with hydrogen are lower than those of the specimens tested in hydrogen gas and are clearly lower than the fracture toughness values determined using the bending specimens precharged with hydrogen. The fracture toughness decreases from 90 MPa $\sqrt{m}$ of the hydrogen-free specimens to 50 MPa $\sqrt{m}$ in specimens with a hydrogen content of 0.5 at.%. The limiting values of the slow crack growth decrease down to 30 MPa $\sqrt{m}$ in the specimens tested in compressed hydrogen and down to 22 MPa $\sqrt{m}$ in the specimens precharged with hydrogen.

Hence, both the type of loading and the type of hydrogen absorption influence the crack behavior. This is attributed to the interaction between the hydrogen concentration, the microstructure and the stress field at the crack tip. Fractographic

Ni	Cr	Co	Ti	Mo	W	V	Al	Fe
33.0–35.0	14.0–16.0	3.0–3,5	2.5–2,7	1.8–2,2	1.5–2,0	0.3–0,5	0.2–0,3	balance

Table 86: Chemical composition of the Fe-Ni super alloy NASA-HR-1 [145]

examinations show that the hydrogen concentration at the crack tip in the specimens precharged with hydrogen and in the specimens tested in hydrogen gas is clearly higher than the equilibrium values in uncharged specimens.

NASA developed an iron-nickel super alloy (NASA-HR-1) exhibiting a comparatively high strength and, in particular, an excellent resistance to compressed hydrogen [145]. Hence, this alloy is suited as a structural material in spaceships. The high resistance to hydrogen embrittlement was achieved by a hydrogen resistant gamma matrix and eta-free grain boundaries. The high strength is the result of finely distributed γ'-precipitates in the γ-matrix and the addition of tungsten and molybdenum. The chemical composition of the Fe-Ni super alloy NASA-HR-1 is reflected in Table 86. The alloy was produced by vacuum induction and vacuum arc melting, homogenized at 1,150 °C, hot rolled (927–1,094 °C) and solution-annealed at 955 °C.

To determine the strength and the ductility slow strain rate tensile tests (8.3×10^{-5} 1/s) were performed in air and 34.5 MPa hydrogen at room temperature. The strength values (yield strength and tensile strength) determined in air are summarized in Figure 130 for the alloy NASA-HR-1 compared to the alloys A-286 (25% Ni, 15% Cr, cf.: 1.4980) and JBK-75 (25% Ni, 15% Cr, 2% Ti). As can be seen, the yield strength of the material NASA-HR-1 is about 25% higher. Fracture elongation serves

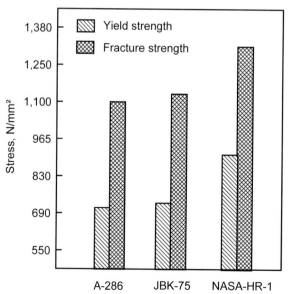

Figure 130: Strength values of the Fe-Ni alloys NASA-HR-1, A-286 and JBK-75 [145]

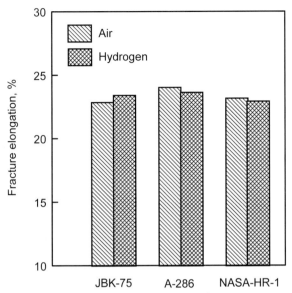

Figure 131: Fracture elongation of the materials NASA-HR-1, A-286 and JBK-75 in air and in 34.5 MPa hydrogen gas [145]

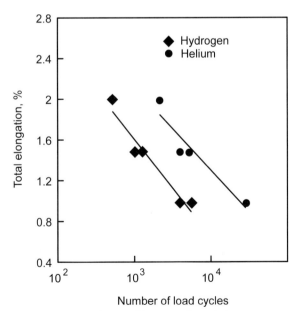

Figure 132: LCF behavior of the Fe-Ni alloy NASA-HR-1 in compressed hydrogen and in helium [145]

as the measure for ductility. Figure 131 shows the fracture elongation in air and in 34.5 MPa hydrogen for the materials NASA-HR-1, A-286 and JBK-75 by comparison. The three tested Fe-Ni alloys hardly differ from each other in terms of their ductility behavior and compressed hydrogen remains without any noticeable influence. Accordingly, all three test materials broke ductilely irrespective of the atmosphere.

Under low-frequency cyclic loading (LCF) a reduction of the fatigue strength of the alloy NASA-HR-1 in 34.5 MPa hydrogen gas is found compared to helium (Figure 132) [145]. The LCF behavior is to be improved by performing a specific heat treatment.

A 26 Nickel

Basically, hydrogen can impair the fracture behavior of nickel and of nickel alloys. Apart from a reduction of the fracture toughness, also changes of the type of fracture – from ductile fracture to brittle, intercrystalline or transcrystalline cleavage fracture – can be observed under the influence of hydrogen. The extent of embrittlement and the fracture appearance depend on the material and the test conditions, e.g. on the microstructure, the hydrogen pressure or the content of dissolved hydrogen, the temperature and the type of mechanical loading. Also the plastic deformation behavior of nickel as well as the behavior under cyclic loading can be impaired by hydrogen [146–149].

Figure 133 shows the example of fatigue test results obtained for pure nickel [150]. Crack growth in hydrogen starts at clearly lower stress intensity values com-

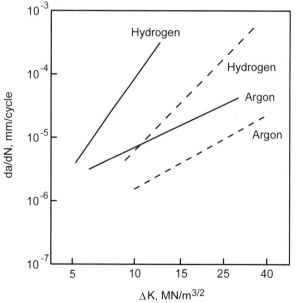

Figure 133: Crack propagation rates of cold-rolled (----) and annealed (----) pure nickel in hydrogen and in argon [150]

pared to argon and in hydrogen higher crack growth rates compared to argon occur at identical stress intensity values. In both test gases annealing of the specimens exerted an unfavorable effect on the fatigue behavior.

Usually, an intercrystalline crack path is found in polycrystalline nickel under the embrittling effect of hydrogen. Monocrystals of nickel charged with hydrogen usually do not show any embrittlement, but a ductile fracture behavior with a transcrystalline crack path. This is indicative of the fact that impurities precipitated at the grain boundaries influence the fracture behavior when the material is charged with hydrogen.

Since the sulfur content and its distribution are considered to play a special role regarding precipitates, the influence of hydrogen on the deformation capacity and the fracture behavior of nickel was examined for specimens with sulfur contents of 0.0044 at.% and 0.00016 at.% [151]. The specimens were subjected to the following different heat treatments:

1) Solution annealing at 1,300 °C for 5 min in flowing gas with 70% CO and 30% CO_2 and rapid cooling in gas at about 80 K/s
2) Same as under 1) with subsequent aging at 900 °C for 1 h in vacuum (2.5 × 10^{-4} Pa) and cooling at 0.03 K/s
3) Solution annealing at 1,300 °C for 20 min in vacuum (1.3 × 10^{-3} Pa) and quenching in silicone oil at 200 K/s
4) Same as under 3) with subsequent aging according to 2).

Slow strain rate tensile tests of the specimens were performed at a strain rate of 2 × 10^{-5} 1/s at 25 °C in air or in hydrogen at a pressure of 100 kPa. Table 87 shows the fracture characteristics of the tensile test specimens.

The specimens tested in air were completely ductile irrespective of their heat treatment and exhibited more than 90% reduction of area. Differences in sulfide precipitates at the grain boundaries had no significant influence on the elongation or fracture behavior.

Sulfur content/ppm heat treatment No.	Test atmosphere	Reduction in area %	Type of fracture
1.6	air	95	ductile
1	hydrogen	55	20% IC; 80% TC
1.6	air	95	ductile
2	hydrogen	14	80% IC; 20% TC
44	air	95	ductile
3	hydrogen	75	20% IC; 80% TC
44	air	90	ductile; 20% IC
4	hydrogen	35	100% IC

IC = intercrystalline, TC = transcrystalline

Table 87: Fracture characteristics of nickel tensile test specimens tested in air and in hydrogen [151]

Tensile tests performed in hydrogen did not only yield clearly lower values obtained for reduction of area, but similar to the fracture pattern they were dependent on heat treatment. Although specimens rapidly cooled down from the solution annealing temperature (heat treatment 1 and 3) to prevent sulfide precipitates exhibited a decrease in the reduction in area, their crack path was still transcrystalline to a major extent. Slowly cooled specimens (heat treatment 2 and 4) for which precipitates are expected to occur at the grain boundaries, exhibited only very low values of the reduction in area and a predominantly intercrystalline crack path. The resulting assumption that the fracture behavior in hydrogen is controlled by the grain boundary precipitates was confirmed by surface analyses during which elevated sulfur contents were found on the fracture surfaces with predominantly transcrystalline crack paths.

The studies described in [152] were continued to examine the influence of the sulfur content and the grain boundary precipitates. Again specimens with sulfur contents of 0.0044 at.% and 0.00016 at.% were examined; in addition, specimens with magnesium contents (0.25 at.% Mg) for sulfur gettering as magnesium sulfide were included in the examinations. The specimens were subjected to the following different heat treatments:

1) Solution annealing at 1,300 °C in vacuum (1.3×10^{-3} Pa)
   quenching in silicone oil at 200 K/s
2) Solution annealing at 1,300 °C in vacuum (1.3×10^{-3} Pa)
   quenching in silicone oil at 200 K/s
   aging at 900 °C for 1 h
   cooling at 0.03 K/s
3) Same as 1;
   but solution annealing and quenching were performed in hydrogen gas at a pressure of 100 kPa.
   The result was a hydrogen content of the specimens of 0.07 at.%.

Slow strain rate tensile tests of the specimens were performed at a strain rate of 2×10^{-5} 1/s either in vacuum, in air, in helium or in hydrogen at a pressure of 100 kPa. Table 88 shows typical examples of values obtained for the reduction in area and the fracture pattern of different specimens in air and in hydrogen.

The results can be summarized as follows:

- Dissolved hydrogen or the hydrogen diffusing into the material during slow deformation lead to a clear reduction of the fluidity.
- Dissolved hydrogen exerts a stronger influence on the deformation behavior than absorbed hydrogen and the embrittlement of specimens with dissolved hydrogen is higher already when they are deformed in air compared to uncharged specimens being deformed in hydrogen.
- The fracture pattern changes from a predominantly transcrystalline to a predominantly intercrystalline crack path with increasing embrittlement.
- Specimens with dissolved hydrogen show a merely transcrystalline crack path also when tested in air.

Sulfur content/ppm heat treatment No.	Test atmosphere	Reduction in area %	Type of fracture
44	air	95	ductile
1	hydrogen	75	20% IC; 80% TC
44	air	90	ductile
2	hydrogen	35	70% IC; 30% TC
1.6	air	95	ductile
1	hydrogen	55	20% IC; 80% TC
1.6	air	26	100% IC
3	hydrogen	-	100% IC

IC = intercrystalline, TC = transcrystalline

Table 88: Examples of fracture characteristics of nickel tensile test specimens tested in air and in hydrogen [152]

- In contrast to the specimens with dissolved hydrogen the crack path of the specimens with a higher sulfur content is influenced by the sulfur distribution in the test in hydrogen.
- Specimens with sulfide precipitates at the grain boundaries predominantly show transcrystalline fractures.
- Higher magnesium contents have a positive effect on the embrittlement behavior since the sulfur is bound as magnesium sulfide within the grains, and hence sulfide precipitates at the grain boundaries are prevented.
- Hydrogen-free specimens always exhibit a ductile behavior during tests in air, inert gas and vacuum and the sulfur content as well as the sulfur distribution remained without any effect on the fracture behavior and the fracture pattern.

However, the statement that the influence of dissolved hydrogen on the deformation behavior is stronger than that of absorbed hydrogen, is contradicts the studies described in [153]. During those studies tensile tests were performed with specimens from pure nickel 200 (Ni 99.2, Mat. No. 2.4066, UNS N02200) in hydrogen (99.995%) at a temperature of 25 °C and a pressure of 101 kPa and the fractographic fracture pattern was compared with that of specimens tested in air and argon. In addition, specimens were aged at 1,000 °C for 1 h in hydrogen gas at a pressure of 100 kPa and then quenched in water of 0 °C. The result of this treatment is a dissolved hydrogen content of about 0.045 at.% in the metal. In air and in argon ductile transcrystalline fractures with marked dimples ("dimples") occurred on the fracture surface. No embrittlement, but ductile fracture behavior was also found in the specimens precharged with hydrogen in high strain rate tensile tests. In contrast, at slow strain rates (6×10^{-5} mm/s) the precharged specimens exhibited signs of embrittlement. Clear hydrogen-induced embrittlement could be found in all specimens tested under pressurized hydrogen. Table 89 summarizes the approximate ratios of the different fracture types determined in the precharged specimens and the specimens tested under compressed hydrogen.

	Intercrystalline fractures	Cleavage fracture content	Ductile fractures
	%	%	%
Compressed hydrogen	5	75	20
Hydrogen-charged	25	15	60

Table 89: Ratio of the fracture types in the specimens tested under compressed hydrogen and specimens precharged with hydrogen [153]

As can be seen, the hydrogen absorbed during a deformation causes a clearly higher embrittlement and crack formation in nickel compared to the hydrogen dissolved in the metal.

In bend test specimens from 99.5% nickel in hydrogen (99.995%) incipient cracks were found only at bend angles above about 45°, whereas in air or in an argon atmosphere no incipient cracks occurred even at larger bend angles [154].

Also the examinations on the influence of compressed hydrogen on the fracture behavior of face-centered cubic metals as described in [155] confirm that nickel becomes brittle during its deformation in hydrogen gas. The tensile tests of specimens from pure nickel (130 ppm C; 68 ppm O; < 10 ppm S) were performed in purified compressed hydrogen with residual contents of oxygen and water vapor at 1 ppm and a tensile rate of 1.27 mm/min. The specimens were solution-annealed (1,000 °C, furnace cooling). Some of the specimens were subjected to 15% or 30% cold forming either in air or in compressed hydrogen of 35 MPa before testing. Table 90 indicates the reduction of area values as a measure of embrittlement measured at different hydrogen pressures.

Test conditions	Air	35 MPa H_2	69 MPa H_2	103 MPa H_2	138 MPa H_2	172 MPa H_2
Solution-annealed	89	47	48	48	46	47
15% predeformed in air		45				
30% predeformed in air		53				
15% predeformed in 35 MPa H_2	91					
15% predeformed in 35 MPa H_2	89					

Table 90: Reduction of area (%) of tensile test specimens from pure nickel in slow strain rate tensile tests in air and in hydrogen [155]

Increasing the test pressure from 35 MPa to 172 MPa had practically no influence on the toughness loss of the specimens. Predeformation also showed the influence on the reduction of area values in the subsequent tensile test in compressed hydrogen. A negative influence of hydrogen absorbed during deformation, was not detected during the test in air following predeformation in compressed hydrogen. During tensile tests in compressed hydrogen at 35 MPa discontinued prior to the occurrence of a fracture, hydrogen-induced crack initiation was found at elongations of about 22%.

A 27 Nickel-chromium alloys

A 28 Nickel-chromium-iron alloys (without Mo)

Nickel-chromium-iron alloys of the NiCrXFe9 type with a value X between 10 und 30% are widely used for compressed hydrogen nuclear reactors. However, in hydrogen-containing water these nickel-based alloys are susceptible to two types of material damage: at lower temperatures below 150 °C a steep decline of the fracture toughness of the material or a rapid subcritical crack growth can be observed, whereas at higher temperatures between 250 °C and 360 °C stress corrosion cracking occurs. This process is characterized by an incubation time of crack formation followed by a relatively low crack growth in the order of magnitude of 5×10^{-10} m/s (0.05 mm/d) at 360 °C in deaerated, hydrogen-containing water depending on the load. Since these Ni–Cr–Fe alloys may be adversely affected in their toughness behavior when absorbing hydrogen, an influence of hydrogen must be assumed also for this crack damage [156, 157].

Examinations regarding the behavior of these alloys compared to oxygen-free high-temperature water with dissolved hydrogen have also shown that the material Alloy 690 (NiCr29Fe, Mat. No. 2.4642) with a nominal content of 27–30% Cr is clearly more susceptible to hydrogen-induced intercrystalline stress corrosion cracking than the material Alloy 600 (NiCr15Fe, Mat. No. 2.4816) with a nominal content of 14–17% Cr. To examine the influence of the chromium content in this alloy type on the behavior towards hydrogen, slow strain rate tensile tests were performed on hydrogen-charged specimens from smelts with chromium contents between 6% and 35% were performed. The alloys and their chemical composition are listed in Table 91 [156].

Melt No.	Ni %	Cr %	Fe %	Ti %	C %	S ppm	P ppm
1	85.3	6.1	8.4	0.21	0.025	< 25	< 25
2	76.7	14.7	8.4	0.21	0.031	< 25	< 25
3	65.1	26.5	8.2	0.21	0.033	< 25	< 25
4	56.7	34.8	8.3	0.21	0.034	< 25	< 25

Table 91: Chemical compositions of the examined melts [156]

The specimens were aged in a high-pressure autoclave at 285 °C for six weeks in a hydrogen atmosphere both at a hydrogen pressure of 34 MPa and a hydrogen pressure of 13 MPa. Following this charging, the hydrogen contents indicated in Table 92 were determined in the specimens.

Melt No.	Chromium content %	Charging at 34 MPa H$_2$ hydrogen content ppm	Charging at 13 MPa H$_2$ hydrogen content ppm
1	6.1	37	25
2	14.7	36	24
3	26.5	55	34
4	34.8	99	60

Table 92: Hydrogen content of tensile test specimens subjected to different charging [156]

The slow strain rate tensile tests were performed at a strain rate of 4×10^{-4} 1/s at 25 °C. Table 93 shows the mechanical properties determined for the uncharged specimens.

Melt No./% Cr	$R_{p0.2}$ MPa	R_m MPa	A %	Z %
1/6	207	462	35	45
2/15	282	586	49	52
3/26	289	593	53	57
4/35	261	606	55	65

Table 93: Mechanical properties of the uncharged specimens [156]

According to the results, the ductility characterized by the values of the elongation (A) and the reduction of area (Z) of the uncharged specimens increases with increasing chromium content.

Table 94 and Table 95 indicate the values for the specimens charged with hydrogen. The values indicated for the loss of elongation (A) and the reduction of area (Z) were calculated according to:

$$\text{Loss A} = \frac{\text{A charged specimens} \times 100}{\text{A uncharged specimens}}$$

and

$$\text{Loss Z} = \frac{\text{Z charged specimens} \times 100}{\text{Z uncharged specimens}}.$$

As can be seen, the hydrogen dissolved in the material exerts only a minor embrittling effect on the alloys with a low chromium content, whereas it causes a clear reduction of the ductility if the chromium content is higher. Hence, the influence of the chromium content on toughness following charging in compressed hydrogen is the same for the crack behavior in case of hydrogen-induced cracking in oxygen-free high-temperature water.

Also the high-strength, precipitation-hardening nickel alloy Alloy X-750 (NiCr15-Fe7TiAl, Mat. No. 2.4669) in power plant applications has turned out to be suscep-

Melt No./% Cr	Hydrogen ppm	$R_{e0.2}$ MPa	R_m MPa	A %	Z %	Loss/A %	Loss/Z %
1/6	24	200	420	31	47	11	0
2/15	22	230	510	43	48	12	8
3/26	35	255	410	22	31	58	46
4/35	60	275	454	25	36	54	45

Table 94: Mechanical properties of the specimens charged with hydrogen at 13 MPa [156]

Melt No./% Cr	Hydrogen ppm	$R_{e0.2}$ MPa	R_m MPa	A %	Z %	Loss/A %	Loss/Z %
1/6	37	220	468	31	42	11	6
2/15	36	255	565	42	45	14	12
3/26	55	282	427	13.5	20	74	63
4/35	99	300	475	18	27	67	57

Table 95: Mechanical properties of the specimens charged with hydrogen at 34 MPa [156]

tible to stress corrosion cracking in hydrogen-containing water at temperatures < 100 °C, attributable to hydrogen-induced cracks. [158] describes fracture-mechanical tests of the behavior of this alloy in a hydrogen atmosphere at 25 °C. The chemical composition of the used test material is indicated in Table 96.

Ni	Cr	Nb	Ti	Al	Fe	C	Mn	Si	S	Co	P	V
71.17	15.46	1.00	2.67	0.76	8.33	0.072	0.10	0.11	0.001	0.07	0.007	0.02

Table 96: Chemical composition of the examined alloy Alloy X-750 [158]

The material was in an aged state subjected to following heat treatment:

annealing at 1,094 °C for 1 to 2 h
rapid cooling in air
aging at 704 °C for 20 h

The fracture-mechanical CT specimens were charged with hydrogen in a high-pressure autoclave using two procedures and the hydrogen content of the specimens obtained was determined with test pieces concurrently subjected to aging. The aging conditions and the hydrogen contents obtained are summarized in Table 97.

Aging time weeks	Temperature °C	Hydrogen pressure MPa	Hydrogen content ppm
3	360	34	65
4	315	21	40

Table 97: Aging conditions and obtained hydrogen contents for fracture-mechanical specimens from Alloy X-750 [158]

To evaluate the fracture toughness of the specimens charged with hydrogen, the stress intensity factor K_{IC} was determined by tests according to ASTM E399-90 at 25 °C, the loading rate being varied between 0.0005 and 0.025 cm/min. For comparison the K_{JC} value for specimens not charged with hydrogen was determined from the J-integral value at 25 °C by performing tests according to ASTM E813-89. Table 98 indicates the results obtained.

Hydrogen content ppm	Loading cm/min	Loading MPa $\sqrt{m}$/min	K_{IC} MPa $\sqrt{m}$	K_{JC} MPa $\sqrt{m}$
0	0.005	7–33	–	147
40	0.0025	3	53	–
65	0.025	33	52	–
65	0.0025	3	45	–
65	0.0005	0.7	42	–

Table 98: Fracture toughness properties of the material Alloy X-750 at 25 °C [158]

Hydrogen charging clearly reduces the fracture toughness of the material and the values of the stress intensity factor decline in the most unfavorable case, i.e. at the highest hydrogen concentration of 65 ppm and the slowest loading rate of 0.0005 m/min, from 147 MPa $\sqrt{m}$ to 42 MPa $\sqrt{m}$. Whereas the loading rate does not influence the value of the stress intensity factor of the non-charged specimens, the decrease of the value of the stress intensity factor in charged specimens is stronger the slower the loading rate is. This is indicative of the fact that the diffusion of hydrogen plays a decisive role for the embrittlement of the material.

Supplementary fracture-mechanical tests were performed using the same alloy (Table 96) also at higher temperatures (260 °C and 338 °C) in air and in hydrogen [157]. The specimens were charged again by aging in hydrogen at a pressure of 13.8 MPa to a hydrogen content of 38 ppm and were again tested in air and in hydrogen at 13.8 MPa to avoid effusion of the hydrogen at high test temperatures.

The influence of hydrogen on the crack propagation in the specimens is illustrated in Figure 134 and the calculated values of the fracture toughness are indicated in Table 99.

At both test temperatures charging with hydrogen leads to a clear decrease of the fracture toughness, the critical stress intensity factor being reduced by more than factor of three. The fractographic fracture pattern changes in the specimens tested in hydrogen from a predominantly ductile, transcrystalline fracture to a predominantly intercrystalline fracture. The fracture patterns of the specimens tested in high-pressure hydrogen and in hydrogen-containing water are similar to a major extent, and hence the conclusion can be drawn that hydrogen plays an important role for the crack damage found in the NiCrFe alloys in high-temperature water.

Although the examinations described in [159] regarding the influence of heat treatment on the susceptibility of Alloy X-750 (2.4669, NiCr15Fe7TiAl) to hydrogen

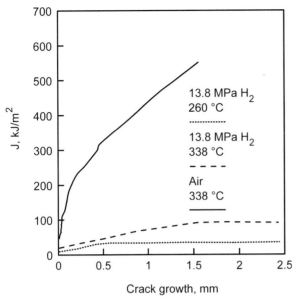

Figure 134: Influence of hydrogen on the fracture toughness of the alloy Alloy X-750 (2.4669, NiCr15Fe7TiAl) [157]

Environment	Test temperature °C	J_{IC} kJ/m²	K_{JC} MPa $\sqrt{m}$	K_{IC} MPa $\sqrt{m}$	$R_{e0.2}$ MPa
Air	260	229	220	–	738
Air	338	252	230	–	728
Hydrogen	260	19	–	68	738
Hydrogen	338	30	–	87	728

Table 99: Fracture toughness of the specimens from Alloy X-750 in air and in hydrogen at higher temperatures [157]

embrittlement were performed using U-type bend test specimens electrolytically charged with hydrogen, the results should be also applicable to the behavior in gaseous hydrogen. The different heat treatments indicated in Table 100 were examined. According to the times to failure determined for the bend test specimens,

Solution annealing	Tempering
982 °C/1 h	704 °C/20 h
	840 °C/24 h + 704 °C/20 h
	735 °C/8 h + 620 °C/10 h
1,066 °C/1 h	704 °C/20 h
1,150 °C/1 h	704 °C/20 h

Table 100: Heat treatments of the examined specimens from Alloy X-750 [159]

the susceptibility to hydrogen embrittlement is influenced by both the solution annealing temperature and the tempering treatment. Compared to the lower temperature, high solution-annealing temperatures improve the resistance to hydrogen. Single-stage tempering (704 °C/1 h) also yielded better resistance compared to multistage tempering treatments.

A 29 Nickel-chromium-molybdenum alloys

The nickel-chromium-molybdenum alloys are slightly less susceptible to hydrogen embrittlement compared to the molybdenum-free nickel-chromium alloys. Examinations regarding the influence of the alloying elements on the resistance to hydrogen embrittlement were performed to compare the behavior of tensile test specimens from alloys of type NiCr18 and NiCr18Mo10 [160]. The chemical compositions of the examined materials are indicated in Table 101.

No.	Type	C	Cr	Mo	Cu	Nb	Al	Fe
1	NiCr18	0.015	18.2					0.3
2		0.092	18.0					0.5
3	NiCr18-Mo10	0.016	18.6	9.9		3.3	1.6	0.1
4		0.09	18.0	9.7			3.1	1.8	17.2
5		0.017	18.0	9.6	4.7		1.5	0.5
6		0.013	18.5	10.0	4.5		1.2	15.7

Table 101: Alloying contents of the examined Ni-Cr and Ni-Cr-Mo materials [160]

The test for susceptibility to hydrogen was carried out by using two methods:

- Tensile tests in air at room temperature using specimens having been charged with hydrogen gas at 500 °C 2 hours before.
- Tensile tests at 20 °C in a pressure chamber in hydrogen with a pressure of 10 MPa.

The susceptibility to hydrogen embrittlement was evaluated on the basis of the loss of plasticity calculated for the hydrogen-influenced tensile test specimens. Figure 135 shows these values. In particular, copper as an additional alloying element proved to have a positive effect on the resistance to hydrogen.

INCONEL® alloy 718 (Alloy 718, UNS N07718, NiCr19Fe19Nb5Mo3, Mat. No. 2.4668) with the nominal reference values indicated in Table 102 is one of the nickel-based super alloys used most frequently. The precipitation-hardening alloy obtained by the addition of aluminum and titanium containing niobium and molybdenum exhibits both a good toughness and strength at temperatures from – 250 °C to 650 °C as well as a good resistance to creep and fatigue. Precipitates of the γ'- and γ''-phases containing considerable amounts of the alloying elements niobium, molybdenum, titanium and aluminum are responsible for the high

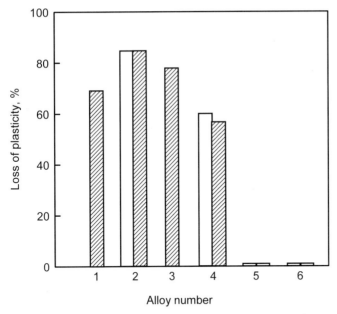

Figure 135: Loss of plasticity of the specimens precharged with hydrogen (not hatched) and of the tensile test specimens tested under compressed hydrogen (hatched) [160]
*For alloys refer to Table 101

Ni	Cr	Mo	Ti	Al	Nb	C	Fe
50–55	17–21	2.8–3.3	0.6–1.2	0.3–0.7	4.7–5.5	0.02–0.08	balance

Table 102: Nominal composition of the material NiCr19Fe19Nb5Mo3 (2.4668) [165]

strength values of this alloy type. The low diffusion rates of these elements and the sluggish precipitation reaction permit welding and tempering without any increased hardening occurring during heating or cooling. Due to its very good mechanical properties INCONEL® 718 is frequently used in gas turbines for temperatures up to 700 °C, however it is also used as a constructional material for hydrogen plants both for low and higher temperatures. The nickel-hydrogen batteries used for the electric power supply of the International Space Station (ISS) are, for instance also, enclosed by pressure vessels consisting of this alloy. The chemical reaction in the battery produces hydrogen gas responsible for the pressure formed in the cells. The material does not only offer the required chemical resistance to the aggressive electrolyte solution, but also to potential hydrogen embrittlement [161].

However, at high temperatures this material can absorb hydrogen and, as a result of its high hardness, becomes susceptible to hydrogen embrittlement then even at room temperature [162–164].

A survey of examinations performed in the United States in connection with the hydrogen problems in the development of space shuttle rockets [166] also reports studies of the influence of "internal" and "external" hydrogen on the deformation capacity of the material INCONEL® alloy 718. The test material was homogenized at 1,050 °C and subsequently annealed at 760 and 650 °C. The specimens were charged by aging in hydrogen at a pressure of 34.5 MPa at 650 °C for 15 minutes. The comparison specimens were aged in helium in the same manner. Tensile tests using smooth and notched specimens as well as fatigue tests using smooth specimens under slow cyclic load were performed at room temperature both in hydrogen and in helium at 34.5 MPa each. To check also the possibility of preventing the hydrogen influence by claddings, appropriate tests were performed including specimens electroplated with copper or gold.

Table 103 summarizes the results obtained for smooth tensile test specimens. Whereas the values for the yield strength and the tensile strength are practically not influenced by precharging with hydrogen ("internal hydrogen") or by the test in compressed hydrogen ("external hydrogen"), the values of fracture elongation and reduction of area under the influence of hydrogen are clearly lower. The effect of the external hydrogen can be seen when comparing the values obtained for the uncharged specimens in helium, on the one hand, and in hydrogen, on the other hand. The effect of the internal hydrogen can be seen when comparing the values obtained for uncharged and charged specimens in helium. Testing of the charged specimens in compressed hydrogen reveals an additive effect of the internal and external hydrogen. Obviously, the toughness properties of the material INCONEL® alloy 718 are influenced by internal hydrogen more strongly than by external hydrogen. Under these test conditions an additive effect of internal and external hydrogen cannot be found. The values of the precharged specimens obtained during testing in a hydrogen atmosphere practically do not differ from those obtained during testing in a helium atmosphere. A reduction of the aging time from 15 to 9 minutes practically remained without any influence of the internal hydrogen at a

		R_p MPa		R_m MPa		A %		Z %	
		He	H_2	He	H_2	He	H_2	He	H_2
IN 718:	a	1,206	1,213	1,427	1,406	18.3	11.9	34.4	15.1
	b	1,275	1,255	1,427	1,372	4,3	3,5	6,9	7,8
IN 718:	c	1,200	1,148	1,400	1,344	4.8	4.3	7.0	6.4
IN 718:	d	1,165	1,144	1,427	1,386	7.3	5.6	8.4	8.8
Copper-clad	a	1,186	1,213	1,434	1,468	19.7	20.1	33.1	32.6
	b	1,220	1,220	1,344	1,351	3,2	3,7	8,1	8,8
Gold-clad	a	1,213	1,172	1,468	1,455	18.9	17.1	23.7	24.7
	b	1,220	1,227	1,482	1,496	12,4	13,2	18,2	16,6

a: uncharged
b: charged 15 min/650 °C/34.5 MPa
c: charged 9 min/650 °C/34.5 MPa
d: charged 9 min/425 °C/34.5 MPa

Table 103: Mechanical properties of tensile test specimens from INCONEL® alloy 718 (2.4668) in helium and in hydrogen at 34.5 MPa and room temperature [166]

hydrogen pressure of 34.5 MPa and a temperature of 650 °C, whereas a reduction of the temperature from 650 °C to 425 °C had a clearly lower damaging effect.

Cladding the specimens with copper does not prevent the absorption of hydrogen and, hence, embrittlement of the material, whereas a gold coating constitutes a good diffusion barrier, although a slight decline of the toughness values by hydrogen can still be found in the gold cladded specimens.

Although a significant additive effect of internal and external hydrogen cannot be detected from the results obtained for the smooth tensile test specimens, the fractographic examinations of the broken specimens reveal that both types of hydrogen contribute to the fracture. The precharged specimens exhibit a predominantly transcrystalline cleavage fracture and both crack initiation and crack propagation follow twin grain boundaries. The cracks caused by external hydrogen clearly start from the surface and take an intercrystalline path. Specimens precharged and thereafter ruptured in a hydrogen atmosphere show a mixed intercrystalline and transcrystalline crack path, without any of the two paths being the dominating process.

In corresponding tensile tests with notched specimens the effect of hydrogen becomes evident as a reduction of the notched ultimate strength as shown in Figure 136.

In the notched specimens the external hydrogen causes the same reduction of the tensile strength in the test of the uncharged specimens in a hydrogen atmosphere as the internal hydrogen in the test of the charged specimens in helium. In addition, the additive effect of internal and external hydrogen becomes evident. This additive effect can be also seen under slow alternating load as shown by the low cycle fatigue test results in Figure 137.

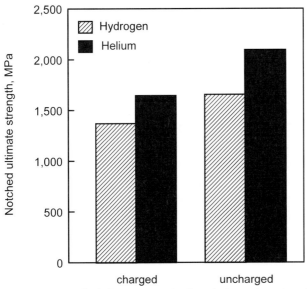

Figure 136: Notched ultimate strength of notched round test bars from INCONEL® alloy 718 (2.4668) during the test in helium and in hydrogen at 34.5 MPa and room temperature [166]

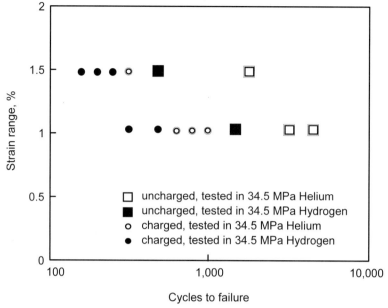

Figure 137: Influence of internal and environmental hydrogen on the low cycle fatigue life of INCONEL® alloy 718 in [166]

The external hydrogen alone clearly reduces the number of cycles withstood, this effect being greater with internal hydrogen and again in combination.

Fracture-mechanical tests of an INCONEL® alloy 718 regarding the influence of the hydrogen pressure on the crack propagation da/dt revealed that under constant load the limiting value of the crack intensity for subcritical crack growth at pressures above about 20 MN/m² is independent from the hydrogen pressure. Under cyclic load the crack growth rate da/dN at low and medium amplitudes of the stress intensity increases with increasing hydrogen pressure. Lowering the frequency from 1.0 Hz to 0.1 Hz leads to a clearly higher crack growth rate both at lower hydrogen pressure (0.7 MN/m²) and at a high pressure (70 MN/m²) [167].

During these examinations the materials were tested in three different heat treatment conditions as indicated in Table 104. The fracture-mechanical tests were performed on WOL (wedge-opening load) and DCB (double cantilever beam) specimens in autoclaves at different hydrogen pressures and in helium. Figure 138

Specimens	Solution annealing		1. Tempering		1. Tempering	
	Temperature °C	Duration Min.	Temperature °C	Duration Min.	Temperature °C	Duration Min.
A	740	60	720	480	620	400
B	1,050	20	760	600	650	530
C	1,040	105	760	630	650	490

Table 104: Heat treatment of the examined specimens from INCONEL® alloy 718 [167]

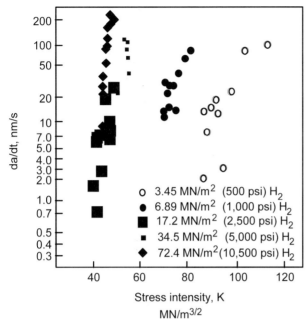

Figure 138: Crack growth rate da/dN as a function of the stress intensity factor K of DCB specimens from INCONEL® alloy 718 with the heat treatment C [167]

shows the results obtained for the crack growth rate da/dt as a function of the stress intensity factor K at different hydrogen pressures.

Table 105 indicates the da/dt values for two specimens with the heat treatment A and one specimen with the heat treatment C at a hydrogen pressure of 34.5 MN/m². Although the load is higher for the specimen C, the values of the crack propagation velocity are clearly higher for the heat treatment A with increasing hydrogen pressure.

Heat treatment	Stress intensity factor MN/m$^{3/2}$	Crack propagation velocity m/s
A	27	4.66×10^{-6}
A	36	3.17×10^{-5}
C	44	1.76×10^{-8}

Table 105: Crack propagation velocity da/dt under constant load of DCB specimens from INCONEL® alloy 718 with heat treatment A and C in compressed hydrogen at 34.5 MN/m² [167]

Figure 139 depicts the critical stress intensity factor K_{tH} for the starting crack growth in compressed hydrogen as a function of the hydrogen pressure for WOL specimens with heat treatment B and the DCB specimens with heat treatment C. At hydrogen pressures above about 20 MN/m² a growth of cracks is not assumed at a stress intensity factor below about 42 MPa $\sqrt{m}$.

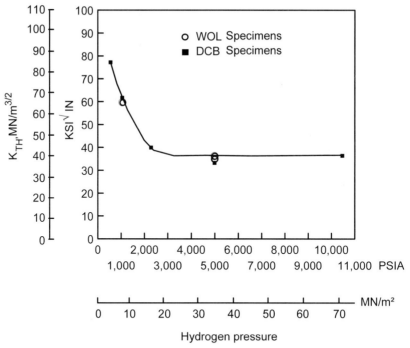

Figure 139: Critical stress intensity factor K_{tH} for WOL and DCB specimens from INCONEL® alloy 718 as a function of the hydrogen pressure at room temperature [167]

Table 106 summarizes further K_I values in hydrogen and in helium for other specimens and temperatures. At room temperature the K_I values are considerably lower in hydrogen compared to helium. Reducing temperatures leads to higher values for the critical crack intensities both in helium and in hydrogen.

Type of specimen	Test gas	Temperature °C	K_{tH} MPa $\sqrt{m}$	K_{IC} MPa $\sqrt{m}$
A	Helium	22	58	78
	Hydrogen		14	–
B	Helium	22	112	119
B + C	Hydrogen		42	
A	Helium	−129	81	98
	Hydrogen		72	
B	Helium	−129	139	122
	Hydrogen		123	
C	Helium	−73	160	149
	Hydrogen		< 47	

Table 106: Critical stress intensity factors K_{tH} and K_{IC} for the crack stop and unstable crack growth of different specimens from INCONEL® alloy 718 (2.4668) in helium and hydrogen at a pressure of 34.5 MN/m² at different temperatures [167]

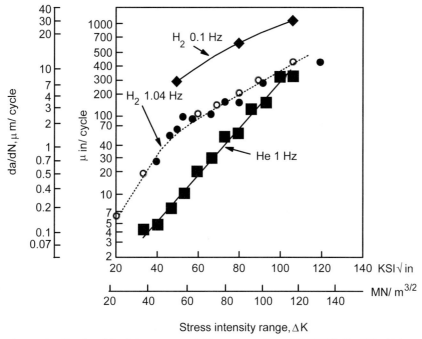

Figure 140: Results of the fatigue tests on DCB specimens from INCONEL® alloy 718 with heat treatment C in hydrogen at 68.9 MN/m² (frequency 0.1 Hz and 1.0 Hz, R = 0.1) as well as in helium (34.5 MN/m²) [167]

The tests under cyclic load were performed on DCB specimens with heat treatment C at hydrogen pressures of 0.069 MN/m², 34.5 MN/m² and 68.9 MN/m² as well as in a helium atmosphere at 34.5 MN/m² for comparison. Figure 140 illustrates the results for the frequencies 0.1 Hz and 1.0 Hz in hydrogen at a pressure of 68.9 MN/m² as well as for 1.0 Hz in helium. Figure 141 summarizes all results of the fatigue tests.

The crack growth da/dN under cyclic loading increases with increasing hydrogen pressure and increasing test frequency. Even at the lowest hydrogen pressure of 0.069 MN/m² and the higher frequency of 1.0 Hz the crack growth is clearly faster than that in helium. The increase of da/dN becomes lower with an increasing amplitude of the stress intensity factor (ΔK) at all hydrogen pressures. At the test frequency of 1.0 Hz da/dN in hydrogen is equal to or smaller than da/dN in helium at ΔK values from 100 MN/m$^{-3/2}$. This effect is no longer found at a test frequency of 0.1 Hz.

Figure 142 shows the influence of the test frequency between 1 Hz to 10 Hz at a load of $\Delta K = 54.7$ MPa $\sqrt{m}$ on the crack growth of the specimens in hydrogen (68.9 MN/m²). For comparison also the value for the specimen tested at 1 Hz in helium (34.5 MN/m²) is indicated. At one second per load cycle, i.e. at a frequency of 1 Hz the da/dN value in hydrogen approximates that in helium.

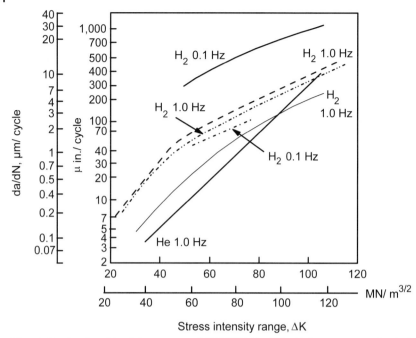

Figure 141: Results of the fatigue tests on INCONEL® alloy 718 with heat treatment C in hydrogen and in helium under different test conditions [167]

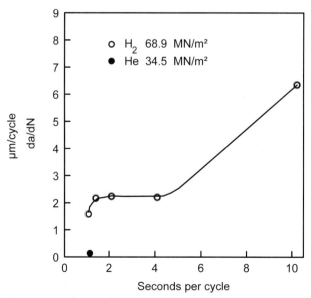

Figure 142: Influence of the test frequency on the crack growth rate of the specimens in hydrogen (p = 68.9 MN/m², ΔK = 54.7 MN/m$^{3/2}$) and in helium (34.5 MN/m², 1.0 Hz) [167]

Cr	Mo	Nb	Fe	C	Ni
20–23	8–10	3.15–4.15	max. 5	0.03–0.1	balance

Table 107: Nominal composition of the material NiCr22Mo9Nb (2.4856) [165]

Also the nickel-based alloy INCONEL® alloy 625 (NiCr22Mo9Nb, Mat. No. 2.4856) with the nominal chemical composition indicated in Table 107 is widely used in the chemical industry, in seawater applications, aviation and in power plants due to its good combination of strength, toughness, weldability and resistance against numerous aggressive atmospheres.

At room temperature the mechanical properties of the material, in particular the toughness behavior, through hydrogen can only be reduced at hydrogen pressures above 7 MPa and the material becomes susceptible to delayed fracture. During the test in compressed hydrogen at 35 MPa and at a temperature of 225 °C the toughness properties were reduced by less than 10% compared to the corresponding test in helium [170].

Comparative examinations of the behavior of both nickel alloys INCONEL® alloy 625 (2.4856) and INCONEL® alloy 718 (2.4668) as well as of the austenitic steel A 286 (X4NiCrTi25-15; Mat. No. 1.4943), towards hydrogen were performed to investigate the materials with their chemical compositions listed in Table 108 [168–170].

	Cr	Ni	Mo	Nb	Ti	Al	V	Fe	Si
2.4668	18.3	balance	3.0	5.07	1.00	0.63	0.008	18.50	0.15
2.4856	20.3	balance	8.22	3.26	0.18	0.19	–	4.49	0.22
1.4943	14.6	24.9	1.20	0.03	2.09	0.16	0.26	balance	0.18
	Mn	Cu	Co	C	N	B	P	S	
2.4668	0.11	0.05	0.42	0.044	0.007	0.0041	0.008	0.001	
2.4856	0.13	0.09	0.07	0.036	0.013	–	0.009	0.0005	
1.4943	0.17	0.08	0.27	0.035	0.0012	0.0065	0.018	0.002	

INCONEL® alloy 625 (NiCr22Mo9Nb; Mat. No. 2.4856)
INCONEL® alloy 718 (NiCr19Fe19Nb5Mo3; Mat. No. 2.4668)
A 286 (X4NiCrTi25-15; Mat. No. 1.4943)

Table 108: Compositions of the examined alloys [168]

To obtain the optimal toughness properties, the alloy 2.4668 was annealed at 1,068 °C for 1 h, cooled in air, aged at 760 °C for 10 h, cooled in the furnace to 650 °C, aged again for 10 h and thereafter cooled in air. The material A 286 was quenched in oil at 980 °C for 1 h, aged at 720 °C for 16 h and quenched in water. 2.4856 was annealed at 1,066 °C for 1 h.

The susceptibility of the alloys to hydrogen embrittlement was determined by means of slow strain rate tensile tests (CERT tests) using notched specimens. All tests were performed at room temperature.

The CERT tests were performed at an extension rate of 8.5×10^{-7} m/s in hydrogen gas at 101 kPa. The influence of the hydrogen was evaluated on the basis of changes of the notch tensile strength and the reduction of area. These CERT tests did not reveal any influence of hydrogen on the mechanical properties mentioned. Also the fractographic images of the fracture surfaces were not indicative of any brittle fracture behavior. These results confirm experience made in the past that higher hydrogen pressures are required to cause embrittlement of these materials in hydrogen gas. During the CERT test hydrogen was cathodically precipitated on the specimens and fine incipient cracks were found on the surface in the area of the notches with a maximum depth of 100 μm, whereas the remaining part of the specimens remained without any influence.

Following cathodic charging with hydrogen in a salt bath also hydrogen-induced crack formation was observed. Figure 143 and Figure 144 illustrate the values of the reduction of area and tensile strength as a function of the hydrogen content of the specimens obtained during slow strain rate tensile tests at a constant strain rate of 8.5×10^{-7} m/s using notched specimens.

As can be seen, the three examined materials exhibit a different susceptibility to hydrogen embrittlement. Susceptibility is clearly more marked in the two precipitation-hardening materials INCONEL® alloy 718 and steel A 286 compared to the single-phase material INCONEL® alloy 625.

The same three materials were also included in the comprehensive examinations in [162]. That study examined three nickel alloys, two steel grades and two titanium grades using different test methods with regard to their behavior in com-

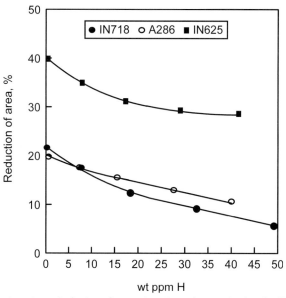

Figure 143: Reduction of area values depending on the dissolved hydrogen content of the specimens [169]

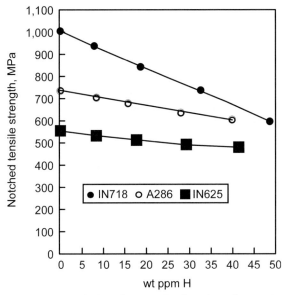

Figure 144: Notched tensile strength values depending on the hydrogen content of the specimens [169]

pressed hydrogen. The materials and their respective heat treatment conditions are indicated in Table 109.

The six different test methods as indicated in Table 110 were used to determine certain mechanical properties of the materials at room temperature and increased temperatures (680 °C for the nickel alloys, 100 °C for the titanium alloys).

Nickel alloys	
INCONEL® alloy 718	annealed at 950 °C
(NiCr19Fe19Nb5Mo3, Mat. No. 2.4668)	annealed at 1,040 °C
INCONEL® alloy 625	annealed at 980 °C
(NiCr22Mo9Nb, Mat. No. 2.4856)	
Hastelloy X alloy	annealed at 1,175 °C
(cf.: NiCr30FeMo, 2.4603)	
(NiCr22Fe18Mo, 2.4665)	
(NiCr21Fe18Mo; Mat. No. 2.4613)	
Steels	
A 286	annealed at 980 °C, tempered
(X4NiCrTi25-15, Mat. No. 1.4943)	
SAE 347	annealed at 1,140 °C, cold-rolled
(Mat. No. 1.4550, X6CrNiNb18-10, S34800)	
Titanium materials	
TiAl6V4, Mat. No. 3.7165	annealed at 750 °C
TiAl5Sn2.5, Mat. No. 3.7115	annealed at 650 °C

Table 109: Examined materials and heat treatment conditions [162]

Test methods	Determined data
Smooth tensile test specimens	0.2% yield strength; tensile strength elongation; necking
Notched tensile test specimen	tensile strength
Low-frequency alternating load (LCF)	elongation against number of load cycles
High-frequency alternating load (HCF)	stress against number of load cycles until fracture
Fracture-mechanical test (CT specimen)	fracture toughness
Creep rupture test	elongation against time stress against time until fracture elongation; reduction of area

Table 110: Test methods used for determining mechanical properties [162]

Identical tests were carried out in helium and in hydrogen at a test pressure of 34.5 MN/m² each. The results obtained during the tensile tests are summarized in Table 111. Impairment of the relevant mechanical properties were evaluated according to equation 18:

Equation 18 $$X = \frac{He - H_2 \times 100}{He}$$

He is the value determined in helium and H_2 is the value determined in hydrogen.

As can be seen, only the fully austenitic steel SAE 347 is insensitive to hydrogen under all test conditions. It is also shown that one test method alone is not sufficient to reveal potential damage by hydrogen. Even if the ductility is not impaired during tensile tests, a negative impact of hydrogen can occur under alternating load or in a creep rupture test.

Examinations regarding the elongation and fatigue behavior of the material INCONEL® alloy 718 in high-pressure hydrogen showed that the fatigue crack growth was increased. Crack initiation started at the carbides and the crack path followed the phase boundaries between the δ-phase and the γ-matrix [171–173]. The influence of hydrogen on the values of elongation and reduction of area during tensile tests of INCONEL® alloy 718 specimens in hydrogen an argon is illustrated in Figure 145.

In [174] tests on the influence of hydrogen on the deformation behavior of INCONEL® alloy 718 under monotonous load were reported. The test material was 0.37 mm thick, cold-rolled strip material with the chemical composition indicated in Table 112.

The tensile test specimens were tested in the fully aged condition using the following heat treatments:

– Solution annealing at 1,060 °C for 1 h
– Quenching in water
– Tempering at 760 °C from 1 h to 50 h.

A 29 Nickel-chromium-molybdenum alloys

	$R_{e0.2}$	R_m	A	Z	R_m/notch	LCF	HCF	Creep rupture	R_m/notch
INCONEL® alloy 718 (annealed at 950 °C)	+/+	+/+	–/+	–/+	–/+	–/+	–/–	–/–	–/+
INCONEL® alloy 718 (annealed at 1,040 °C)	+/+	+/+	–/+	+/+	+/+	–/–	–/–	+/–	+/+
INCONEL® alloy 625 (NiCr22Mo9Nb, Mat. No. 2.4856)	+/+	+/+	–/+	–/+	+/+	–/–		+/–	+/+
Hastelloy X alloy (cf.: 2.4603, 2.4613, 2.4665)	+/+	+/+	+/+	+/+	+/+	–/			+/+
A 286 (X4NiCrTi25-15, Mat. No. 1.4943)	+/+	+/+	+/–	+/–	+/+	–/+		+/–	+/+
SAE 347 (Mat. No. 1.4550, X6CrNiNb18-10, S34800)	+/+	+/+	+/+	+/+	+/+	+/+	+/	+/+	+/+
TiAl6V4, Mat. No. 3.7165	+/+	+/+	–/–	–/–	–/–	+/+		+/–	–/–
TiAl5Sn2.5, Mat. No. 3.7115	+	+	–/–	–/–	–/–	–/–	+/	+/–	–/–

– = X > 10%, + = X < 10% (calculated according to equation 18)

Table 111: Impairment of the mechanical properties of the examined materials by hydrogen at room temperature (1st character) and elevated temperature (2nd character) [162]

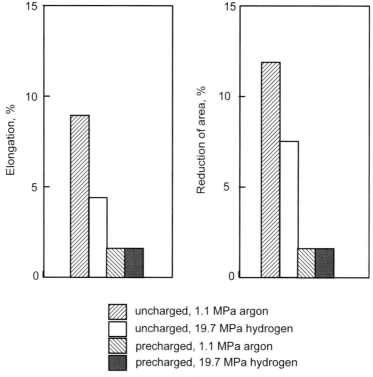

Figure 145: Tensile properties of INCONEL® alloy 718 during the tensile test in 1.1 MPa argon and 19.7 MPa hydrogen [172]

The specimens were charged with 11 MPa hydrogen in a pressure chamber at 300 °C for 72 h. The subsequent tensile tests were performed at an initial strain rate of 3.3×10^{-5} 1/s in air at ambient temperature.

Figure 146 shows the influence of hydrogen on the path of the stress-strain curve of the specimens tempered between 8 and 16 h. Hydrogen leads to clearly reduced values of the fracture stress and, in particular, elongation.

The effect of the aging time duration on the elongation of the tensile test specimens in air and in hydrogen is shown in Figure 147.

Although the strain rates reduce both in the specimens charged with hydrogen and the hydrogen-free specimens with increasing aging time, the values of the specimens charged with hydrogen are always lower by about a factor of 3. However, the ratio of the values remains practically constant at about 0.38 attributable to the

C	Mn	Si	P	S	Cr	Co	Mo	Nb + Ta	Ti	Al	B	Cu	Ni	Fe
0.05	0.21	0.12	< 0.006	< 0.002	18.1	0.4	3.01	5.08	1.02	0.55	0.004	0.05	52.7	balance

Table 112: Chemical composition of the examined INCONEL® alloy 718 [174]

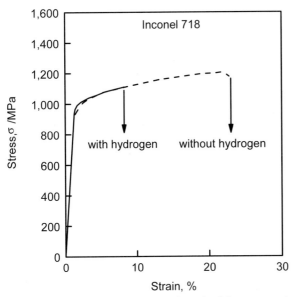

Figure 146: Influence of hydrogen on the path of the stress-strain curve of INCONEL® alloy 718 [174]

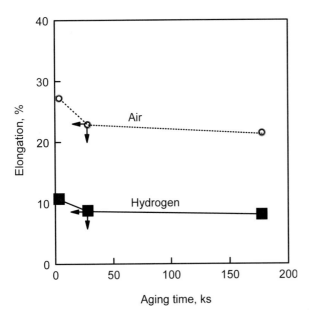

Figure 147: Influence of the aging time duration at 760 °C on the elongation of the tensile test specimens from INCONEL® alloy 718 in air and in hydrogen [174]

fact that the susceptibility to hydrogen embrittlement is independent of the aging time and, hence, also independent of the morphology of the precipitates.

Whereas the hardness in the precipitation-hardened condition of INCONEL® alloy 718 amounts to about 450 HV, it is only 250 HV in the solution-annealed condition. Therefore, it can be assumed that the susceptibility to hydrogen embrittlement can be reduced by solution annealing. On the other hand, the suitability of this alloy as a constructional material is also reduced due to the lower strength values as a result of this treatment. However, since hydrogen-induced cracks most frequently occur at sites of increased stress, susceptibility to hydrogen embrittlement can be also reduced due to the fact that only in these areas the hardness is reduced by local solution annealing.

To check this assumption, the temperature of the surface area of tensile test specimens from two commercial melts of the material INCONEL® alloy 718 was increased to a solution-annealing temperature by performing a laser-beam treatment without causing this area to melt [175]. Thereafter the specimens were subjected to a strain rate of 1.67×10^{-3} mm/s in a slow strain rate tensile test in a hydrogen atmosphere at a pressure of 29.4 MPa and at room temperature until fracture. Both INCONEL® alloy 718 melts A and B differed only slightly in terms of their chemical composition (refer to Table 113). For comparison specimens were also tested in the solution-annealed state. In addition, specimens were included in the examination, having been precharged with hydrogen prior to the tensile test in compressed hydrogen by aging in compressed hydrogen at 25 MPa and 527 °C for 2 h.

	Ni	Cr	Fe	Nb + Ta	Mo	Ti	Al	Co	Mn	Si	Cu	C
A	52.55	18.50	18.40	5.12	3.02	0.98	0.42	0.43	0.15	0.08	0.04	0.04
B	52.57	18.47	18.40	5.14	2.95	0.99	0.39	0.34	0.16	0.08	0.04	0.04

Table 113: Chemical compositions of the examined INCONEL® alloy 718 melts in wt% [175]

For the heat treatment of the specimens including solution annealing, hardening and two-stage tempering reference is made to Figure 148.

The hardness distributions illustrated in Figure 149 show how hardness in the surface area of the specimens could be reduced by different laser treatments.

The results of the tensile tests of the specimens not precharged with hydrogen in air and in compressed hydrogen are shown in Figure 150.

The results reflected in Figure 150a show that hydrogen has a minor influence on tensile strength only in the solution-annealed specimens and has practically no influence at all in hardened and surface-annealed specimens. The tensile strength values of the surface-annealed specimens being about 13% lower than those of the hardened specimens are attributed to the influence of the softer surface areas.

In contrast to the tensile strength, the ductility of all specimens is clearly reduced in the hydrogen atmosphere compared to air as can be seen from the values for the reduction of area in Figure 150b. Whereas the reduction of area of the surface-annealed specimens assumes the same value in air as the hardened specimens,

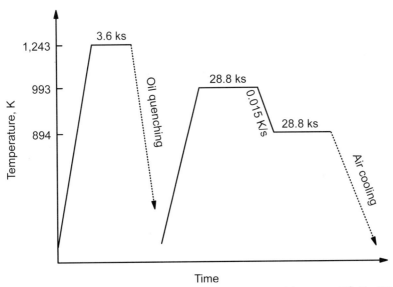

Figure 148: Temperature-time diagram for the heat treatment of the INCONEL® alloy 718 specimens [175]

this value is higher in compressed hydrogen by about a factor of 1.5 and corresponds to the solution-annealed specimens. Hence, the susceptibility of the surface area to hydrogen embrittlement determines the fracture behavior of the specimens such that the resistance to hydrogen-induced crack formation can be increased by a softer surface.

Also in the specimens precharged with hydrogen the susceptibility to hydrogen embrittlement was clearly reduced by the laser treatment of the surface compared to the hardened specimens as shown by the values of the reduction of area in Table 114.

Apart from containing chromium and molybdenum, polycrystalline and single crystal nickel based superalloys such as PWA 1480 (Pratt & Whitney Alloy 1480) also contain other alloying elements such a aluminium, cobalt and tungsten. They are considered to be suitable materials for turbine blades, e.g. in the space shuttle main engine unit, due to their good mechanical properties both at ambient temperatures and at higher temperatures as well as due to their good creep and fatigue strength. The alloy contains about 70% coherent, ordered precipitates of the γ-phase (Ni_3Al,Ti) in a γ-matrix (face-centered cubic nickel). Usually, it contains a larger number of pores as well as a eutectic γ-γ′-phase acting as hydrogen traps in the alloy when hydrogen is absorbed. Therefore, it can be an efficient measure to reduce the number of these traps to improve the hydrogen resistance of the material. Since the fatigue strength in a hydrogen environment is an important factor in space technology, the influence of the porosity and the content of γ-γ′-eutectic on the mechanical properties and on the fatigue behavior of the alloy charged with hydrogen was investigated (chemical composition in Table 115) [176–179].

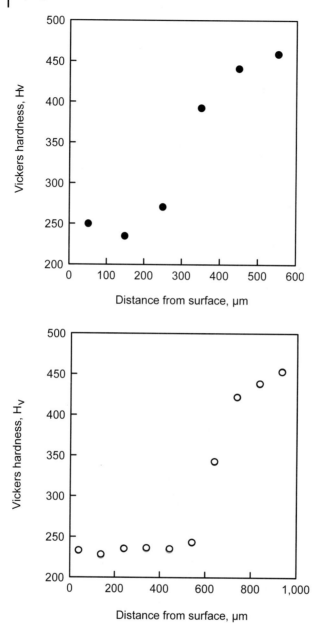

Figure 149: Hardness distributions in the surface area of the tensile test specimens from IN-CONEL® alloy 718 following laser treatment [175]
focal distance a) 20 mm; feed rate 16.7 mm/s/b) 30 mm; feed rate 6.7 mm/s

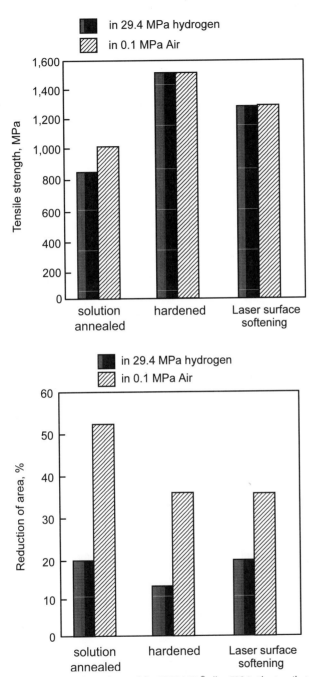

Figure 150: Values obtained for INCONEL® alloy 718 in the tensile test in air and in compressed hydrogen for the tensile strength (a) and the reduction of area (b) depending on the heat treatment of the specimens [175]

	Reduction of area, %
Hardened specimen	10.0
Surface-annealed specimen	19.5

Table 114: Reduction of area of the hardened and surface-annealed specimens from INCONEL® alloy 718 in a hydrogen atmosphere following hydrogen precharging [175]

Cr	Al	Ti	Ta	W	Co	Ni	References
10.4	4.8	1.3	11.9	4.1	10.4	balance	[176]
9.4	4.7	1.0	11.0	5.2	4.8	balance	[177, 178]

Table 115: Chemical composition (wt%) of the examined PWA 1480 alloy

Different contents of pores and eutectic were yielded by hot isostatic pressing (HIP) and varying the thermomechanical treatment.

Following their manufacture, the specimens were aged in compressed hydrogen (138 MPa) at 300 °C [177] for 15 days or at 350 °C and 103.5 MPa for 15 days [176, 178]. The hydrogen content was then 400 ppm. The subsequent tensile tests and the strain-controlled fatigue tests (frequency 0.1 Hz; R = −1) were performed at room temperature in air.

Table 116 and Table 117 indicate the mechanical properties determined for uncharged specimens and specimens charged with hydrogen in the tensile test (strain rate 0.1%/s). The unfavorable influence of hydrogen becomes visible, in particular, by the values for elongation and the reduction of area.

Heat treatment	Yield strength MPa	Tensile strength MPa	Reduction of area %
Standard, not charged	1,138	1,151	3.3
Standard, charged	1,081	1,086	0.35
HIP, not charged	1,147	1,179	5.9
HIP, charged	1,071	1,081	1.2

Table 116: Mechanical properties of specimens subjected to different treatment [176]

Heat treatment	Yield strength MPa	Tensile strength MPa	Elongation %	Reduction of area %
	Uncharged specimens	< 5 ppm hydrogen		
Standard	1,041	1,241	7.6	8.4
	Charged specimens	400 ppm hydrogen		
Standard	936	1,048	2.6	7.3
Eutectic annealed	1,060	1,083	7.7	6.9
HIP	960	1,069	3.0	6.8
Eutectic annealed + HIP	952	1,117	3.2	4.2

Table 117: Mechanical properties of specimens subjected to different treatment [177]

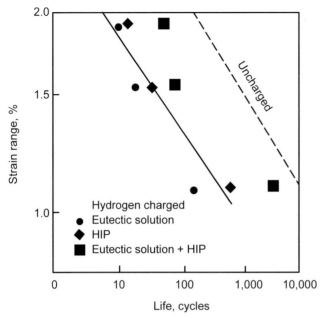

Figure 151: Results of the fatigue tests of the alloy PWA 1480 [177]

The results of the fatigue tests are summarized in Figure 151. Hydrogen charging had a considerably detrimental effect on the fatigue behavior of the alloy PWA 1480. Specimens subjected to a standard heat treatment having therefore normal pore and eutectic contents exhibited times to failure which were reduced by more than one order of magnitude. This detrimental effect of the hydrogen can be diminished by a double thermomechanical heat treatment consisting of solution annealing at 1,290 °C followed by hot isostatic pressing yielding the strongest decrease of the size and number of pores and the eutectic. The influence of hydrogen could be reversed by aging of the charged specimens at 426 °C for 48 h. Thereafter, the typical fatigue values of the materials were obtained again.

Also the single crystal super alloy CMSX-2 with a similar composition (Table 118) was examined regarding its behavior following hydrogen charging [180].

The tensile test specimens for the examinations were manufactured from supplied rods in the solution-annealed state (at 1,315 °C for 3 h). The tensile test specimens were encapsulated in dry argon and subjected to the following heat treatment to obtain a uniform size and distribution of the γ'-precipitates:

- annealing at 1,050 °C, 16 h cooling in air
- aging at 850 °C, 48 h cooling in air.

Cr	W	Ta	Al	Co	Ti	Mo	Fe	S	Ni	C	S	N	O
%	%	%	%	%	%	%	%	%i	%	ppm	ppm	ppm	ppm
8.0	8.0	6.0	5.6	4.06	1.07	0.6	0.08	0.015	balance	15	10	4	2

Table 118: Chemical composition of the nickel-based alloy CMSX-2 [180]

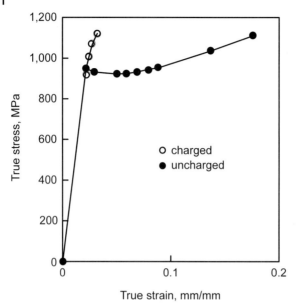

Figure 152: True Stress-true strain curves of slow strain rate tensile tests on specimens from the CMSX-2 alloy charged with hydrogen and non-charged [180]

Some of the specimens were tested in the non-hydrogen charged state, whereas the rest were aged in compressed hydrogen (138 MPa) at 300 °C for 15 days. Then, the hydrogen content of the specimens was between 0.50 and 0.55 at.%. The charged and uncharged specimens were used thereafter to perform slow strain rate tensile tests at an initial strain rate of 1.2×10^{-3} 1/s. Figure 152 shows the true stress-true strain curves obtained for the charged and uncharged specimens. Table 119 contains the mechanical properties determined during the tensile tests.

	0.2% Yield strength MPa	Tensile strength MPa	Uniform elongation %	Elongation loss %
uncharged	962	988	16.2	–
H$_2$-charged	935	1,123	1.2	92.6

Table 119: Results of the slow strain rate tensile tests on specimens from the CMSX-2 alloy charged with hydrogen and uncharged [180]

The uncharged specimens exhibit a high yield strength value in combination with a good plastic behavior and, although the dissolved hydrogen does not cause major changes in the yield strength value, it does reduce the elongation behavior by more than 90%.

Other nickel-based super alloys frequently tested for their suitability as a turbine material and behavior towards hydrogen and their chemical compositions are listed in Table 120.

Table 120: Chemical composition of frequently tested nickel-based super alloys (balance Ni) [171, 181]

Alloy	Cr	Co	W	Mo	Ta	Nb	Ti	Al	Hf	B	Zr	C
Waspaloy®	19.5	13.5		4.25			3.0	1.25				0.07
INCONEL®100	12.4	18.4		3.25			4.35	5.0				0.06
MAR-M 247	8.4	10.0	10.0	0.6	3.0	–	1.0	5.5	1.4	0.015	0.05	0.15
Udimet® 720	18.0	14.5	1.2	3.2	–	–	5.1	2.5		0.030	0.05	0.04

As shown in Table 121 for two melts of the alloys Waspaloy® (2.4654) and INCONEL® 100 each, the materials absorb different amounts of hydrogen when aged in compressed hydrogen (10.3 MPa and 34.4 MPa) for 24 h depending on the temperature [181].

Temperature/pressure °C/MPa	Waspaloy® ppm		INCONEL® 100 ppm	
	A	B	A	B
315/10.3	31	44	33	69
538/10.3	47	46	132	138
760/10.3	50	47	94	104
760/34.4	101	94	157	157

Table 121: Solubility of hydrogen in nickel-based alloys [181]

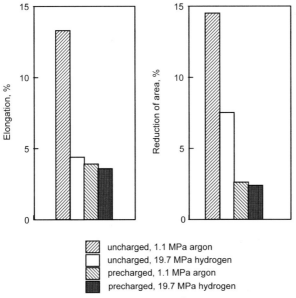

Figure 153: Tensile properties of Udimet® 720 during the tensile test in 1.1 MPa argon and 19.7 MPa hydrogen [172]

The diagrams in Figure 153 illustrate the negative influence of compressed hydrogen on the values of reduction of area and elongation of the tensile test specimens from Udimet® 720.

Examinations regarding the fatigue behavior of the alloys Udimet® 720 and MAR-M 247 exhibit a clearly stronger influence of hydrogen compared to the INCONEL® alloy 718 [171, 173]. Figure 154 summarizes the results of fatigue tests in compressed hydrogen for various nickel alloys [182].

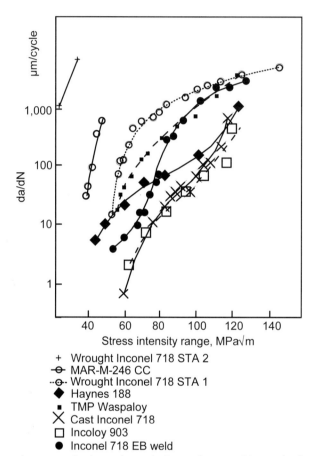

+ Wrought Inconel 718 STA 2
⊖ MAR-M-246 CC
⊙ Wrought Inconel 718 STA 1
◆ Haynes 188
■ TMP Waspaloy
✕ Cast Inconel 718
☐ Incoloy 903
● Inconel 718 EB weld

Figure 154: Crack growth rate da/dN as a function of the amplitude of the stress intensity factor ΔK for various nickel alloys in compressed hydrogen at 34.5 MN/m^2 and 48.3 MN/m^2 at room temperature [182]

A 30 Nickel-copper alloys

Since nickel-copper alloys contain both a metal sensitive to hydrogen (nickel) and a metal insensitive to hydrogen (copper), the susceptibility to hydrogen-induced embrittlement of the alloy depends on the ratio of both metals. From a nickel content of about 50% the alloys become susceptible to hydrogen embrittlement as shown by the examinations described under A 9 [70].

A 32 Other nickel alloys

Also the nickel-based materials from the intermetallic compounds of type A_3-B are sensitive to hydrogen. [183] describes examinations of the behavior of the intermetallic compound $Ni_3(Si, Ti)$ under the following atmospheres:

Vacuum
Air
Oxygen
Distilled water
Hydrogen
Argon + 10% hydrogen.

These examinations were performed using singlecrystals of this intermetallic compared with no boron addition at room temperature in slow strain rate tensile tests. The composition of both compounds is indicated in Table 122.

	Ni mole%	Si mole%	Ti mole%	B wt%	H wt%	O wt%	N wt%
Ni_3(Si, Ti)	78	10.8	10.6		0.00011	0.0059	0.0013
Ni_3(Si, Ti) + B	78	11.2	10.5	0.005	0.00017	0.0072	0.0018

Table 122: Chemical composition of the examined monocrystals [183]

The slow strain rate tensile tests were performed at three strain rates

$\dot{\varepsilon} = 6.4 \times 10^{-3}$ 1/s

$\dot{\varepsilon} = 6.4 \times 10^{-4}$ 1/s

$\dot{\varepsilon} = 6.4 \times 10^{-5}$ 1/s.

The two diagrams in Figure 155 show the nominal stress-strain curves obtained for the tensile test specimens

The tests on the boron-free Ni_3(Si, Ti) specimens in oxygen, air and vacuum revealed the highest fracture elongation values. In water and, in particular, in the presence of hydrogen considerably lower strain rates are reached. In H_2 and in Ar/H_2 the specimens broke already in the elastic range without plastic deformation. Although the addition of boron to the single crystal alloy increases the fracture

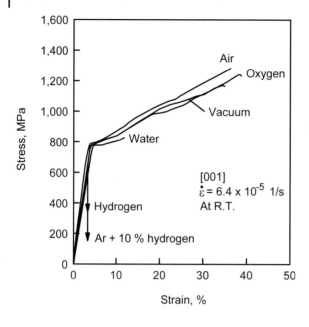
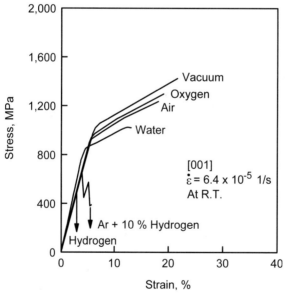

Figure 155: Stress-strain curves of the boron-free $Ni_3(Si, Ti)$ alloy (a) and the $Ni_3(Si, Ti)$ alloy with boron addition (b) in different test environments at $\dot{\varepsilon} = 6.4 \times 10^{-5}$ 1/s [183]

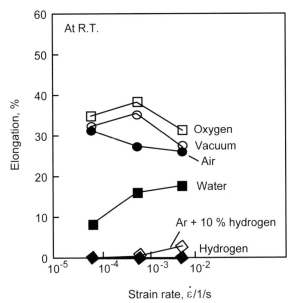

Figure 156: Values of elongation as a function of the strain rate of the boron-free alloy $Ni_3(Si, Ti)$ [183]

stress in oxygen and in vacuum, the strain values are, in general, reduced and negatively impacted in the same manner through the presence of hydrogen. Figure 156 shows the influence of the strain rate on the elongation values of the alloy not doped by boron in slow strain rate tensile tests.

Whereas the fracture elongation in vacuum, oxygen and air slightly increases with decreasing strain rate, the values in water, hydrogen and argon-hydrogen gas mixture decrease more strongly at a slower strain rate since the hydrogen has more time to penetrate the material and exert its embrittling effect. The addition of boron has only a very minor effect on the embrittlement behavior in hydrogen.

Similar examinations were also performed for intermetallic materials on the basis of Ni_3Al [184]. The chemical compositions of the tested alloys are indicated in Table 123.

The mechanical parameters determined during the slow strain rate tensile test ($\dot{\varepsilon} = 2 \times 10^{-4}$ 1/s) in the different test atmospheres are summarized in Table 124.

Alloy	Al at.%	Cr at.%	Zr at.%	Mn at.%	B wt ppm	Ni
$Ni_3Al + B$	23	–	–	–	120	balance
$Ni_3(Al,Cr,Zr) + B$	17.7	7.9	1.1	–	80	balance
$Ni_3(Al,Mn)$	10.2	–	–	13.7	–	
$Ni_3(Al,Mn) + B$	9.2	–	–	14.2	400	balance

Table 123: Chemical compositions of the tested intermetallic alloys [184]

Test atmosphere	Yield strength MPa	Tensile strength MPa	Elongation %
Air	173	433	10.1
Vacuum (1.3×10^{-3} Pa)	171	762	25.5
Ar + H_2O (0.04 MPa)	177	422	10.6
H_2 (0.1 MPa)	174	216	0.6
O_2 (0.1 MPa)	182	581	20.0
a) O_2 (0.1 MPa)	153	663	20.6
a) H_2 (0.1 MPa)	151	203	1.5
b) O_2 (0.1 MPa)	163	524	16.5
b) H_2 (0.1 MPa)	158	173	0.5

a) oxidized before hand at 900 °C for 24 h
b) oxidized before hand at 900 °C for 0.5 h

Table 124: Mechanical properties of the material Ni_3Al + B in different test atmospheres [184]

The yield strength values are practically not influenced by the test atmosphere. Whereas for the elongation as a measure of the ductility comparable values were obtained in air and in water vapor-saturated argon, the value increases by more than the double during the test in vacuum and in oxygen gas with a water content below 2.6 ppm. The tensile strength decreases in accordance with the increase of the elongation. In hydrogen the elongation value reached amounts to 0.6% only. Also preliminary oxidation of the specimens cannot prevent the negative influence of hydrogen. Annealing in air at 900 °C for half an hour does not show any effect and an extension of the annealing time to 24 h only leads to a very minor improvement in the elongation.

Table 125 shows the results of the respective tests for the material $Ni_3(Al,Cr,Zr)$ + B.

Test atmosphere	Yield strength MPa	Tensile strength MPa	Elongation %
Air	422	610	6.2
Vacuum (1.3×10^{-3} Pa)	412	744	11.7
Ar + H_2O (0.04 MPa)	426	584	4.2
H_2 (0.1 MPa)	424	500	2.4

Table 125: Mechanical properties of the material $Ni_3(Al,Cr,Zr)$ + B during the slow strain rate tensile test in different test atmospheres [184]

Basically, the result is the same as that of the alloy Ni_3Al + B, and toughness decreases in the sequence vacuum – air – hydrogen.

The influence of the boron addition on the mechanical properties in the tensile test were only investigated for the alloy $Ni_3(Al,Mn)$ by performing tests in air and in dry oxygen. As shown in Table 126, the values of the material with boron addi-

Alloy	Test atmosphere	Yield strength MPa	Tensile strength MPa	Elongation %
$Ni_3(Al,Mn) + B$	Air	201	703	49
	Oxygen (0.1 MPa)	198	718	51
$Ni_3(Al,Mn)$	Air	192	456	15.5
	Oxygen (0.1 MPa)	192	692	46.9

Table 126: Mechanical properties of the material $Ni_3(Al,Mn)$ with and without boron during the slow strain tensile test in air and in dry oxygen [184]

tion in air and in oxygen are practically identical and the material is very ductile with elongation values of about 50%, whereas the boron-free alloy exhibits a comparably high elongation in dry oxygen which, however, drops to 15.5% during the test in air.

In general, it can be concluded from the examination results that the Ni_3Al alloys are not only highly susceptible to embrittlement in hydrogen gas, but are also susceptible to hydrogen embrittlement in a humid atmosphere, the hydrogen being formed by the reaction of the water vapor with the aluminum of the alloy according to Equation 19.

Equation 19 $1\ Al + 3\ H_2O \rightarrow Al_2O_3 + 6\ H^+ + 6e^-$

The effect in a humid atmosphere, however, is clearly lower than that in hydrogen such that here a positive influence of the boron addition can be detected (refer to Table 126).

Similar results were also obtained by the examinations described in [185], which were performed with the alloy $Ni_3(Al,Ti)$ with 0.05% boron addition. The nominal stress-strain curves obtained from the slow strain rate tensile test in vacuum, air, distilled water and hydrogen for specimens from this material (Figure 157) clearly reveal the damaging effect of hydrogen again.

Again, the elongation values decrease in the sequence vacuum – air – water; in hydrogen the specimens fail already in the elastic range and elongation is no longer measured. Figure 158 shows the influence of the strain rate on the elongation values measured in the various test atmospheres. Whereas virtually no deformation is measured in hydrogen irrespective of the strain rate, a transition from ductile to brittle fracture behavior can be observed with decreasing strain rate in moist air and in hydrogen. The explanation is that more hydrogen can be formed in the reaction according to Equation 19 when the strain rate decreases and the dwell time of the specimens in the test atmospheres is longer.

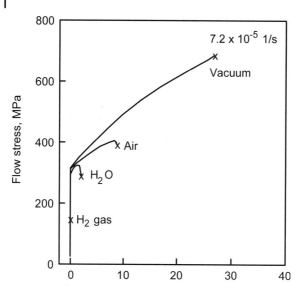

Figure 157: Influence of the testing environment on the nominal stress-strain curves of the alloy Ni$_3$(Al,Ti) + B (room temperature, strain rate $\dot{\varepsilon}$ = 7.2 × 10^{-5} 1/s [185])

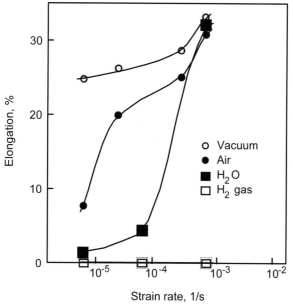

Figure 158: Influence of the strain rate and the test environment on the elongation values of the alloy Ni$_3$(Al,Ti) + B [185]

A 35 Platinum metals (Ir, Os, Pd, Rh, Ru) and their alloys

It is known that hydrogen can be easily absorbed by palladium and diffuses through palladium. Compact metallic palladium is able to dissolve hydrogen of an amount of about 600 times its volume at room temperature, finely dispersed palladium dissolves 850 times, an aqueous suspension of microdispersed palladium 1,200 times and colloidally dissolved palladium even 3,000 times its volume.

Palladium belongs to the metals in which dissolved hydrogen leads to the formation of two phases ($\alpha + \beta$) with different hydrogen concentrations. Both phases are solid solutions of hydrogen in palladium, have the same crystal structure with a face-centered cubic lattice, however with different lattice parameters, and different solubility for hydrogen (in [51] p. 183–185). At room temperature the lattice constants a of pure palladium are 3.886×10^{-10} m for the α-phase and 4.020×10^{-10} m for the β-phase. The hydrogen solubility limit is H/Pd $\approx$ 0.03 in the α-phase and 0.65 in the β-phase.

Figure 159 shows the lattice constants for the palladium-hydrogen system as a function of the temperature [186] (quoted in [51]). As the temperature increases, the lattice constants approximate the two parallel phases and reach the same value at the critical temperature of about 300 °C. Then, the corresponding values for the concentration and the pressure of hydrogen are H/Pd $\approx$ 0.27 and $p_{H2} \approx$ 22 bar.

Tests regarding the solubility of hydrogen in palladium and palladium-silver alloys are described in [187]. The examined alloys had a silver content of 10, 20, 23, 26, 29 and 40%. The tests were performed in a temperature range from 100 °C to

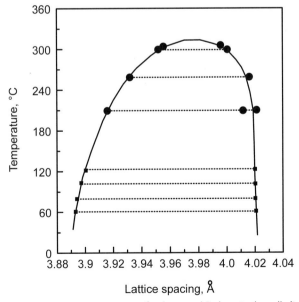

Figure 159:   Lattice parameters for the α- and β-phase in the palladium-hydrogen system [186]

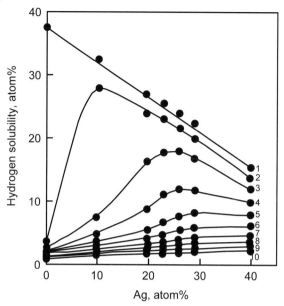

Figure 160: Solubility of hydrogen in palladium–silver alloys with different silver contents at different temperatures in °C 1) 100, 2) 140, 3) 180, 4) 220, 5) 260, 6) 300, 7) 350, 8) 400, 9) 450, 10) 500 [187]

500 °C at a hydrogen pressure of 760 mm Hg. The results of the examinations are summarized in Figure 160.

The solubility for hydrogen decreases with increasing temperature. In the temperature range from 300 °C to 500 °C the solubility slightly increases with increasing silver content of the alloys up to the highest examined content of 40% silver. At temperatures below 300 °C the solubility maximum shifts to lower silver contents and is reached at 140 °C for the alloy with 10% silver. The highest solubility in pure palladium is found at 100 °C. Since in the silver-hydrogen system at temperatures below the range from 140 °C to 160 °C the β-phase is more stable with the highest solubility for hydrogen, it is assumed that with an increasing silver content of the alloy the tendency to form the β-phase and, hence, the solubility from about 25% silver decreases.

Ready diffusion of hydrogen through palladium and palladium-silver alloys makes this material interesting as a diffusion cleaner for hydrogen and as a hydrogen electrode in fuel cells. The dependence of hydrogen diffusion through the palladium membranes on the partial hydrogen pressure on the entry and exit sides is of special importance to the separation of hydrogen from mixtures with other gases. The studies described in [188] examined the relationship between the partial pressure, temperature, diffusion rate and permeability of technically pure hydrogen through 0.015 mm thick membranes from palladium and a palladium-25% silver alloy. The non-dried hydrogen had the nominal composition indicated in Table 127 and contained about 2.5% moisture.

% H₂	% CO₂	% N₂	% CH₄
98.5	0.3–0.5	0.5–0.8	0.5–0.6

Table 127: Nominal composition of hydrogen [188]

In the examined temperature range from 250 °C to 700 °C and in the pressure range up to 3 bar the diffusion rate follows the Sievert's Law and is linearly dependent on the square root of the partial pressure and increases with increasing temperature as the curves in Figure 161 and in Figure 162 show as examples for the two metals at various temperatures.

Since these examinations did not reveal any influence of the small contents of carbon dioxide, methane and nitrogen on the permeability of hydrogen, subsequent tests were also performed with converter gas obtained from reforming hydrocarbons with water [189]. The nominal composition of the gas is indicated in Table 128.

As in the previous examinations, the same laws and regularities were found for this gas. The higher contents of gaseous contaminants had no influence on the permeability of the hydrogen.

[190] describes examinations of the influence of hydrogen on the behavior of alloys of the palladium-silver system. The chemical compositions of the examined binary alloys are indicated in Table 129. The susceptibility of the alloys to hydrogen embrittlement was checked in slow strain rate tensile tests.

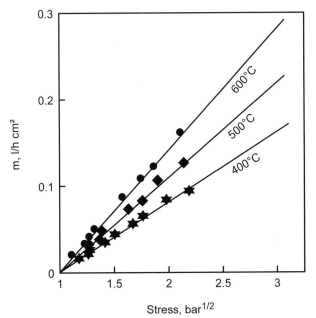

Figure 161: Dependence of the diffusion rate m on hydrogen through a palladium membrane at different temperatures [188]

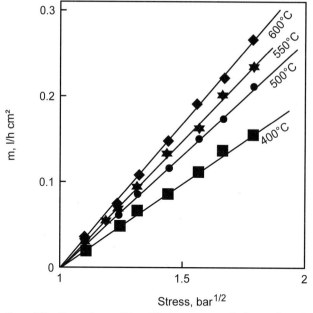

Figure 162: Dependence of the diffusion rate m on hydrogen through a palladium-silver membrane at different temperatures [188]

% H_2	% CO_2	% CO	% N_2	% CH_4
75.2	13.0	9.6	0.4	1.8

Table 128: Nominal composition of the hydrogen from the converter gas [189]

Alloy	Pd	Pd-10Ag	Pd-20Ag	Pd-25Ag	Pd-30Ag	Pd-50Ag	Ag
at.% Ag	0.0	9.7	19.3	24.3	29.5	49.6	100

Table 129: Chemical composition of the examined palladium-silver alloys [70]

However, the specimens were not charged or tested in hydrogen gas, but the tensile tests were performed in air at a strain rate of 3×10^{-4} 1/s following cathodic charging. Figure 163 illustrates the elongation values obtained for charged and uncharged specimens as a function of the silver content of the alloys.

As can be seen, the alloys with a silver content from 30% are not susceptible to hydrogen embrittlement, whereas at silver contents up to 10% strong embrittlement occurs and the elongation values approach zero.

As the studies in [191] show, the values of the strength and ductility of palladium-silver alloys with silver contents up to 30% at lower hydrogen contents can be also improved. For these examinations the test material was obtained by melting palladium (99.99%) and palladium alloys with silver contents of 5, 20, 25 and 30% in an argon atmosphere, forging at temperature from 1,000 °C to 1,200 °C to

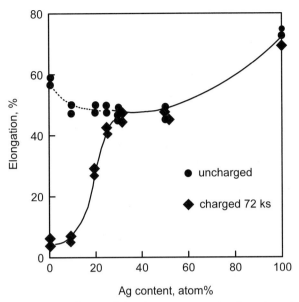

Figure 163: Influence of the silver content on the elongation values obtained during the slow strain tensile tests [70]

produce plates of 8 to 10 mm thickness and then rolling these plates to a thickness of 0.5 mm. This material was used for performing slow strain rate tensile tests at a strain rate of 3×10^{-3} 1/s both in the deformed state and following an annealing treatment at 800 °C in vacuum for 1 h. The influence of hydrogen on the mechanical properties was again checked following electrolytic charging of the specimens in a 4% sodium fluoride solution with an electric current density of 300 A/m². Figure 164 illustrates the values obtained with the palladium specimens for the 0.2% yield strength, tensile strength and elongation as a function of the hydrogen content of the specimens for the rolled and heat treated state.

The results show that absorption of up to 0.25% to 0.30% hydrogen leads to an increase both of the strength values and the ductility characterized by the elongation values. For instance, elongation of the heat treated specimens increases from 20% of the hydrogen-free specimens to 40% up to 45% at a hydrogen content of about 0.2%. Whereas in hydrogen-free specimens in the rolled state the elongation is practically zero, the value increases to 20% to 25% with 0.2% hydrogen, corresponding to the value of the non-charged, annealed specimens.

From a hydrogen content of 0.3% the strength values slightly decrease again and the strong decline of the elongation values down to almost zero is indicative of strong embrittlement. The performed X-ray fine structure investigations show that at the beginning of hydrogen absorption mainly the β-phase is formed in which the overwhelming amount of the dissolved hydrogen is bound. This leads to the observed increase of the values for the mechanical properties.

The influence of the β-phase on the behavior in hydrogen is confirmed by the examination of the palladium-silver alloys. Since the formation of the β-phase is

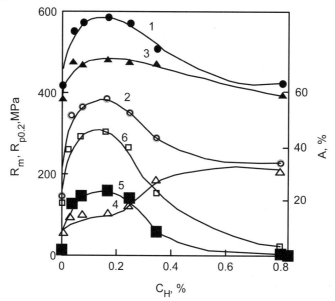

Figure 164: Influence of the hydrogen content on the mechanical properties of the palladium specimens in the rolled and heat treated state [191]
1) fracture stress, 2) fracture stress (tempered) 3) 0.2% yield strength 4) 0.2% yield strength (tempered) 5) elongation 6) elongation (tempered)

inhibited with increasing silver content, this should also have an effect on the behavior of the examined alloys. Relevant examination results are summarized in Figure 165. The relative elongation values K, i.e. the ratio of the elongation difference of charged and uncharged specimens, are plotted:

$$K = \frac{A_H - A}{A}.$$

Here, A = elongation of the uncharged specimens and A_H = elongation of the charged specimens.

With an increasing silver content the positive effect of low hydrogen contents on the elongation values is only low or disappears completely. If the hydrogen contents increase, again embrittlement occurs as has been observed during the examinations at low silver contents of the palladium-silver alloys described above. Apart from the amount of the hydrogen content also the content of the β-phase in the microstructure reducing with increasing silver content is another important factor of the influence exerted by hydrogen on the ductility of the alloys.

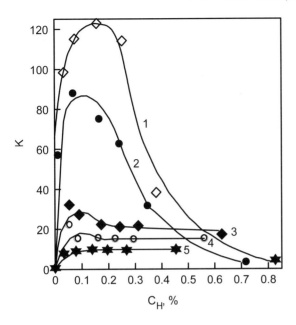

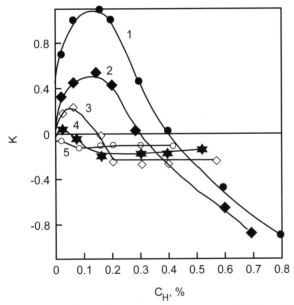

Figure 165: Relationship between hydrogen content (C$_H$) in palladium-silver alloys and the relative elongation values K in the rolled (a) and in the heat treated (b) state [191]
1) Pd
2) Pd-5 Ag
3) Pd-20 Ag
4) Pd-25 Ag
5) Pd-30 Ag

A 37 Tantalum, niobium and their alloys

The refractory metals of the 4th and 5th subgroup of the Periodic Table of Elements exhibit an excellent corrosion resistance against numerous atmospheres. However, they are extremely susceptible to embrittlement by hydrogen formed, for instance, during the cathodic partial reaction of corrosion on the surface in atomic form and absorbed by the metals. Whereas metallic materials are usually considered sufficiently resistant at a corrosion rate of ≤ 0.1 mm/a, a limit value of ≤ 0.01 mm/a must be applied to the mentioned metals to obtain sufficient resistance to the impacts by hydrogen.

The refractory metals vanadium, niobium and tantalum usually exhibit sufficient ductility even over a wide low-temperature range up to the temperature of liquid nitrogen. However, if the metals contain dissolved hydrogen, their mechanical properties are strongly influenced and during the tensile test an increase of the yield stress and a decrease of the fracture elongation and reduction of area can be observed when the temperature decreases.

Hydrogen embrittlement of the refractory metals vanadium, niobium and tantalum may be caused either by dissolved hydrogen or the hydrides formed. Both mechanisms presuppose that the hydrogen is transported either to a crack tip or to sites with increased stresses in the presence of a triaxial stress field. Depending on the material and the test conditions, the toughness behavior at decreasing temperature is characterized by

– continuous decrease of ductility
– sharp transition from ductile to brittle
– sharp transition from ductile to brittle with the ductility subsequently increasing again after having passed its minimum.

To this end, Figure 166 shows the influence of the test temperature on the yield stress in slow strain rate tensile tests at a strain rate of 8.3×10^{-5} 1/s for the pure metals niobium and tantalum as well as an alloy with 50 at.% Nb and 50 at.% Ta [192]. All three materials show the known increase of the yield stress with decreasing test temperature.

If these specimens are charged with 0.3 at.% hydrogen, this increase in the yield stress is intensified as can be seen from Figure 167. This figure indicates the additional increase of the yield stress compared to the corresponding value of the pure materials for the different test temperatures. The specimens were charged in the Sievert's apparatus described in more detail under A 38 (Titanium), the hydrogen having been obtained from the thermal decomposition of uranium hydride.

The influence of dissolved hydrogen on toughness can be seen in Figure 168. The reduction of area as a measure of toughness changes only slightly in the hydrogen-free materials at decreasing test temperature as shown by the upper array of curves. The lower array of curves for the specimens charged with hydrogen illustrates the strong decline of the ductility in the temperature range between about −40 °C and −70 °C (depending on the material). Having passed a minimum, the reduction of area values increase again as the temperatures further decrease, with almost room temperature values being reached in the case of pure tantalum.

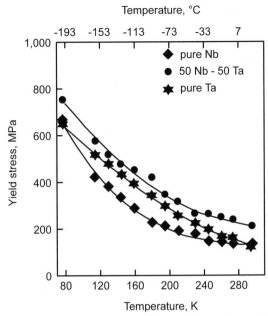

Figure 166: Temperature dependence of the yield stress of the pure metals niobium, tantalum and a 50Nb-50Ta alloy [192]

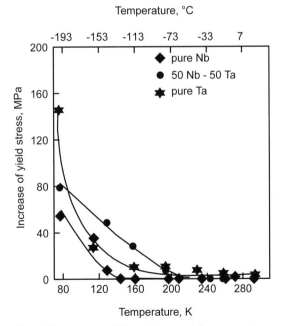

Figure 167: Increase of the yield stress at low temperatures in specimens from niobium, tantalum and the 50Nb-50Ta alloy with a hydrogen content of 0.3 at.% compared to the specimens not charged with hydrogen [192]

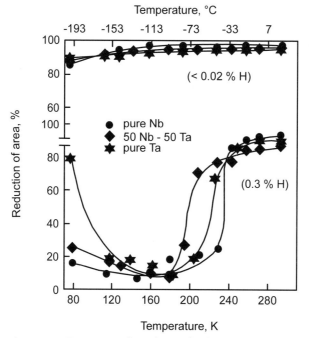

Figure 168: Temperature dependence of reduction of area for the hydrogen-free specimens (upper array of curves) and for the specimens charged with 0.3 at.% hydrogen (lower array of curves) from niobium, tantalum and the 50Nb-50Ta alloy [192]

The sharp decline of toughness is not attributable to the precipitation of hydrides since the relevant temperatures are clearly above the equilibrium temperature for the formation of hydrides (about – 200 °C for Nb and Ta as well as about – 150 °C for the alloy).

To examine the influence of the contents of the gases hydrogen and nitrogen on the yield stress and the reduction of area of the three mentioned metals, slow strain rate tensile tests were performed at a strain rate of 8.3×10^{-5} 1/s in the temperature range from –223 °C to 27 °C [193]. Table 130 lists the analytical values of the three examined metals.

The specimens with the elevated nitrogen content were manufactured by the addition of the respective amounts of nitrogen containing alloys during melting of the high-purity materials in an electric arc furnace. The finished specimens were annealed in a vacuum chamber at 850 °C to expel the contained hydrogen and activate

	Si	Ta	W	C	O	N	H
V	< 0.0001	0.010	0.024	0.012	0.020	0.0006	0.015
Nb	0.001	0.029	0.0017	0.035	0.035	0.015	0.028
Ta	< 0.0001		0.010	0.035	0.035	0.006	0.018

Table 130: Analytical values (at.%) of the three examined metals [193]

	H-charged		N-alloyed		N-alloyed + H-charged	
	N	H	N	H	N	H
V	0.00073	0.31	0.16	0.025	0.15	0.33
	0.0073	0.43			0.16	0.45
Nb	0.013	0.23	0.16	0.034	0.14	0.54
	0.017	0.56	0.27	0.030	0.15	0.91
	0.016	0.93			0.27	0.21
Ta	0.008	0.36	0.18	0.018	0.17	0.62
	0.006	0.63	0.49	0.090	0.18	1.30
	0.005	1.36			0.47	0.34

Table 131: Contents of hydrogen and nitrogen (at.%) of the examined tensile test specimens [193]

the surface. After cooling of the chamber to 600 °C, hydrogen with the desired pressure was added. The high-purity hydrogen was produced by the thermal decomposition of uranium hydride. Following a holding time of 2 hours, the specimens were slowly cooled down to 100 °C within 5 hours and then removed from the chamber. Table 131 contains the contents of hydrogen and nitrogen of the specimens.

Figure 169 to Figure 171 show the results obtained for the influence of hydrogen on the temperature dependence of the yield stress. The relevant results for the combined influence of hydrogen and nitrogen are shown in Figure 172 to Figure 174.

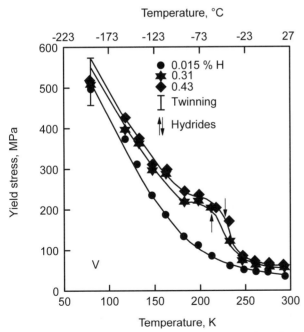

Figure 169: Influence of hydrogen on the temperature dependence of the yield stress of vanadium [193]

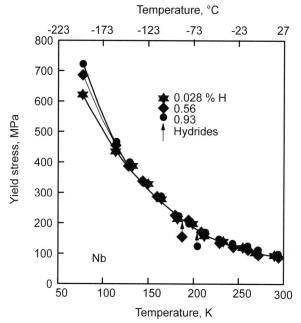

Figure 170: Influence of hydrogen on the temperature dependence of the yield stress of niobium [193]

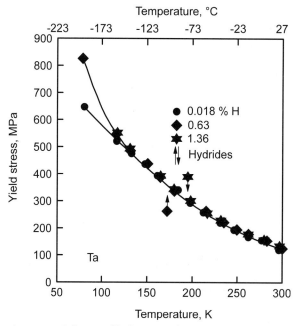

Figure 171: Influence of hydrogen on the temperature dependence of the yield stress of tantalum [193]

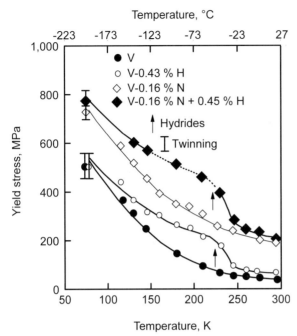

Figure 172: Influence of hydrogen and nitrogen on the temperature dependence of the yield stress of vanadium [193]

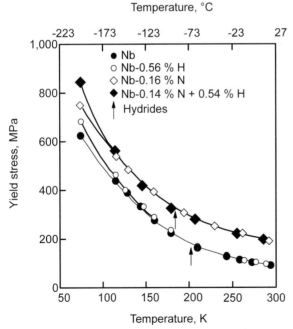

Figure 173: Influence of hydrogen and nitrogen on the temperature dependence of the yield stress of niobium [193]

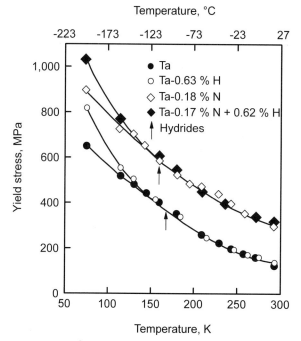

Figure 174: Influence of hydrogen and nitrogen on the temperature dependence of the yield stress of tantalum [193]

For vanadium the curve paths for the specimens with 0.31 at.% and 0.43 at.% hydrogen are identical to a major extent and in the temperature range from −23 °C to −123 °C a quick increase of the yield stress can be observed with decreasing temperature, with a plateau at about −73 °C, i.e. near the temperature at which hydrides are formed.

In the case of niobium and tantalum the higher hydrogen contents of up to 0.93 at.% and 1.36 at.% exert an influence on the yield stress only at temperatures below about −140 °C, i.e. far below the temperature of the hydride formation.

The hydrogen content in the nitrogen-containing specimens does not change the curve path of the temperature dependence of the yield stress, but shifts the curve path of the temperature dependence by a certain amount towards higher stress values. Obviously, this shift is attributable alone to nitrogen.

The toughness behavior characterized by the reduction of area is shown for the vanadium specimens with hydrogen, with nitrogen and with hydrogen plus nitrogen as a function of the temperature in Figure 175.

The ductility of the specimens not charged with hydrogen is practically not changed by the addition of 0.16 at.% N down to very low temperatures (bottom picture). Hydrogen strongly lowers the ductility independently of the nitrogen content at decreasing temperature with a transition from ductile to brittle fracture (DBTT = ductile-brittle transition temperature). In the range of the brittle fracture behavior the toughness is independent of the hydrogen content to a major extent. At tem-

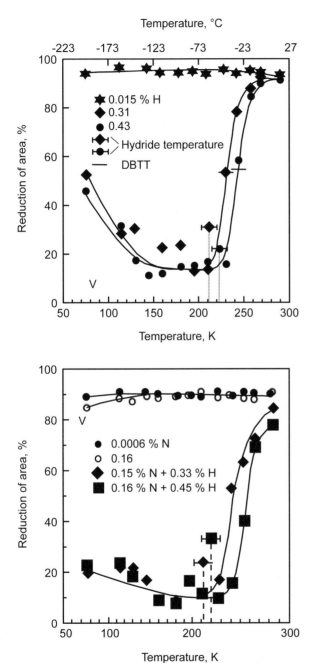

Figure 175: Influence of hydrogen, nitrogen and hydrogen plus nitrogen on the temperature dependence of the reduction of area of vanadium [193]

peratures below −73 °C it can be observed that the reduction of area values increase again, this increase being more marked in the nitrogen-free specimens. The corresponding results obtained for niobium and tantalum are indicated in Figure 176 and in Figure 177.

Although the toughness of niobium is reduced by the addition of nitrogen throughout the examined temperature range, the material remains ductile with the reduction of area values of not less than 60%. The hydrogen-containing specimens exhibit a steep reduction of the reduction of area values when the temperature decreases. In contrast to vanadium, toughness does not increase again at lower temperatures.

The influence of nitrogen on the toughness behavior of tantalum corresponds to the influence on niobium. Although the reduction of area values constantly decrease as the temperature decreases, the material remains ductile. The reduction of area values of the nitrogen-free specimens decrease already at room temperature with increasing hydrogen content and the steep decline occurs at about 22 °C. However, toughness recovers clearly to about 70% to 90% of the initial values at temperatures below −173 °C depending on the hydrogen content. Compared to vanadium and niobium hydrogen causes a very strong reduction of the reduction of area values of the nitrogen-containing specimens also in the upper temperature range and the increase of the values occurring again at lower temperatures is clearly lower than that of the nitrogen-free specimens.

Results regarding the influence of hydrogen on the temperature dependence of the yield stress which where similar to a major extent for the three metals vanadium, niobium and tantalum were also determined by performing relevant examinations of single crystal materials [194].

The different location of the maxima of the temperature dependence of the yield stress of vanadium, on the one hand, and niobium and tantalum, on the other hand, is explained in [194] that in vanadium hydride particles, which are aligned almost parallel to the stress axis, follow an orientation perpendicular to the direction of the tensile stress if an external tensile stress is applied in the temperature range of the yield stress increase. This stress-induced orientation may also explain the abnormal strength increase observed in hydrogen-charged tantalum and niobium crystals [194]. However, from other examinations the conclusion is made that the different behavior of the three metals is attributable to the different diffusion of hydrogen in the metals [195]. The diffusion coefficients at −130 °C amounting to 4×10^{-7} cm^2/s for niobium and 1×10^{-7} cm^2/s for tantalum are lower than that of vanadium (8×10^{-6} cm^2/s). Since the largest part of the dissolved hydrogen is bound by the formation of the hydrides in vanadium, there is only little or no hydrogen left in solution to facilitate further hardening at lower temperatures. On the other hand, enough hydrogen is left following the lower diffusion of the hydrogen in niobium and tantalum even below the hydride formation temperature to increase the yield stress in solution.

Examinations regarding the mechanical behavior of hydrogen-charged and uncharged, high-purity niobium crystals in the temperature range from −196 °C to 107 °C confirm the strength increase by dissolved hydrogen at temperatures below

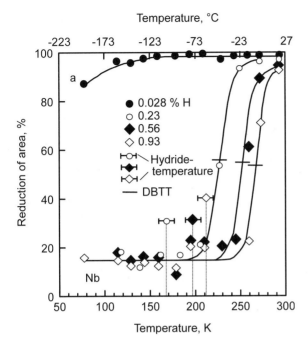
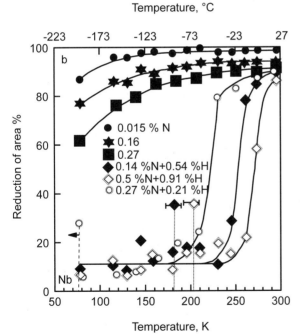

Figure 176: Influence of hydrogen, nitrogen and hydrogen plus nitrogen on the temperature dependence of the reduction of area of niobium [193]

Hydrogen

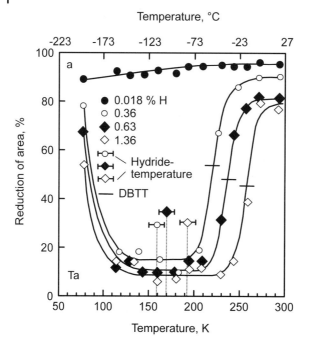

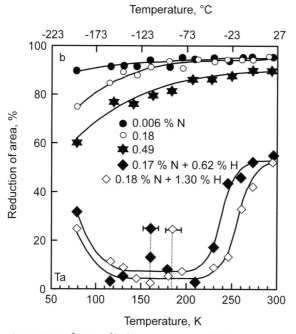

Figure 177: Influence of hydrogen, nitrogen and hydrogen plus nitrogen on the temperature dependence of the reduction of area of tantalum [193]

about −123 °C [196]. As a result, hydride precipitates, which are probably facilitated by elongation in the mechanical test, contribute to a major extent to the decline of the toughness of alloys with hydrogen contents up to 0.07 at.%.

The temperature dependence of the reduction of area of the two metals niobium and vanadium does not only depend on the hydrogen content, but also on the deformation rate as shown by the examinations described in [197]. The tests were performed in high-purity materials, such as those used in the reactor technology. The specimens were charged with hydrogen by aging in a vacuum chamber at 600 °C for 4 h with the hydrogen – obtained from the thermal decomposition of zirconium hydride – added with the necessary pressure to yield hydrogen contents of 12 ppm and 25 ppm in the vanadium specimens as well as 25 ppm, 50 ppm and 100 ppm in the niobium specimens. For the slow strain rate tensile tests strain rates between 10^{-1} 1/s and 3.4×10^{-4} 1/s were chosen. The values of fracture elongation and the reduction of area served as parameters for the ductility of the specimens. Figure 178 shows the values of reduction of area in the temperature range from 25 °C to −196 °C for different strain rates obtained for uncharged specimens and specimens charged with 25 ppm hydrogen. The corresponding values for uniform elongation are shown in Figure 179.

Whereas the hydrogen-free specimens exhibit only a very low reduction of the ductility with decreasing temperature irrespective of the strain rate, a steep transi-

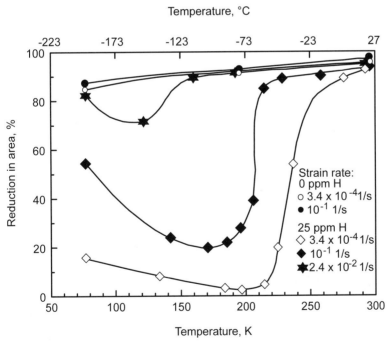

Figure 178: Influence of the test temperature on the reduction of area of niobium tensile test specimens charged with 25 ppm hydrogen compared to hydrogen-free specimens at different strain rates [197]

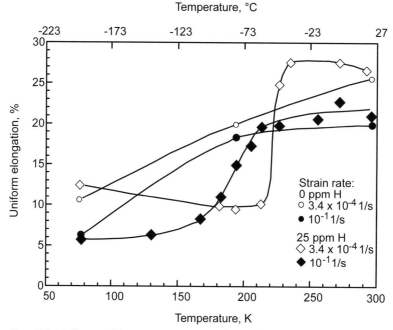

Figure 179: Influence of the test temperature on the uniform elongation of the niobium tensile test specimens charged with 25 ppm hydrogen compared to hydrogen-free specimens at different strain rates [197]

tion from ductile to brittle behavior with a minimum toughness at medium temperatures can be found in the specimens charged with 25 ppm hydrogen. The location and the value of this minimum depend on the strain rate. The ductility increases again by further reducing temperatures. As known also from other metals, the negative influence of hydrogen is stronger and turns out more clearly at increasing temperatures the slower the deformation of the specimens is.

The fracture elongation values of the specimens charged with hydrogen only show a ductile-brittle transition, however no subsequent increase in ductility. This is attributed to incipient cracks competing with each other in the area of the necking making it difficult to obtain a reading near the minimum.

If the hydrogen contents are higher, the toughness behavior is different as shown in Figure 180 and Figure 181 for the specimens charged with 100 ppm hydrogen.

As can be seen, the transition from ductile to brittle behavior and the minimum in ductility during cooling occur much earlier and, in addition, there is only a very low subsequent increase of the toughness values at further decreasing test temperature followed by a second ductile-brittle transition at about − 60 °C after which the toughness values do not increase again during further cooling. This change in the toughness behavior at higher hydrogen contents is attributed to the precipitation of niobium hydrides in this lower temperature range. The specimens with

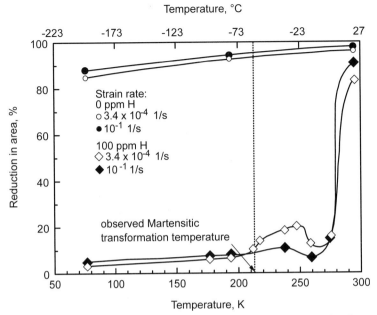

Figure 180: Influence of the test temperature on the reduction of area of niobium tensile test specimens charged with 100 ppm hydrogen compared to hydrogen-free specimens at different strain rates [197]

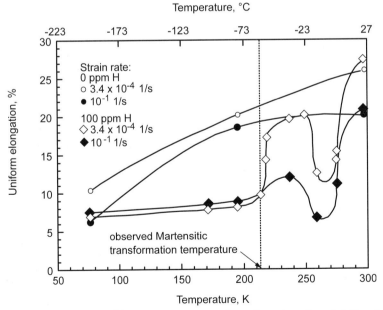

Figure 181: Influence of the test temperature on the uniform elongation of niobium tensile test specimens charged with 100 ppm hydrogen compared to hydrogen-free specimens at different strain rates [197]

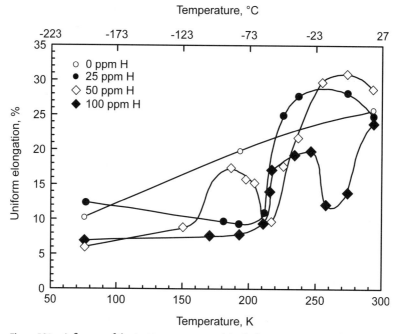

Figure 182: Influence of the test temperature and the hydrogen content on the uniform elongation of niobium tensile test specimens at a strain rate of 3.4×10^{-1} 1/s [197]

50 ppm exhibited a toughness behavior which ranked roughly between that of the specimens with the lower and the higher hydrogen contents (Figure 182).

The toughness behavior of the vanadium specimens with 12 ppm hydrogen is almost identical to that of the niobium specimens with 25 ppm hydrogen as shown by comparing Figure 178 and Figure 179 with Figure 183 and Figure 184. A higher hydrogen content in vanadium also leads to a shift of the ductile-brittle transition towards higher temperatures, attributed to the precipitation of hydrides. However, in contrast to niobium here the toughness values increase again as the temperature continues to decrease. The explanation is that the higher diffusion of hydrogen in vanadium masks the embrittling influence of the hydride precipitates.

The subsequent increase in the toughness values of the refractory metals after the minimum has been passed is not determined by the precipitated hydrides and the diffusion behavior of the hydrogen, but can be also influenced by the microstructure of the material, in particular by the grain size [198]. To this end, Figure 185 shows the example of a 25V-75Nb alloy with a hydrogen content of 0.3 at.% and a grain size of 25 μm where a marked subsequent increase of the ductility with further decreasing temperatures can be observed after the ductility minimum has been passed, whereas this subsequent increase does not occur in the same alloy with a grain size of 175 μm.

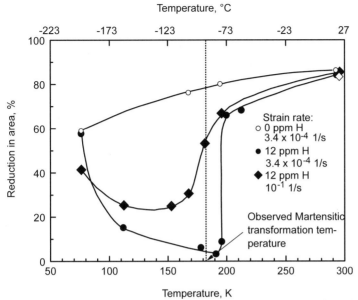

Figure 183: Influence of the test temperature on the reduction of area of vanadium tensile test specimens charged with 12 ppm hydrogen compared to hydrogen-free specimens at different strain rates [197]

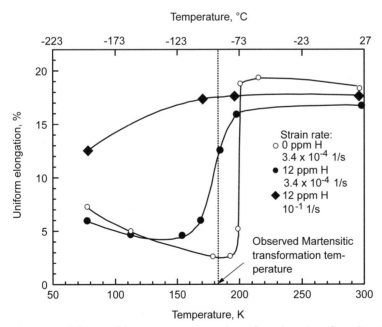

Figure 184: Influence of the test temperature on the uniform elongation of vanadium tensile test specimens charged with 12 ppm hydrogen compared to hydrogen-free specimens at different strain rates [197]

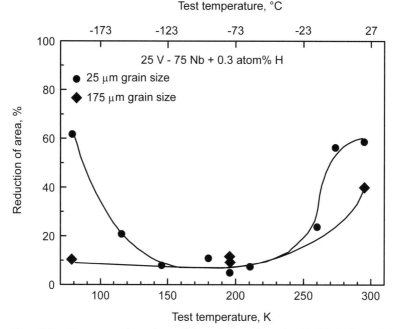

Figure 185: Temperature dependence of the reduction of area in a 25V-75Nb alloy with 0.3 at.% hydrogen at different grain size [198]

Systematic examinations of a 25V-75Nb alloy to investigate the influence of the grain size and the hydrogen content revealed that the subsequent increase of the toughness at low temperatures only occurs under the condition of a small grain size and low hydrogen contents. Grain sizes between 30 µm and 220 µm were obtained following an appropriate heat treatment. The hydrogen contents were between 0.1 at.% and 1.4 at.%. Figure 186 shows the results of the slow strain rate tensile tests (strain rate 8.3×10^{-5} 1/s) for the alloy with 0.4 at.% hydrogen. Figure 187 illustrates the influence of higher hydrogen contents on the uniform elongation and the reduction of area for two temperatures and two grain sizes.

Also the behavior of the refractory metals under cyclic loading is influenced by dissolved hydrogen or hydrogen bound as a hydride as shown by using the example of niobium in the examinations described in [199]. The tests on the crack propagation under cyclic load were performed at room temperature using CT specimens (Figure 188) from a 1 mm thick niobium sheet (99.85).

First, the specimens had been annealed in vacuum at 900 °C to remove the hydrogen resulting from the manufacturing process (15 ppm) and to activate the surface before it was charged with purified hydrogen at the specified temperature and pressure until the equilibrium content was reached. Thus, the hydrogen contents of the examined specimens indicated in Table 132 were obtained.

The fatigue tests were performed with a sinusoidal load at a frequency of 4 Hz and with the two R values of 0.05 and 4 (R = ratio of minimum stress and max-

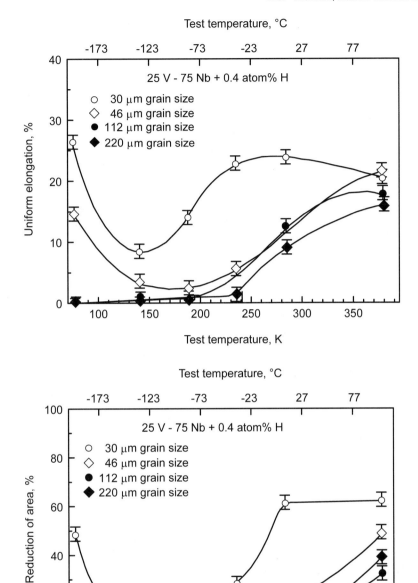

Figure 186: Temperature dependence of the uniform elongation and reduction of area of the 25V-75Nb alloy with 0.4 at.% hydrogen at different grain sizes [198]

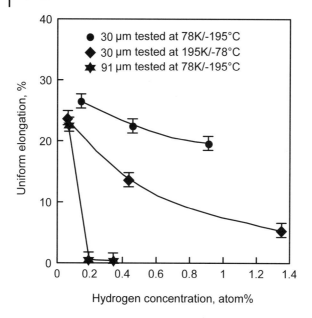
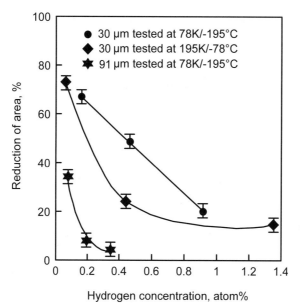

Figure 187: Influence of the hydrogen content in the 25V-75Nb alloy with different grain sizes at two different temperatures [198]

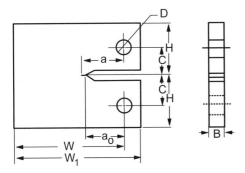

$W_1 = 6.78$ cm, $W = 5.46$ cm, $a_0 = 2.21$ cm
$H = 3.24$ cm, $C = 0.95$ cm, $D = 0.64$ cm
$B = 0.1$ cm

Figure 188: Shape and dimensions of the CT specimens (compact tension) for determining the crack propagation under cyclic load [199]

| 0 | 70 | 116 | 180 | 210 | 250 | 360 | 535 | 660 | 700 | 1,025 |

Table 132: Hydrogen contents (ppm) of the examined niobium specimens [199]

imum stress). Figure 189 and Figure 190 show the fatigue crack propagation rates da/dN (m/load cycle) for the two R values, measured on the specimens charged with different hydrogen contents depending on the stress intensity amplitude ΔK (MPa $\sqrt{m}$).

Frequently, the diagrams obtained during fatigue tests, in which the propagation velocity of fatigue cracks (da/dN) is plotted as a function of the cyclic stress intensity factor ΔK effective on the crack front, reveal a characteristic path which can be divided into three steps. At low stress intensity values there is a permanently steep increase in the crack propagation velocity with increasing stress intensity (area I) followed by a plateau area (area II) at medium K values and a subsequent steep increase in the area III with further increasing K values until the specimen finally breaks by unstable crack propagation after the critical value K_{IC}, i.e. the so-called fracture toughness has been reached. In practice, in particular area I is of interest as an evaluation criterion since, if the curves are sufficiently steep, a limiting value K_I can be defined for the stress intensity factor in this area. below which the crack propagation rate practically approaches zero (cf. also Figure 258 in A 40).

In Figure 189 and Figure 190 the overwhelming part of the curves shows a crack growth behavior according to the areas I and II. At lower crack propagation values a strong dependence of the da/dN values both on ΔK and the hydrogen content of the specimens can be detected. In the subsequent area II the da/dN values increase only gradually with increasing ΔK and the influence of hydrogen is less marked. The threshold stress value K_I is illustrated in Figure 191 for both R values depending on the hydrogen content of the specimens. This figure also shows the

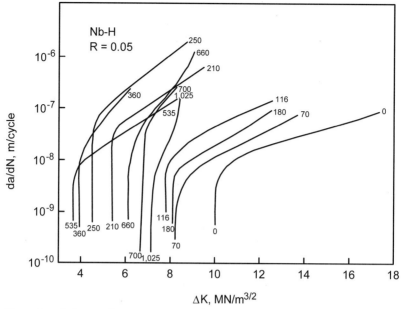

Figure 189: Fatigue crack propagation rates as a function of the amplitude of the stress intensity for different hydrogen contents (refer to Table 132) in niobium at R = 0.05 [199]

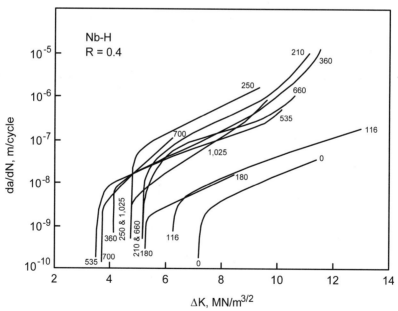

Figure 190: Fatigue crack propagation rates as a function of the amplitude of the stress intensity for different hydrogen contents (refer to Table 132) in niobium at R = 0.4 [199]

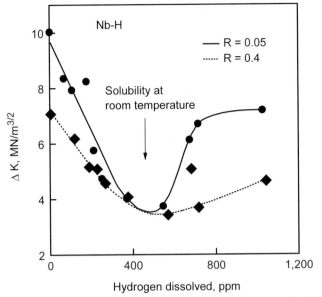

Figure 191: Threshold stress intensity range ΔK depending on the hydrogen content (refer to Table 132) for R = 0.05 and R = 0.4 [199]

threshold value for the solubility of hydrogen in niobium at room temperature (400 ppm). The highest ΔK limit values are reached in hydrogen-free niobium specimens. These values show a relatively sharp decline with increasing hydrogen content. This decline in the ΔK threshold values follows the hydrogen content practically linearly until a minimum is reached at the solubility threshold value. Thereafter, the ΔK threshold values increase again with increasing hydrogen content probably as a result of hydrides being formed.

A 38 Titanium and titanium alloys

Titanium

As all refractory metals titanium does not only exhibit high corrosion resistance, but also a high affinity for gas absorption, such as hydrogen, oxygen and nitrogen. Whereas the build up of smaller amounts of these gases in the metal lattice causes an increase in strength, larger amounts may initiate massive embrittlement [200].

Titanium is offered in four strength groups, the second and third strength grades being preferably used in practice. Due to its low oxygen content Ti 1 has a comparatively low strength and, hence, exhibits good ductility and is suited for deep drawing. Ti 2 is most frequently used in practice since it offers the optimal combination of strength, weldability and ductility. Ti 3 is suited, for instance, for heat exchangers. Table 133 lists the chemical composition of the different titanium

Hydrogen

Material	Mat. No.	Fe max.%	O max.%	N max.%	C max.%	H max.%	Others Σ max.%
Ti 1	3.7025	0.15	0.12	0.05	0.06	0.013	0.4
Ti Grade 1		0.20	0.18	0.03	0.08	0.015	0.4
Ti 2	3.7035	0.20	0.18	0.05	0.06	0.013	0.4
Ti Grade 2		0.30	0.25	0.03	0.08	0.015	0.4
Ti 3	3.7055	0.25	0.25	0.05	0.06	0.013	0.4
Ti Grade 3		0.30	0.35	0.05	0.08	0.015	0.4
Ti 4	3.7065	0.30	0.35	0.05	0.06	0.013	0.4
Ti Grade 4		0.50	0.40	0.05	0.08	0.015	0.4

Table 133: Comparison of the chemical compositions of Ti 1 to Ti 4 according to DIN 17850 and Ti Grade 1 to Grade 4 according to ASTM B 265-99

grades from titanium 1 to titanium 4 according to DIN 17850 compared to Grade 1 to Grade 4 titanium according to ASTM B 265-99. It should be noted that the maximum permissible alloying contents, in particular the oxygen and iron contents of both specifications (DIN 17850 and ASTM B 265-99) are not identical. Consequently, also the mechanical properties are different (Table 134). The physical properties are summarized in Table 135.

The coating-free, clean titanium surfaces reversibly absorb hydrogen according to the overall reaction:

Equation 20 $H_2(\text{gas}) \leftrightarrow 2\,H$ (in metal)

Material	Mat. No.	$R_{p0.2}$ N/mm²	R_m N/mm²	A_5 %
Ti 1	3.7025	min. 180	290–410	min. 30
Ti Grade 1		170–310	min. 240	min. 24
Ti 2	3.7035	min. 250	390–540	min. 22
Ti Grade 2		275–450	min. 345	min. 20
Ti 3	3.7055	min. 320	460–590	min. 18
Ti Grade 3		380–550	min. 450	min. 18
Ti 4	3.7065	min. 390	540–740	min. 16
Ti Grade 4		483–655	min. 550	min. 15

Table 134: Comparison of the mechanical properties of Ti 1 to Ti 4 according to DIN 17850 and Ti Grade 1 to Grade 4 according to ASTM B 265-99

Atomic number	Atomic mass	Lattice type	α/β-conv. °C	Density g/cm³	Melt. point °C	Lin. expansion coefficient 10^{-6}/°C	Thermal conductivity W/cm	Spec. electr. resistance. Ω cm²/m
22	47.9	α hex β bcc	≈ 885	4.51	1,668 ± 5	8.5	22	45

Table 135: Physical properties of titanium [200]

T in °C	319	100	60	20
wt% H	0.18	0.025	0.006	0.001

Table 136: Influence of the temperature on the hydrogen dissolution capability of α-titanium [200]

The capability to dissolve hydrogen in α-titanium is highest at 319 °C, the capacity decreasing with increasing or decreasing temperature (Table 136) [200].

Below 80 °C the absorbed amount of hydrogen from the gaseous phase is described to be safe for the devices and equipment used in practice. The hydrogen solubility of high-purity α-titanium ranges between 44 and 300 °C:

Equation 21 $\quad c\,(\text{ppm}) = 8.60 \times 10^4 \exp(-2.52 \times 10^3 / T)$

The hydrogen solubility in titanium also depends on the microstructure. Figure 192 depicts the H-Ti phase diagram with the α-, β- and δ-areas [201]. It should be noted that the designation of the γ-phase used before 1992 was changed into δ-phase. As can be seen, the β-phase decomposes into α- and β-phase at 319 °C. This phase diagram also shows the maximum hydrogen solubility of the individual phases; the values are summarized in Table 137. The maximum hydrogen solubility is reached in α-Ti at 319 °C, in β-Ti at 640 °C and in δ-Ti at about 600 °C.

If the hydrogen solubility limit in the refractory metals is exceeded, hydrides are formed (refer to Figure 192). Often, if not as a rule, the formed hydrides are hypostoichiometric compounds ($TiH_{1.5}$ to $TiH_{1.94}$) attributable to vacancies in the hydride lattice.

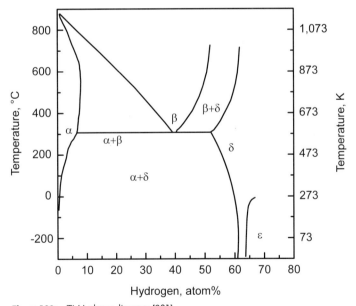

Figure 192: Ti-H phase diagram [201]

238 | Hydrogen

Temp. in °C	α-phase	β-phase	δ-phase
320	max. 7.9%		48–63%
600			max. 60–63%
640		max. 49%	

Table 137: Influence of the microstructure on the maximum solubility of hydrogen in titanium (figures in at.%) [201]

The diffusion rate of hydrogen in α-titanium is low due to its hexagonal lattice type; it amounts to about 10^{-11} cm²/s at 40 °C. The diffusion rate clearly increases only at temperatures above 80 °C.

In aircraft construction technically pure titanium is used because of its low density, good mechanical properties and corrosion resistance. Since the strength depends on the oxygen content of the material to a major extent, the question arises whether the oxygen content also influences the creep behavior of titanium when charged with hydrogen. To this end, titanium with different oxygen contents – Ti Grade 1 with 0.064% and Ti Grade 2 with 0.15% – was tested [202]. Some of the specimens were charged with hydrogen in a hydrogen-containing argon atmosphere at a pressure of 1,333 Pa at 750 °C for 6.5 h and quenched in silicone oil. Subsequently, creep tests were performed at room temperature using uncharged specimens and specimens charged with hydrogen under loads between the yield strength and the tensile strength. The results are indicated for Ti Grade 1 in Fig-

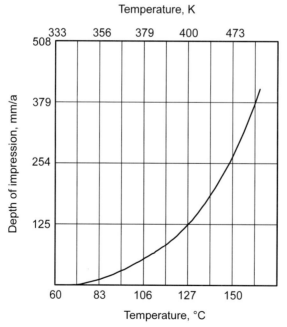

Figure 193: Diffusion rate of hydrogen in titanium as a function of the temperature (24 h test) [200]

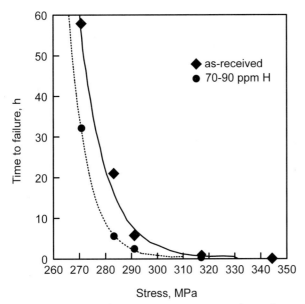

Figure 194: Results of creep rupture tests on Ti Grade 1 without and with hydrogen charging at room temperature [202]

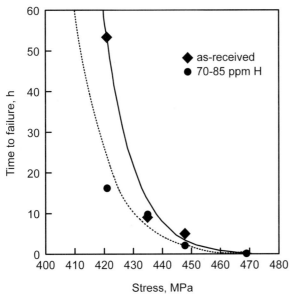

Figure 195: Results of creep rupture tests on Ti Grade 2 without and with hydrogen charging at room temperature [202]

ure 194 and for Ti Grade 2 in Figure 195. As can be seen, for titanium with both the lower and high oxygen contents small concentrations (70–90 ppm) of absorbed hydrogen are sufficient to cause a noticeable reduction in the time to failure at room temperature.

In another test series creep rupture tests were performed at different temperatures (–173 to –73 °C for Ti Grade 1 and –73 to +31 °C for Ti Grade 2) subjected to the same exposure of an Ar/H$_2$ mixture at 1,333 Pa compared to pure Ar at atmospheric pressure. The constant load over time amounted to 241 MPa for Ti Grade 1 and 276 MPa for Ti Grade 2. The results are summarized in Table 138 [202]. Whereas the absorption of hydrogen causes a clear reduction of the time to failure at lower temperatures, even a slight increase of the time to failure can be observed at higher temperatures. This behavior is attributed to the activation of different slip systems [202].

| | Ti Grade 1 | | |
| | Loading: 241 MPa | | |
Temperature °C	Ar time to failure, h	Ar/H time to failure, h	Ratio (Ar/H)/Ar
+27	0.001	n.d.	n.d.
–73	0.015	n.d.	n.d.
–123	0.38	0.44	1.13
–148	0.95	1.02	1.07
–161	4.36	2.36	0.541
–173	20.51	4.26	0.208
	Ti Grade 2		
	Loading: 276 MPa		
Temperature, °C	Ar Time to failure, h	Ar/H Time to failure, h	Ratio (Ar/H)/Ar
31	0.019	n.d.	n.d.
–23	0.34	n.d.	n.d.
–58	1.26	2.74	2.17
–68	15.90	3.26	0.205

Table 138: Influence of the temperature on the time to failure of Ti Grade 1 and Ti Grade 2 in argon and an Ar/H-gas mixture

To further clarify the influence of oxygen, hydrogen and the temperature on the embrittlement of technical-purity titanium notched-bar impact tests were performed in a temperature range from –196 to 102 °C [201]. The test materials were Ti Grade 1 and Ti Grade 2 containing 26 and 40 ppm hydrogen in the as-received state. According to the different oxygen contents (0.15% and 0.25%, respectively) of the two materials, Ti Grade 1 exhibits a much more ductile behavior, in particular, at temperatures ≤ room temperature compared to Ti Grade 2 and the temperature influence on Ti Grade 2 is clearly more marked (Figure 196). When the hydrogen content is increased to 67 ppm in Ti Grade 1 and 70 ppm in Ti Grade 2, the notch bar impact work decreases to different extents. Whereas the higher hydro-

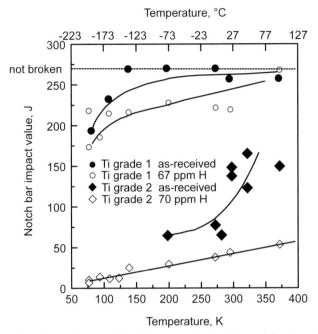

Figure 196: Influence of the hydrogen content on the notch bar impact value of Ti Grade 1 and Grade 2 in the temperature range from 127 to –223 °C [201]

gen content has almost no effect on Ti Grade 1 at temperatures of about 77–97 °C, the strongest decline in the notched bar impact work is found in Ti Grade 2 within this temperature range. In general, Ti Grade 2 exhibits a more brittle behavior throughout the investigated temperature range (Figure 196).

Also the fatigue behavior of the titanium materials Ti Grade 1 and Grade 2 differs [201]. Ti Grade 1 specimens with a hydrogen content of 70–90 ppm exhibited a strong deformation under rotating bending load testing at room temperature and did not break. In the Ti Grade 2 specimens with an identical hydrogen content (70–90 ppm) high load amplitudes shifted the finite life fatigue strength to lower numbers of load cycles (Figure 197). Regarding fatigue strength (about 280 MN/m²) hydrogen remains without any noticeable influence. In pure titanium hydrogen contents of up to 140 ppm remain without any influence on the fatigue behavior under tension-compression loading [201].

Technically pure titanium (IMI 115, IMI 130 and IMI 155 corresponding to Ti Grade 1, Ti Grade 3 and Ti Grade 4) was used to investigate the hydrogen absorption in pure hydrogen and H_2S-containing hydrogen (H_2/5% H_2S, $p_{S_2} = 10^{-14}$ bar at 500 °C) [203]. The chemical compositions of the materials are shown in Table 139. The specimens were charged with hydrogen in a furnace at 300–500 °C for a maximum duration of up to 480 h, the gas flow rate amounting to about 7.5 l/h. Figure 198 shows the mass increase using the example of the material Ti Grade 3, corresponding to an elevated hydrogen absorption, depending on the time of char-

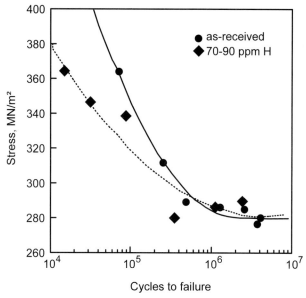

Figure 197: Influence of hydrogen (70–90 ppm H) on the fatigue behavior of Ti Grade 2 under rotating bending load. The material in the as-received state contained 40 ppm H [201]

ging for 400 and 500 °C. As the results show, a temperature increase of 100 °C leads to an approximately tenfold increase in the hydrogen absorption.

H_2S in hydrogen gas (e.g. H_2/5% H_2S) exerts a strongly inhibiting influence on the hydrogen pick-up in the temperature range from 300–500 °C. The mass increase, for instance, of the materials Ti Grade 3 and Grade 4 following charging in pure hydrogen at 500 °C for 120 h amounts to 7 mg/cm² and 1 and 1.8 mg/cm², respectively, in H_2/H_2S gas (Figure 199). Shot-blasting of the titanium surface had a slightly negative effect since this treatment causes iron to penetrate the surface.

Thermal oxidation treatment in air (15 min at 600 °C) reduces the hydrogen absorption. The average weight gain is lower by a factor of 10 compared to the material without pretreatment (Figure 200). The oxide layer acts as a diffusion barrier.

In the presence of wet hydrogen gas, the formation of a protective oxide surface layer prevents the absorption of hydrogen. It was proven that, at a temperature of

	IMI 115 Ti Grade 1	IMI 130 Ti Grade 3	IMI 155 Ti Grade 4
O	0.07	0.20	0.28
N	0.0075	0.009	0.01
C	0.02	0.02	0.02
Fe	0.025	0.03	0.04
H	0.002–0.004	0.002–0.004	0.002–0.004

Table 139: Chemical analysis of technically pure titanium test materials

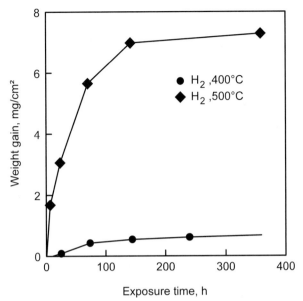

Figure 198: Weight gain as a function of the exposure time charging time of the material Ti Grade 3 at 400 and 500 °C [203]

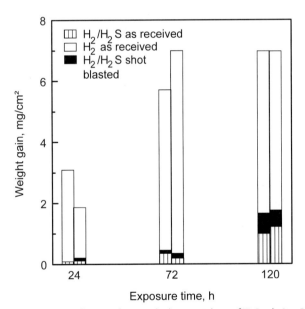

Figure 199: Influence of H_2S on hydrogen pick-up of Ti Grade 3 at 500 °C [203]

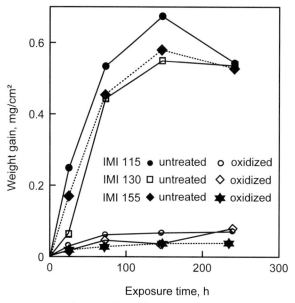

Figure 200: Influence of thermal oxidation on weight gain in hydrogen at 400 °C [203]

315 °C and a hydrogen pressure of 5 MPa, a moisture content of only 2% is sufficient to prevent the absorption of hydrogen [200].

To examine the embrittlement notched CT specimens were charged with hydrogen at 500 and 600 °C at 1 bar H_2 pressure for 6 h. Subsequently, vibration tests were performed in air at 100 Hz with R = 0.2 in a ΔK range from 3 to 15 MN/m$^{3/2}$ to determine the crack growth [203]. The result was that with a hydrogen absorption of about 200 ppm at 500 °C and > 2,.000 ppm at 600 °C with hydride precipitates compared to the comparison specimen with about 20 ppm hydrogen no major differences were found in the crack growth over the entire loading range.

The absorption of hydrogen can cause hydrogen embrittlement in titanium. To clarify the question, at which H concentration and which temperature embrittlement starts, examinations were performed using pure titanium and technically pure titanium, ASTM Grade 2 8chemical composition in Table 140) [204].

The specimens of pure titanium were charged with high-purity hydrogen at 825, 890 and 945 °C for two hours and at 975 °C for 2, 4 and 6 hours. Technically pure titanium (ASTM Grade 2) was charged with hydrogen at 710, 770, 825, 895, 955 and 1,010 °C for 2 h each. All specimens were cooled in the furnace. The hydrogen content amounted to values between 32 and 48 ppm. To determine embrittlement,

Fe	C	N	O	H
0.11	0.026	0.012	0.10	0.004

Details in wt%

Table 140: Chemical composition of technically pure titanium, ASTM Grade 2 [204]

H-charging °C	H-content ppm	Lateral expansion mm	Crack propagation energy J	Total fracture energy J
825	39	1.65	26.44	93.01
890	36	1.65	32.94	86.00
945	26	1.73	41.38	98.16
975	38	1.14	18.82	56.18
975-4 h	48	0.20	3.29	6.14
975-6 h	46	0.20	1.90	4.62

2 h tests, if not indicated otherwise

Table 141: Fracture energy of pure titanium after hydrogen charging

instrumented notched-bar Charpy impact tests [205] were performed during which the force and deformation paths were measured during the impact process. This makes it possible to determine the crack formation energy and the crack propagation energy in addition to the total fracture energy. The results for pure titanium are summarized in Table 141. As can be seen, the ductility of the pure titanium decreases abruptly during the hydrogen charging for 2 h between 945 and 975 °C and, hence, exhibits the transition to brittle behavior. Following hydrogen charging for 4 and 6 hours at 975 °C a brittle fracture occurs. Additional fractographic examinations revealed that the fracture after charging at 945 °C is ductile with numerous deformation twins and first indications of a hydride formation can be found at 975 °C. With a longer charging time the hydride formation increases and the ductility further decreases. This means that hydrogen embrittlement in pure titanium starts with the initiating hydride formation and the degree of embrittlement increases with the advancing hydride growth.

In technically pure titanium (ASTM Grade 2) embrittlement by hydrogen decisively depends on the temperature-dependent microstructure [α-, (α + β-) or β-phase], refer to Figure 192, present during hydrogen charging [204]. The results of the notched-bar impact tests after hydrogen charging are summarized in Table 142.

Microstructure	Temperature °C	H-content ppm	Fracture toughness MJ/mm²
α	710	32	131.7
	770	42	103.5
	825	48	20.8
α + β	895	40	19.9
β	955	41	28.5
	975	39	23.4
	1,010	39	17.0

2 h H-charging

Table 142: Fracture toughness of technically pure titanium (ASTM Grade 2) after hydrogen charging

In the area of the α-phase, embrittlement starts at 770 °C. In contrast to pure titanium, the ductile-brittle transition in technically pure titanium is not initiated by the initial hydride formation since no hydride could be detected [204]. However, on the fracture surface both inside the grain and at the grain boundaries isolated agglomerates are found at the β-phase. At 825 °C a coherent network can be found at the β-phase at the grain boundaries, leading to complete embrittlement. This means that the initial embrittlement is initiated by the residual content at the β-phase in the α-matrix since this phase exhibits a higher hydrogen solubility than the α-phase. The embrittlement of the (α + β)-alloy is enhanced by the α/β-phase boundary. As proven by the fracture toughness values (Table 142), the extent of hydrogen embrittlement of the β-phase is similar to that of the (α + β)-phase.

Following a review of the literature [204] the critical limiting concentration for initiating the hydrogen embrittlement in titanium is indicated with different values. On the one hand it is reported that the strength properties of pure α-titanium up to 1.0 at.% hydrogen (200 ppm) are not affected to a major extent. Elsewhere in the literature it is shown that at 1.0 at.% hydrogen in α-titanium the notched bar impact strength is considerably lower. Also the cooling rate of H-charged titanium seems to exert an influence. It was observed that, following quenching of α-titanium with 1.0 at.% hydrogen, the notched bar impact strength is similar to that of uncharged titanium, however complete embrittlement occurred during slow cooling of α-titanium with 2.0 at.% hydrogen. Also the purity of the α-titanium plays a role. Technically pure α-titanium needs only half the amount of hydrogen (1.0 at.%) compared to the pure metal to initiate embrittlement. If the material contains a residual content at the β-phase, embrittlement is also possible without hydride formation.

Comparative experiments were performed to study the hydrogen influence under constant load on the notched bar impact strength and the crack propagation of technically pure metals of the IVb group titanium, zirconium and hafnium [206]. Table 143 shows the chemical composition of the semi-finished products (manufacturer analyses).

For the notched bar impact bend tests subsized specimens according to DIN 50115 were manufactured and for the tensile tests LT-oriented flat tensile specimens (direction of stress = rolling direction) were taken, notched on one side and

Trace element wt%	Ti (Mat. No. 3.7025)	Zr (ASTM R60702)	Hf
Fe	0.03	0.04	0.01
Cr	< 0.03	0.05	< 0.002
Ta	–	–	< 0.01
Nb	–	< 0.02	< 0.005
C	0.04	< 0.05	0.004
N	< 0.05	< 0.02	0.002
O	0.08	0.1	0.024
H	< 0.011	0.003	< 0.0005
Zr	–	balance	3.1
Hf	–	4.2	balance

Table 143: Chemical compositions of the materials (specifications of manufacturer) [206]

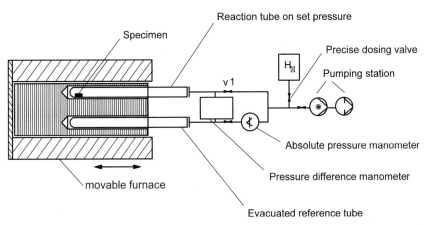

Figure 201: Schematic diagram of the UHV charging system according to Sievert [206]

provided with a sharp fatigue crack. First, the specimens were vacuum-annealed at 800 °C for 30 minutes. The reaction tube of the UHV charging apparatus filled with the specimens (Figure 201) was inserted into hydrogen gas of the quality grade 6.0 (99.9999) at RT. The partial H_2 pressure could be determined by measuring the differential pressure compared to an evacuated reference tube. Solution annealing was performed under isochoric test conditions (V1 closed) at 800 °C. The amount of absorbed hydrogen can be determined from the residual pressure in the recipient after cooling down to RT.

During the notched bar impact bend tests hydrogen charging was 50 atomic ppm up to 1.4 at.%. After cooling down to room temperature in the furnace the notch bar impact values were determined. Figure 202 shows the notch bar impact values determined as a function of the hydrogen content. It should be noted that the examined titanium grade exhibited much higher toughness compared to Hf and Zr.

Figure 203 shows the impact force-displacement paths registered with an instrumented impact machine [205] for uncharged specimens from titanium, zirconium and hafnium as well as the influence of hydrogen in the case of titanium [206]. The maximum impact force ranged between 0.8 and 0.9 kN for all examined materials and is only slightly influenced by the H-content. The impact work and the notched bar impact strength, on the other hand, are strongly influenced by the deformation component decreasing with increasing H-content.

Due to the low hydrogen solubility in titanium at RT, the decreasing deformability can be attributed to the hydride precipitates preferably congregating at grain boundaries in the form of hydride plates at lower hydrogen contents, which is fractographically proven [206]. Apart from H-induced cleavage fracture areas also intercrystalline fracture parts can be found on the fracture surfaces, whereas in uncharged specimens the transcrystalline, cleavage-type fractures are the dominant fracture type.

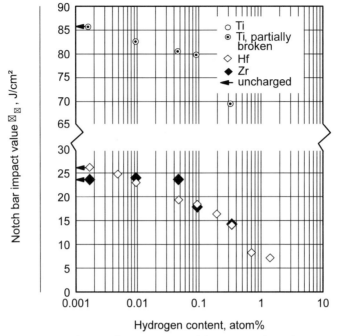

Figure 202: Hydrogen influence on the notch bar impact values of titanium, zirconium and hafnium [206]

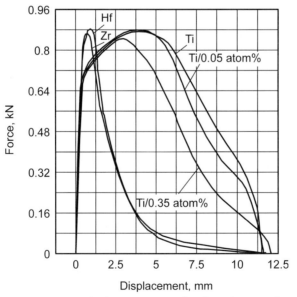

Figure 203: Force-displacement curves of uncharged Ti, Zr, Hf specimens and H-containing Ti specimens [206]

Hydrogen content, at.%	K_C, MPa $\sqrt{m}$		
	Ti	Zr	Hf
0	56 ± 2	55 ± 3	51 ± 2
1	56 ± 3	53 ± 2	42 ± 2
5	52 ± 2	45 ± 2	34 ± 1
10	50 ± 2	39 ± 1	22 ± 1

Table 144: Fracture toughness K_C at 25 °C [206]

To determine the influence of hydrogen on fracture toughness, creep rupture tests were performed on notched LT-oriented flat tensile test specimens (direction of stress = rolling direction). The fracture toughness values obtained from 2 tests each at room temperature are summarized in Table 144 [206]. A significant reduction of K_C can be found in titanium and zirconium only at c_H > 1 at.%, whereas hafnium exhibits a clear reduction of the fracture toughness already at 1 at.%. In contrast to titanium the reduction of K_C is much more marked in zirconium and hafnium at higher hydrogen contents.

Summarizing the above, it was proven that the notched bar impact strength is determined, first of all, by the influence of hydride precipitates on the deformation capacity and that the susceptibility to notched-bar impacts increases in the presence of hydrides in the sequence titanium – zirconium – hafnium. Examinations to investigate the crack growth under constant mechanical load revealed that titanium and zirconium do not show any indication of hydrogen-induced, delayed crack propagation at room temperature. Apart from lower fracture toughness, in particular grain boundary embrittlement as a result of the segregation of hydride precipitates contributes to the crack propagation in hafnium.

For storing nuclear waste also titanium tanks are used. Since titanium can contain up to 50 ppm hydrogen from the production process, it is vital for safety reasons to know whether this may cause embrittlement effects. To this end, examinations on the influence of the hydrogen content on the fracture behavior were carried out using notched titanium specimens [207]. The specimens were produced from rolled Ti Grade 2 (sheet B) and Ti Grade 12 (sheet C) as well as from rod material Ti Grade 12 (rod E). The chemical analyses of the test materials (Table 145) and the mechanical properties (Table 146) show considerable differences. In addition, differences were found in the texture determined in X-ray tests.

Following heating to 400 °C the prenotched specimens were charged with pure hydrogen up to a maximum of 1,800 ppm in a UHV plant (< 10^{-5} torr) and homo-

			Chemical analysis, %						
Material	Ti grade	Fe	Ni	Mo	C	O	N	H
Sheet B	2	0.03	–	–	0.082	0.13	0.020	0.005
Sheet C	12	0.10	0.65	0.29	0.070	0.15	0.013	0.004
Rod E	12	0.05	0.85	0.235	0.116	0.15	0.009	0.002

Table 145: Chemical analysis of the titanium test materials [207]

Hydrogen

Material	Grade	0.2% yield stress MPa	Tensile strength MPa	Reduction of area %
Sheet B	2	470	535	61
Sheet C	12	495	652	28
Rod E	12	368	588	26

Table 146: Mechanical properties of longitudinal specimens from sheet and rod material

genized for 4 days. Subsequently, the CT specimens were ruptured in the slow strain rate tensile test at 10^{-3} and 10^{-5} mm/s. The force-time diagram obtained (Figure 204) was used to determine that force by using the 5% secant method, which is necessary to yield 5% deviation from the straight line. As metallographically proven, this point P_s shows the beginning of the slow crack growth and the fracture P_f occurring soon after as by a rapid decline of the force.

The critical stress intensity factors determined on that basis depending on the hydrogen concentration are summarized for the materials B, C and E in T-L direction in Figure 205 to Figure 207. Additional results for the S-L, S-T and L-T directions can be found in the original literature [207].

The results can be summarized as follows:

Ti Grade 2:

Ti Grade 2 is characterized by a comparatively high toughness at hydrogen contents up to 400 ppm. A content above 500 ppm hydrogen causes already brittle fracture. It is of vital importance to applications in practice that Ti Grade 2 in its as-received state breaks exclusively ductilely with strong necking during the slow strain tensile test (10^{-3} mm/s). It is typical for this material that it breaks merely

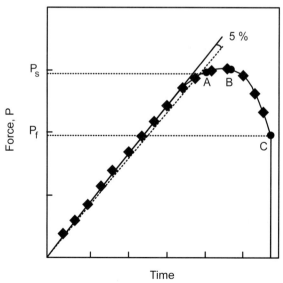

Figure 204: Force-time curve with images of the fracture surfaces showing that the beginning of the slow crack growth can be determined with the 5% secant method [207]

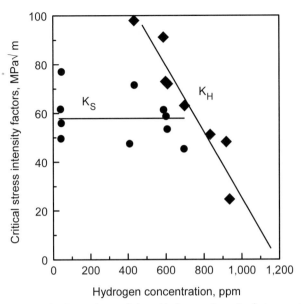

Figure 205: Dependence of the critical stress intensity factors on the hydrogen content for Ti Grade 2 in the T-L direction:
● for the slow crack growth (K_S) and ◆ for the H-induced unstable crack propagation (K_H) [207]

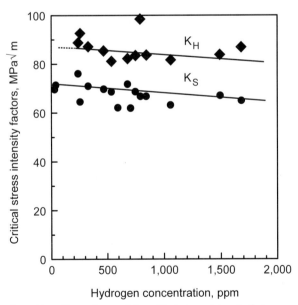

Figure 206: Dependence of the critical stress intensity factors on the hydrogen content for Ti Grade 12, sheet C, in T-L direction:
● for the slow crack growth (K_S) and ◆ for the H-induced unstable crack propagation (K_H) [207]

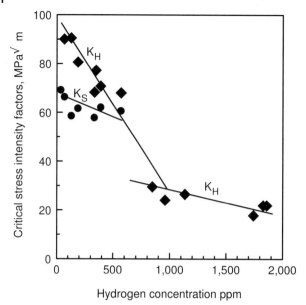

Figure 207: Dependence of the critical stress intensity factors on the hydrogen content for Ti Grade 12, rod material C, in T-L direction:
● for the slow crack growth (K_S) and ◆ for the H-induced unstable crack propagation (K_H) [207]

ductile at hydrogen contents < 500 ppm and the fracture is merely brittle at hydrogen contents > 700 ppm.

Ti Grade 12:
The typical feature of Ti Grade 12 is that the brittle fracture is always preceded by slow crack growth in both sheet and rod material (Figure 206 and Figure 207). In addition, the individual values also depend on the specimen orientation.

All examined titanium materials (sheets Ti Grade 2 and Ti Grade 12, rod material Ti Grade 12) with hydrogen contents of 20–80 ppm in their as-received state exhibited a exclusively ductile behavior in the slow strain rate tensile test, and hence hydrogen embrittlement is not expected. However, the risk of embrittlement by hydrides or of a delayed fracture cannot be completely excluded if additional hydrogen is absorbed by the materials when used in practice [207].

Further examinations were performed to investigate also the susceptibility of other technically pure titanium materials to hydrogen embrittlement [208]. They differed noticeably in their chemical composition (Table 147), microstructure and mechanical properties (Table 148).

Again, the prenotched CT specimens were charged with pure hydrogen of different concentrations at 400 °C, homogenized for 4 d and subsequently deformed until fracture in the slow strain rate tensile test (1.7×10^{-5} mm/s) at RT. The crack growth was measured across the potential drop at the notch. These examinations revealed that the plastic deformation required for H-induced crack initiation decreases more or less linearly with the hydrogen content (Figure 208). For details

Material	Designation	Chemical analysis, wt%				
Sheet		Fe	C	O	N	H
SC-38	A	0.14	0.019	0.19	0.017	0.001
BASL	B	0.03	0.082	0.13	0.020	0.010
AT-7248	F	0.04	-	0.12	0.012	0.005
P-2030	G	0.10	0.014	0.19	0.015	0.005

Table 147: Analysis of the examined titanium sheets [208]

Material	0.2% Yield strength MPa	Tensile strength MPa	Necking %	Vickers hardness
A	380	495	57	220
B	470	535	61	245
F	276	439	51	183
G	385	468	44	219

Table 148: Mechanical properties of longitudinal specimens from titanium sheet: strain rate = 10^{-3} 1/s [208]

regarding the influence of the hydrogen content on the critical stress intensity factor reference is made to the original literature [208]. Figure 209 shows the influence of the 0.2% flow stress on the critical hydrogen content of the various titanium materials examined to initiate a brittle fracture. From these findings it can be concluded:

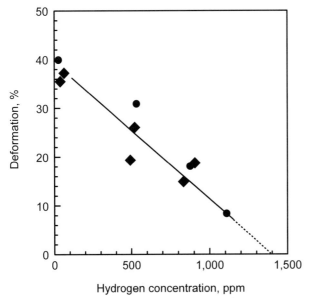

Figure 208: Influence of the hydrogen content on the plastic deformation required for crack initiation in the technically pure Ti material F [208]
(◆ L-T and ● T-L direction)

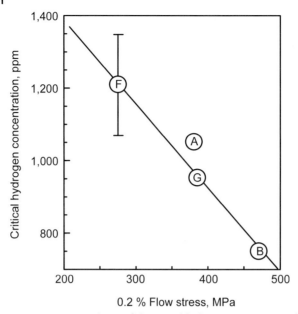

Figure 209: Dependence of the critical hydrogen concentration for initiating a brittle fracture from the 0.2% flow stress (at 10^{-3} 1/s) in the examined technically pure titanium materials A, B, F and G (T-L direction) [208]

The higher the strength of the titanium (the lower the 0.2% yield strength), the lower is the critical hydrogen concentration for causing a brittle fracture.

In addition, both the distribution of the residual β-phase and the texture of the material exert an influence on the susceptibility to H-induced brittle fracture.

A number of failure analyses in operating plants were investigated, which were attributed to H-induced embrittlement in various, mostly liquid environments [209]. In a plant for molten urea (200 °C) comprising titanium components including a Ti cladded steel vessel, hydrogen absorption was detected after an operating time of 10 months. In contrast to the titanium components exposed to the liquid phase or gas-liquid interface and suffered a noticeable reduction of the notched bar impact strength, no embrittlement was found at those parts having only been exposed to the gaseous phase (Table 149).

For further examinations of the hydrogen absorption behavior a number of Ti coupons were provided at many locations of the plant, which were removed and tested at regular intervals during operation. The results of the corrosion rate and hydrogen absorption are summarized in Table 150.

Whereas the corrosion rate in the gaseous phase was very low and only a low amount of hydrogen ($\leq$ 36 ppm) was absorbed, the material consumption rate in the liquid phase and at the gas-liquid interface was just below 0.1 mm/a and the hydrogen absorption continuously increased during the aging time and reached 1,150 ppm within 3 years.

Titanium component	H-content* ppm	Notched bar impact strength kg m/cm²
Gas discharge pipe (gas phase)	42	14.3
Feed pipe (gas-liquid interface)	926	0.4
Baffle plate (liquid phase)	856	1.1

* H-content prior to use in operation: 20–50 ppm

Table 149: Hydrogen content and notched bar impact strength of titanium components following the use in molten urea at 200 °C for 10 months [209]

Phase	Corrosion rate, mm/a and hydrogen content*, ppm	Aging time, years		
		1	2	3
Gas	Corrosion rate	< 0.01	< 0.01	< 0.01
	Hydrogen content	13	13	36
Gas-liquid interface	Corrosion rate	0.084	0.078	0.084
	Hydrogen content	210	505	947
Liquid	Corrosion rate	0.083	0.085	0.089
	Hydrogen content	224	560	1,150

* Before test: 12 ppm hydrogen

Table 150: Corrosion rate and hydrogen absorption of titanium coupons provided in a titanium vessel for urea melts at 200 °C [209]

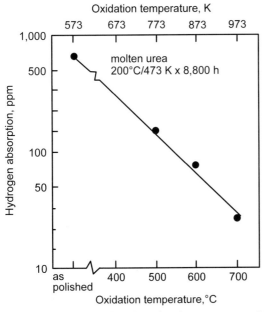

Figure 210: Influence of the thermal oxidation treatment of titanium on the hydrogen absorption in molten urea [209]

Thermal oxidation proved to be a suitable measure [209] to inhibit the absorption of hydrogen. Figure 210 illustrates that the absorption of hydrogen during aging in molten urea is clearly reduced by increasing temperature of a thermal oxidation treatment of the titanium. According to these findings, the new titanium vessel was thermally oxidized before use in the urea melt with the result that after an operating time of two years no damage was found.

[210] confirmed that both an oxide layer and an oxygen-rich α-Ti-layer can protect the base material titanium from hydrogen embrittlement. Examinations of the Ti-O-H system show that hydrogen solubility is lower at a specified hydrogen pressure if oxygen is dissolved in the material. Therefore, a higher hydrogen content and a higher pressure are required for the hydride formation in oxygen-containing α-titanium compared to oxygen-free titanium [210]. Also results from hydrogen permeation tests on preoxidized α-titanium are available [211]. For preliminary oxidation the specimens were first degassed in vacuum at 600 °C for 4 h, purged with helium at room temperature and then oxidized in oxygen at 1 atm at 270 °C for 4 h. Thereafter, the thickness of the oxide film is about 50 angstroms (5 nm). Hydrogen permeation tests were performed at hydrogen pressures of 2.67, 5.34, and 26.7 N/m². The influence of the temperature and the pressure on the permeation rate F of non-preoxidized and preoxidized specimens is shown in Figure 211 and Figure 212. A comparison of the results reveals that even an oxide film of a thickness of only 5 nm considerably inhibits the permeation of hydrogen. The experimental results correspond to the temperature dependence of the Arrhenius equation.

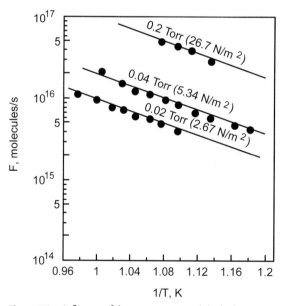

Figure 211: Influence of the temperature and the hydrogen pressure on the permeation rate F of non-preoxidized α-titanium. The solid lines are calculated figures [211]

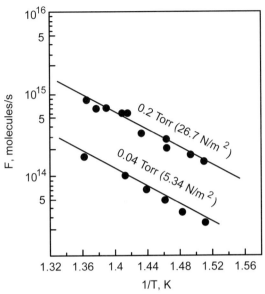

Figure 212: Influence of the temperature and the hydrogen pressure on the permeation rate F of preoxidized α-titanium. The solid lines are calculated data [211]

Occasionally, crack damage occurred in welded titanium pipes of aircraft exhaust systems, attributable to hydrogen embrittlement [212]. The pipes consisted of Ti Grade 2 and Ti Grade 3 (Table 151) and had a hydrogen content of 30–50 ppm and, in exceptional cases, of about 60–65 ppm. The operating conditions were 120–175 °C at a mean pressure of 0.31 MPa (Figure 213). Crack formation mainly occurred in the heat-affected zone and sometimes also in the weld seam itself. Metallographic examinations proved pronounced hydride formation in the area around the cracks. It is assumed that under the present operating temperatures and during the thermal cyclic loading when switching the exhaust system on and off, hydrogen accumulates at the sites of the highest stress concentrations, i.e. in the area of the weld seam, in particular of defective weld seams with undercuts.

Simulations of the failure analysis were conducted in laboratory tests. To this end, welded tensile test specimens notched at the weld seam were charged with hydrogen (60 and 150 ppm) and creep rupture tests were performed at 260 and 315 °C under constant load of 140 MPa. Figure 214 shows the result, i.e. the dependence of the elongation on the test duration [212]. As can be seen, the ductility

Material	Ti min.%	O max.%	H max.%	Fe max.%
Ti Grade 2	99.2	0.15	0.006	0.2
Ti Grade 3	99.0	0.4	0.013	–

Table 151: Chemical compositions of the titanium materials used [212]

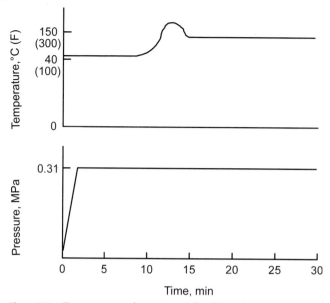

Figure 213: Temperature and pressure profile of the exhaust system after commissioning [212]

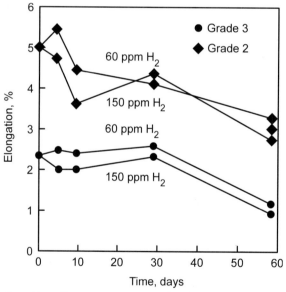

Figure 214: Elongation-time curves of H-charged, notched specimens from Ti Grade 2 and 3 following constant load over time at 140 MPa [212]

noticeably decreases after a prolonged test duration, attributable to local hydrogen enrichment with hydride formation. Local hydride formation, initially in the area of the notch, could be also shown metallographically.

In welded joints of technically pure titanium to higher-alloyed titanium materials (e.g. TiAl6V4,5) migration of hydrogen from one material to the other may cause local hydrogen enrichment with hydride formation in the transition zone, causing embrittlement there [213].

Titanium alloys

Based on the microstructure titanium alloys are divided into:

- α-alloys
- (α + β)-alloys
- β-alloys.

Starting from pure titanium via α-, (α + β)- up to β-alloy the strength of the alloys clearly increases (Figure 215) [214]. The alloying elements Al, O, N, C stabilize the α-phase and Mo, V, Ta, Nb stabilize the α-phase [215]. A selection of common titanium materials is listed in Table 152, including their chemical composition and strength properties [215]. A complete list of the titanium materials is contained in

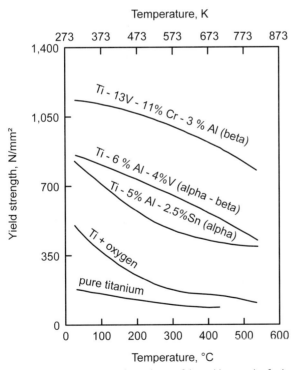

Figure 215: Temperature dependence of the yield strength of selected Ti-alloys [214]

Chemical composition wt%	E min-max GPa	R_p min-max MPa	R_m min-max MPa	A min-max %	K_{IC} min-max MPa m$^{1/2}$
α-Ti-alloys					
Ti-6Al-2Sn-4Zr-2Mo-0.1Si	114	990	1,010	13	70
Ti-6Al-2.7Sn-4Zr-0.4Mo-0.4Si	112	900–950	1,010–1,050	10–16	60–75
Ti-6Al-5Zr-0.5Mo-0.25Si	120	850–910	990–1,020	6–11	68
Ti-5.8Al-4Sn-3.5Zr-0.5Mo-0.7Nb-0.35Si-0.06C	120	910	1,030	6–12	45
(α+β)-Ti-alloys					
Ti-6Al-4V	110–140	800–1,100	900–1,200	13–16	33–110
Ti-6Al-6V-2Sn	110–117	950–1,050	1,000–1,100	10–19	30–70
Ti-6Al-2Sn-2Zr-2Mo-2Cr-0.25Si	110–120	1,000–1,200	1,100–1,300	~15	65–110
Ti-6Al-2Sn-2Zr-6Mo	114	1,000–1,100	1,100–1,200	13–16	30–60
Ti-5Al-2Sn-2Zr-4Mo-4Cr	112	1,050	1,100–1,250	8–15	30–80
β-Ti-alloys					
Ti-4.5Al-3V-2Mo-2Fe	110	900	960	8–20	60–90
Ti-11.5Mo-6Zr-4.5Sn	83–103	800–1,200	900–1,300	8–20	50–100
Ti-3Al-8V-6Cr-4Mo-4Zr	86–115	800–1,200	900–1,300	6–16	50–90
Ti-10V-2Fe-3Al	110	1,000–1,200	1,000–1,400	6–16	30–100
Ti-15V-3Cr-3Al-3Sn	80–100	800–1,000	800–1,100	10–20	40–100

Table 152: Chemical compositions and mechanical properties of selected titanium materials [215]

the ASME specifications SB-265 of 2002 corresponding to the ASTM specifications B 265-99.

It is fundamental for the absorption of hydrogen that hydrogen solubility also increases in the sequence α-, (α + β)-, β-alloy. Whereas the H solubility in the α-phase at room temperature amounts to about 20 to 200 ppm only, it is > 4,000 ppm in the β-phase [216]. In α-alloys embrittlement does not start until after the solubility limit has been exceeded and hydrides are formed.

Ti-Al-V and Ti-Al-Sn alloys

The material TiAl6V4 is one of the most frequently used titanium alloys. It is used in aircraft construction, in chemical plants, for condensers and as a bio material. It is known that titanium materials are extremely susceptible to hydrogen embrittlement and hydrogen fosters crack growth.

H-induced crack growth was examined in close detail using the technical material TiAl6V4 (Table 153) [217]. First, the sheet specimens were heat treated in an argon atmosphere at 954 °C for 4, thereafter cooled to 760 °C at 0.8 °C/min and to 480 °C at 7 °C/min and then cooled to room temperature in air. The microstructure consists of globular α-grains with β-phase along the grain boundaries. The hydrogen content is 47 ± 2 ppm after heat treatment. The strength values determined in 202 kPa Ar are listed for 4 different test temperatures in Table 154.

Fe wt%	Al wt%	V wt%	N ppm	C ppm	O ppm	H ppm
0.19	6.29	3.85	1,300	100	1,800	57

Table 153: Chemical composition of the test material TiAl6V4 (3.7165) [217]

	20 °C	45 °C	70 °C	95 °C
R_p	1,605	1,580	1,502	1,440
R_m	1,814	1,925	1,782	1,689

Table 154: Strength values for TiAl6V4 (3.7165) determined on smooth specimens in 202 kPa Ar [217]

The specimens were notched and provided with a 1 mm long incipient fatigue crack. Prior to the start of the test the autoclave was evacuated, flushed with high-purity argon (99.9995%) at a pressure of 202 kPa, evacuated again and then flushed with hydrogen twice. Following these operations, the specimens were exposed to high-purity compressed hydrogen (> 99.9995%) at a pressure of 101 and 505 kPa and the crack growth was microscopically analyzed under constant load at temperature of 20, 45, 70 and 95 °C. If the initial load K_A was too low and the crack growth was less than 10 µm within 24 hours (< 1.2×10^{-10} m/s), the load was increased by 30 kg.

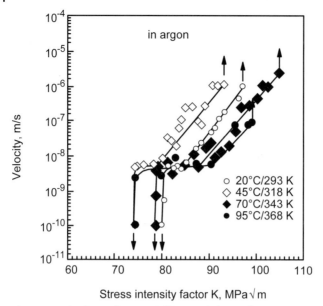

Figure 216: Crack growth rate in TiAl6V4 in 202 kPa Ar from 20 to 95 °C [217]

The influence of the stress intensity factor K_I on the crack growth rate da/dt in the material TiAl6V4 is shown in Figures 216 to Figure 218 for 202 kPa Ar, 101 kPa H_2 and 505 kPa H_2 for four different temperatures.

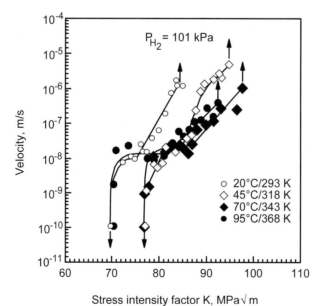

Figure 217: Crack growth rate in TiAl6V4 in 101 kPa H_2 from 20 to 95 °C [217]

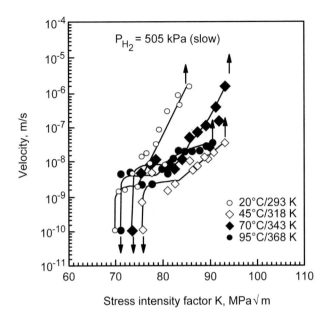
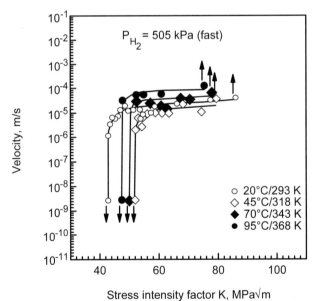

Figure 218: Crack growth rate in TiAl6V4 (3.7165) in 505 kPa H_2 at 20 to 95 °C [217]
a) $K_A > 55$ MPa $\sqrt{m}$, b) $40 < K_A < 50$ MPa $\sqrt{m}$
K_A = initial load

All ln v-K curves illustrate the characteristic behavior in the areas I, II and III. In argon the crack propagation velocity in the plateau area (area II) amounting to 6×10^{-9} m/s is independent of the temperature to a major extent, however in area II propagation increases with the temperature and the same holds true for the K_c value. The K_0 value (threshold value for initial crack gowth) is between 74 and 80 MPa $\sqrt{m}$ (Figure 216).

In 101 kPa H_2 the plateau velocity $(da/dt)_{II}$ is again independent of the test temperature to a major extent and, amounting to about 10^{-8} m/s, is only slightly higher than that in Ar (Figure 217). The expansion of area II increases with the temperature. The K_0 value is between 70 and 77 MPa $\sqrt{m}$ and, hence, slightly lower than that in Ar.

In 505 kPa H_2 the crack growth rates were measured in TiAl6V4 (3.7165) under a higher and lower initial load K_A. As the results in Figure 218a and b show, this has a clear effect on the plateau area. Whereas the plateau velocity $(da/dt)_{II}$ under a higher initial load ($K_A > 55$ MPa $\sqrt{m}$) amounts to $2-5 \times 10^{-9}$ m/s, it is considerably higher under a lower initial load ($40 < K_A < 50$ MPa $\sqrt{m}$) amounting to about 10^{-5} m/s, with no influence of K being detected. Also the threshold values for the initial crack growth K_0 and the accelerated crack growth K_H differed clearly from each other. To clarify this behavior, an additional experiment was conducted in 505 kPa H_2 at 45 °C and under a lower initial load. After the crack had started to continuously propagate, the hydrogen pressure was lowered to 202 kPa. The crack stopped growing and crack advancement could be observed again only after the pressure had been increased to 505 kPa. In addition, the fracture mode changed at 505 kPa H_2. At $K_A > 55$ MPa $\sqrt{m}$, $K_0 > 70$ MPa $\sqrt{m}$ and $(da/dt)_{II} = 2 \times 10^{-9}$ m/s the fracture surface exhibited only individual smooth cleavage fracture areas and was similar to that of the specimen tested in argon. In contrast, a complete brittle fracture with microcracks on the fracture surface occurs at $40 < K_A < 50$ MPa $\sqrt{m}$, $41 < K_0 < 53$ MPa $\sqrt{m}$ and $(da/dt)_{II} = 10^{-5}$ m/s. An autocatalytic process of hydride formation and crack growth is assumed. The effect of the initial load is attributed to the different size of the plastic zone and the hydrogen-influenced zone.

Slow strain rate tensile tests were performed on the material TiAl6V4 with two different microstructural states (globular and acicular) in compressed hydrogen at room temperature [218]. The microstructure called "globular" consists of a primary α-phase with a finely dispersed β-phase at the grain boundaries and is obtained by solution annealing at 830 °C 40 min/H_2O and aging at 510 °C for 12 h. The microstructure called "acicular" consists of a β-matrix with embedded α-needles (Widmannstätten structure) and is obtained by solution annealing at 1,038 °C for 40 min, stabilization at 704 °C for 1 h and 593 °C 1 h/air cooling. First, the influence of the strain rate at a hydrogen pressure of 9.04×10^4 N/m^2 on the embrittlement of both microstructural states of TiAl6V4 was examined (Figure 219). As the measure for embrittlement the K_0/K_H ratio was used, i.e. the ratio of the stress intensity factor for the initiating measurable crack growth K_0 and the critical stress intensity K_H for the fracture. As illustrated in Figure 219, embrittlement of the globular α-microstructure is lower compared to the acicular Widmannstätten structure.

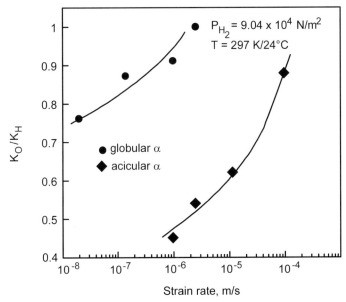

Figure 219: Influence of the strain rate on the embrittlement of two microstructural states (globular and acicular) of the TiAl6V4 alloy in compressed hydrogen at room temperature [218]

Apart from the effects of the strain rate also the influence of the hydrogen pressure from $p_{H2} = 1$ N/m² to 9×10^4 N/m² on embrittlement and the fracture type of the different microstructures of the material TiAl6V4 were examined [218]. The obtained results are summarized in Table 155. Whereas, the embrittlement of the Widmannstätten structure of TiAl6V4 (3.7165) strongly depends on the pressure of the surrounding hydrogen gas and no fracture can be detected at $p_{H2} = 1$ N/m², embrittlement of the globular α-phase microstructure is largely independent from the hydrogen pressure and always moderate. Also the type of fracture, i.e. transcrystalline cleavage fracture or intercrystalline fracture, depends on the microstructure.

The materials TiAl6V4 (3.7165) and TiAl6Nb2Ta1Mo0,8 were examined in high-pressure hydrogen at room temperature [219]. Following heat treatment 1 h 870 °C/air + 2 h 95 °C/air the 2.5 cm thick plates exhibited a α-phase microstructure with about 5–10 vol% continuous β-phase along the grain boundaries. The chemical composition of the alloys and the mechanical properties are shown in Table 156. WOL specimens were used to determine the threshold values for the subcritical crack growth in ultra-pure hydrogen at pressures of 14 and 52 MPa. The specimens were loaded in the linear-elastic range and tested over a period between 65 and 105 days. The results in Table 157 show that both materials are susceptible to subcritical crack growth. The fracture was predominantly intercrystalline along the α/β-phase boundary.

To clarify the question from which hydrogen concentration the ductility of the material TiAl6V4 is impaired, examinations were performed to study the influence of the hydrogen content from 10 ppm up to 3,000 ppm on the mechanical proper-

Conditions		Continuous α-structure (globular α)	Continuous β-structure (acicular α)
Time dependence	decreasing strain rate	embrittlement increases with decreasing strain rate	embrittlement increases with decreasing strain rate
H_2 pressure	$p_{H2} = 9.04 \times 10^4$ N/m², slow strain rate	moderate embrittlement	strong embrittlement
	$p_{H2} = 100$ N/m², slow strain rate	moderate embrittlement	moderate embrittlement
	$p_{H2} = 1$ N/m², slow strain rate	moderate embrittlement	negligible embrittlement
Crack branching	at all pressures and strain rates	strong	negligible
Type of fracture	$p_{H2} = 9.04 \times 10^4$ N/m²	transcrystalline cleavage fracture	intercrystalline fracture
	$p_{H2} = 100$ N/m²	transcrystalline cleavage fracture	transcrystalline cleavage fracture
	$p_{H2} = 1$ N/m²	transcrystalline cleavage fracture	no fracture

Globular = primary α-phase with finely dispersed β-phase at the grain boundaries, acicular = β-matrix with embedded α-needles (Widmannstätten structure)

Table 155: Embrittlement of TiAl6V4 by compressed hydrogen [218]

	TiAl6V4	TiAl6Nb2Ta1Mo0.8
Chemical composition, wt %		
Al	6.0	6.1
V	4.0	–
Nb	–	1.90
Ta	–	0.88
Mo	–	0.88
Fe	0.05	0.05
C	–	0.02
N	0.009	0.009
O	0.075	0.088
H	–	0.0036
Ti	balance	balance
Mechanical properties		
$R_{p0.2}$, MPa	710	772
R_m, MPa	875	854
A, %	12	15
Z, %	27	29

Table 156: Chemical compositions and mechanical properties of the test materials [219]

ties [220]. Round test bars were produced from rod material (6.65 Al, 4.10 V, 0.17 Fe, 0.01C, 0.009 N and 0.003 H in wt%). Following preannealing in vacuum (10^{-4} mbar) the specimens were volumetrically charged with purified hydrogen at 600 °C for 2 h and subsequently deformed until fracture in a slow strain rate tensile test (strain rate: 1.1×10^{-3} to 2.5×10^{-6} 1/s). The decrease in the reduction of area Z (Figure 220a) and the fracture elongation A (Figure 220b) served as the measure for hydrogen embrittlement. The results clearly show that from a hydrogen content of 1,500 ppm embrittlement of the material TiAl6V4 starts and continues to increase with increasing hydrogen concentration. At 3,000 ppm H both reduction of area and fracture elongation are virtually zero.

The influence of the hydrogen content on the crack growth rate in TiAl6V4 (3.7165) was determined using notched CT specimens provided with an incipient fatigue crack [220]. The results in Figure 221 show that both the threshold value K_0

Material	K_H MN/m$^{3/2}$	Test duration days
Test atmosphere: 14 MPa H_2		
TiAl6V4	36	105
	46	67
TiAl6Nb2Ta1Mo0.8	34	63
	72	67
Test atmosphere: 52 MPa H_2		
TiAl6Nb2Ta1Mo0.8	< 29	91

Table 157: Results for subcritical crack growth [219]

268 | Hydrogen

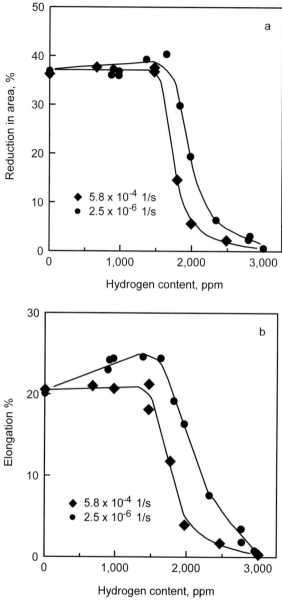

Figure 220: Influence of the hydrogen content on the ductility of TiAl6V4 determined in the slow strain rate tensile test at 5.8×10^{-4} and 2.5×10^{-6} 1/s [220]
(a) reduction of area and (b) elongation

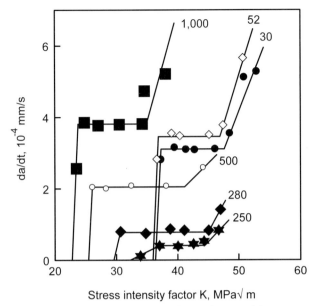

Figure 221: Relationship between crack growth rate and the stress intensity factor for CT specimens from TiAl6V4 at a hydrogen content from 30 to 1,000 ppm. Results from the slow strain rate tensile test at a strain rate of 2.8×10^{-5} 1/s. da/dt $\approx$ 0 is assumed for K_0 [220]

for the initiating crack growth and the K_H value for the fast crack growth decrease successively as the hydrogen content decreases. In addition, a clear difference can be found in the crack velocity in the plateau area below and above about 100 ppm hydrogen. The explanation for this finding is that at low hydrogen contents the crack growth is determined by the creep behavior under the influence of hydrogen and at high hydrogen contents it is predominantly the hydride formation at the α/β-phase boundary, which controls the fracture process. From these findings it is understandable that a hydrogen content of < 125 ppm is required and demanded for the application of TiAl6V4 in aircraft construction.

The influence of hydrogen on the fracture toughness was examined using the α-β-alloy TiAl6V4 (5.8 Al, 3.8 V, 0.3 Fe, 0.2 O, 0.05 N) [221]. The measurements were performed on a CNSB specimen (chevron notched short bar specimen, Figure 222). This specimen is advantageous in that it is much smaller than the CT specimen (only about 10 vol%) and does not require to be precracked. Under constant load the crack grows continuously until the critical crack length is reached. Hydrogen absorption took place in a Sievert's apparatus. The test parameters for the absorption of hydrogen are summarized in Table 158 and the mechanical properties obtained from the tensile tests are contained in Table 159. Both the strength and the ductility of the material TiAl6V4 are hardly influenced up to the maximum examined hydrogen content of 200 ppm. However, the fracture toughness turned out to be strongly dependent on the hydrogen concentration (Figure 223). At hydrogen contents below 50 ppm the fracture toughness strongly decreases with in-

270 | Hydrogen

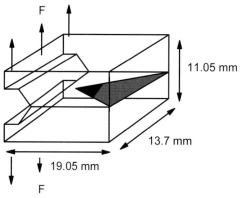

Figure 222: Schematic diagram of the CNSB specimen [221]

H-content in ppm	Test conditions
10	4 h, 800 °C in vacuum
25	4 h, 800 °C in argon
30	4 h, 800 °C in argon
50	4 h, 800 °C in hydrogen
60	4 h, 800 °C in hydrogen
115	4 h, 800 °C in hydrogen
200	4 h, 800 °C in hydrogen

Table 158: Test parameters for hydrogen absorption in the TiAl6V4 specimens [221]

creasing hydrogen concentration and seems to approach a limit value above 50 ppm H. By means of metallographic examinations and ultrasonic measurements it could be demonstrated that the crack propagation is discontinuous at high hydrogen concentrations and rather continuous at low hydrogen concentrations [221].

As early as in 1973 the influence of lower hydrogen concentrations (< 120 ppm) on the critical stress intensity factor K_{I0} for the start of subcritical crack growth and the fracture toughness K_{Ix} was examined for titanium materials [222]. TiAl8-Mo1V1, TiAl6V4 (3.7165) and TiAl6V6Sn2 (3.7175) in the as-received state and following different heat treatments (Table 160) served as test materials. The thickness of the specimens was 0.67 cm for TiAl8Mo1V1, about 2.6 cm TiAl6V4 and 0.6 cm for TiAl6V6Sn2. All experimentally determined values are summarized in Table 160. The influence of the hydrogen content on K_{I0} and K_{Ix} is also illustrated using the example of the TiAl6V4 material (Figure 224). As can be seen, the K_{I0} value in the TiAl6V4 material reaches a minimum at about 50 ppm hydrogen and crack growth occurs at higher hydrogen contents.

A more recent study examined the influence of the microstructure and the heat treatment on the fracture behavior of smooth and notched, hydrogen-containing TiAl6V4 tensile test specimens [223]. Two TiAl6V4 alloys with a similar composition (Table 161), but different microstructures served as test materials. Tensile test

H-content ppm	R_p MPa	R_m MPa	A %	Z %
*	912.9	981.2	18.6	41.5
	906.7	971.5		
	903.2	985.3		
30	921.2	982.6	20.4	39.5
30	906	985.3		
30	912.9	985.3		
62	921.2	982.6	20.8	43
62	930.1	999.1		
62	939.1	1,005		
115	935.6	994.3	21.5	43.8
115	943.9	1,004		
115	948.1	1,005		
200	937	1,011	20.8	41.5
200	941.9	1,020		
200	930.1	989.5		

* Delivery state

Table 159: Influence of the hydrogen concentration on the mechanical properties of the material TiAl6V4 (3.7165) [221]

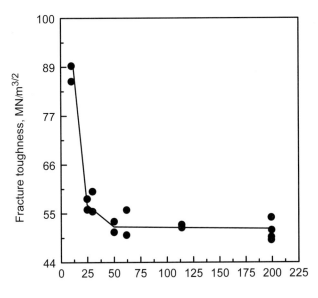

Figure 223: Influence of the hydrogen concentration on the fracture toughness tested on CNSB specimens [221]

Material	Heat treatment	H ppm	R_m MN/m²	R_p MN/m²	K_{Ix} MN/m^{3/2}	K_{I0} MN/m^{3/2}
TiAl8Mo1V1 (R7)	as-received	14–21	1,000	930	102	45
	as-received	14–21	1,000	930	72	38
	871V8OA	5			143	118
	871V8OA (2 x)	7			148	–
	871V8OA + 871H8OA	39			154	44
	1065V4OA	21			198	46
TiAl6V4 (3.7165) (R23B)	as-received	35	1,030	992	66	44
	927L7OA	64	1,030	978	96	66
	927V7OA	8	1,000	909	145	92
	927V7OA + 927H8OA	36	1,013	965	118	65
	927V7OA + 927H8OA	53	978	930	104	67
	927V7OA + 927H8OA	122	1,030	978	100	64
	927V7OA + 927H8OA	215	1,020	971	100	66
TiAl6V4 (3.7165) (R14)	927L7OA	38			112	84
	927V7OA	9			133	89
	927V7OA + 927H8OA	36			125	–
	927V7OA + 927H8OA	50			96	88
	927V7OA + 927H8OA	125			101	90
	as-received	35		785	91	74
	927V7OA	10			123	88
TiAl6V4 (3.7165)	0.11% O; 927 rolled	45		937	81	55
	0.11% O; 960 rolled	38		937	79	47
	0.15% O; 927 rolled	67		937	74	59
	0.15% O; 960 rolled	65		937	86	50
TiAl6V6Sn2 (3.7175)	as-received	60		1,068	55	31

927V7OA = at 927 °C vacuum annealing 7 h/furnace cooling, OA = furnace cooling, V = vacuum, L = air, H = hydrogen, O = oxygen

Table 160: Influence of the hydrogen content on the mechanical properties of titanium alloys [222]

specimens in the L direction and CT specimens with a TL orientation were produced from rod material. Following vacuum treatment (< 10^{-4} mbar), the specimens were volumetrically charged with purified hydrogen at 600 °C. At this temperature the microstructure is not changed.

Following hydrogen charging, the smooth TiAl6V4 specimens were subjected to slow strain rate tensile tests at a strain rate of 2.5×10^{-6} 1/s and 5.8×10^{-4} 1/s. Fracture elongation served as the measure for embrittlement. Figure 225 shows the results for the materials B and C in the as-received state and for heat treatments at 860 and 920 °C. As can be seen, the microstructure of the material C in the as-received state exhibits a higher resistance to hydrogen embrittlement compared to material B. If the alloy B is subjected to suitable heat treatment (e.g. 860 or 920 °C/furnace cooling) prior to hydrogen charging, hydrogen embrittlement

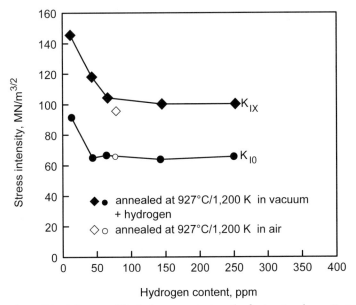

Figure 224: Influence of the hydrogen content on the fracture toughness K_{Ix} and the threshold value for the initiating crack propagation K_{I0} in the material TiAl6V4 (3.7165) [222]

only starts above 2,000 ppm H. The reason for this behavior is considered to be the microstructure obtained as a result of heat treatment with primary α-crystals and a discontinuous β-phase along the grain boundaries.

Moreover, it could be proven that, in particular, the solution treatment temperature and also the type of cooling (furnace cooling, air cooling, quenching in water) can have a considerable influence on the ductility of hydrogen-containing TiAl6V4 alloys (Figure 226). The results are indicated for a hydrogen content of 1,800 and 2,500 ppm compared to the heat treated state without H-charging. As can be seen, solution annealing above 950 °C basically has a negative impact. Following solution treatment between 800 and 920 °C hydrogen contents of 1,800 ppm do not have any negative impact on the ductility of TiAl6V4 (3.7165). However, at hydrogen contents of 2,500 ppm embrittlement can be clearly detected depending on the solution annealing temperature and the type of subsequent cooling. Quenching in water is much more unfavorable than cooling in the furnace. The reason for this behavior is the existing condition of the microstructure illustrated in images in the original paper [223]. For comparison the measured values are also presented in a table. The findings show that, although solution treatment at 1,030 °C increases the mechanical strengths, this heat treatment exerts a negative influence if hydro-

Specimen	Al	V	Fe	O	C	N	H
B	6.55	4.10	0.17	0.200	0.01	0.09	0.003
C	6.55	4.19	0.19	0.185	0.02	0.07	0.001

Table 161: Chemical compositions of the TiAl6V4 test materials in wt% [223]

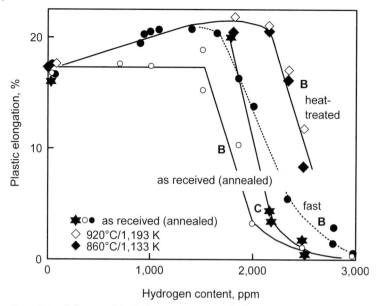

Figure 225: Influence of the hydrogen content on the fracture elongation of TiAl6V4 (3.7165) in the delivery state (test materials B and C) and of the specimen B following heat treatment at 860 and 920 °C/furnace cooling. Strain rate 2.5 × 10⁻⁶ 1/s, open symbols at 8.5 × 10⁻⁴ 1/s [223]

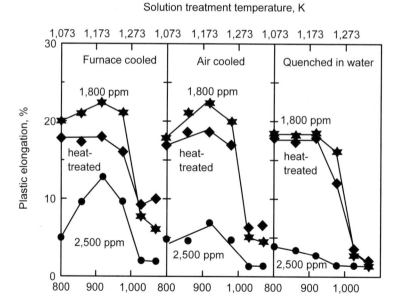

Figure 226: Influence of the solution treatment temperature and cooling (furnace cooling, air cooling, quenching in water) on the ductility of TiAl6V4 specimens with and without hydrogen charging [223]

Heat treatment	$R_{p0.2}$	R_m	Z	A	A at 1,800 ppm H	A at 2,500 ppm H
	MPa	MPa	%	%	%	%
Delivery state	806	1,119	34.1	15.7	19.5	1.4
920 °C/furnace cooling	795	1,125	29.3	11.7	21.7	11.9
1,030 °C/furnace cooling	931	1,171	28.6	10.4	17.3	1.3
920 °C/H_2O	801	872	13.9	5.9	6.7	0.6
1,030 °C/H_2O	1,116	1,225	4.4	2.1	1.3	0

Table 162: Influence of heat treatment on the mechanical properties of smooth TiAl6V4 specimens [223]

gen is absorbed (Table 162). Here, heat treatment at 920 °C/furnace cooling is clearly advantageous.

The influence of the heat treatment temperature and the type of cooling (cooling rate) on embrittlement can be derived from the embrittlement index indicated in Figure 227 for smooth specimens from TiAl6V4 (test material B). This index is the ratio of the fracture elongation of hydrogen-charged and uncharged specimens with identical heat treatment. The embrittlement index depends on the solution annealing temperature, on the one hand, and decisively on the cooling rate following heat treatment, on the other hand. The higher the cooling rate, the higher is

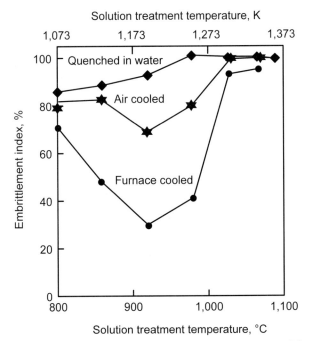

Figure 227: Influence of the solution treatment temperature and the cooling rate on the embrittlement index of smooth tensile test specimens from TiAl6V4 (test material B) [223]

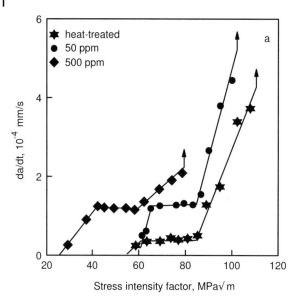

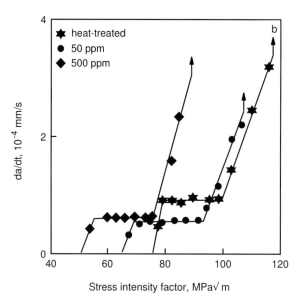

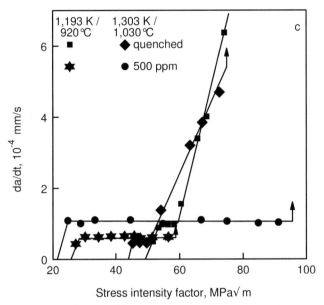

Figure 228: Influence of heat treatment on the crack growth rate-stress intensity curves of TiAl6V4 (3.7165, test material C) with and without hydrogen charging [223]
a: 920 °C/furnace cooled
b: 1,030 °C/furnace cooled
c: 920 and 1,030 °C/ quenched in water

also the embrittlement. This behavior can be explained through the microstructure. Whereas a α-microstructure with a discontinuous β-phase is formed at the grain boundaries during slow cooling, fast quenching leads to a higher retained β-fraction and hence to coherent β-phase areas favoring hydrogen embrittlement.

The influence of the heat treatment on the crack propagation velocity da/dt in TiAl6V4 (3.7165, test material C) was examined using notched CT specimens with incipient fatigue cracks [223]. The results for different heat treatments (920 and 1,030 °C) as well as different cooling rates (furnace cooling and quenching in water) are summarized in Figure 228.

The results show that at a hydrogen content of 500 ppm the threshold value K_0 for initial crack growth is considerably reduced by all heat treatments. At 50 ppm hydrogen the K_0 values in the area of the uncharged specimens are even slightly higher. Also the K_H values of the initial accelerated crack growth with subsequent fracture are considerably reduced at 500 ppm hydrogen. At 50 ppm H the K_H value in the area of the uncharged specimens is slightly lower depending on the heat treatment. The determined critical stress intensity factors K_0 and K_H for the crack growth in TiAl6V4 (test material C) for the various heat treatments are summarized in Table 163.

Due to the favorable strength/weight ratio and the good toughness at low temperatures, the material TiAl5Sn2.5 (3.7115) is frequently used for liquid hydrogen tanks, e.g. in spacecrafts [28]. Pipes and fittings typically consist of titanium or a

Heat treatment	10 ppm H		50 ppm H		500 ppm H	
	K_0	K_H	K_0	K_H	K_0	K_H
as-received	38.9	90.0	46.2	83.6	29.5	84.4
920 °C/furnace cooling	54.8	111.0	59.7	102.9	25.7	78.9
1,030 °C/furnace cooling	65.5	108.9	76.1	120.1	50.2	90.0
920 °C/H$_2$O	47.8	73.6		50.3	24.6	56.8
1,030 °C/H$_2$O	42.5	71.6		61.0	21.2	92.5

K_0 = threshold value for initial crack growth K_H = threshold value for the accelerated crack growth with subsequent fracture

Table 163: Critical stress intensity factors for the crack growth in CT specimens from TiAl6V4 (test material C) [223]

titanium alloy. These materials are exposed to hydrogen at different pressures and temperatures from −253 °C up to ambient temperature. To minimize the damage in hydrogen storage tanks and clarify whether titanium materials are basically susceptible to embrittlement in hydrogen gas, tensile tests were performed on technically pure titanium, TiAl5Sn2.5 and TiAl6V4 (Table 164) in air as well as at 70 MPa gas pressure in helium and hydrogen [28]. The results are summarized in Table 165 and Table 166. Changes of the mechanical properties were only detectable in notched specimens. However, it should be noted that the tests were only short-time tests and, hence, do not permit any conclusion regarding the long-time behavior, in particular since hydride formation depends on time.

The material TiAl6V4 (3.7165) was used to examine the influence of the temperature on the fatigue behavior in hydrogen, nitrogen and helium [28, 224]. The alloy contained 6.0% Al, 4.1% V, 0.23% C, 0.08% Fe, 0.010% N, 0.12% O and

	Al	Sn	V	Fe	C	H	O
TiAl5Sn2,5	5.47	2.70		0.09	0.011	0.0122	0.09
TiAl6V4	6.0		4.1	0.15	0.24	0.008	0.019
TiAl6V4	6.4		4.2	0.15	0.023	0.008	0.14

Table 164: Chemical compositions of the test materials [28]

Material			R_p MPa	R_m MPa	Elongation, %	A %
Titanium	rolled	u	392	490	25	52
		n		987		11
TiAl5Sn2.5	heat treated	u	812	847	18	46
		n		1,435		3.7
TiAl6V4	heat treated	n	973	1,008	14	

n: notched, u: unnotched

Table 165: Mechanical properties of unnotched and notched titanium materials in air [28]

			R_m, MPa		Elongation, %		Z, %	
			He	H_2	He	H_2	He	H_2
Titanium	rolled	u	441	427	32	31	61	61
		n	882	840			10	7,3
TiAl5Sn2.5	heat treated	u	791	798	20	18	45	39
		n	1,407	1,134			3,1	1,8
TiAl6V4	heat treated	n	1,701	1,092			2.2	1.9
		n	1,589	1,162			2,2	1,7

u: unnotched, n: notched

Table 166: Mechanical properties of unnotched and notched titanium materials in helium and hydrogen at 70 MPa gas pressure [28]

0.004% H. The notched specimens were additionally provided with an incipient fatigue crack (2.5 mm). The specimen chamber was first evacuated and subsequently filled with the desired gas at an overpressure of 0.14 MPa. The maximum oxygen content was 1 ppm. Axial loading was between $\sigma_{min} = 87.5$ and $\sigma_{max} = 437.5$ MPa with reference to the unnotched specimen cross-section. The influence of the temperature on the cycles to failure in the different atmospheres is shown in Figure 229.

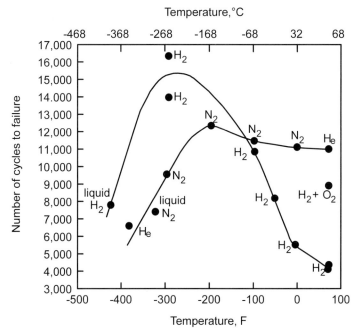

Figure 229: Influence of the temperature on the fatigue behavior of the material TiAl6V4 in hydrogen, helium and nitrogen [28]
$\sigma_{min} = 87.5$ and $\sigma_{max} = 437.5$ MPa

280 | Hydrogen

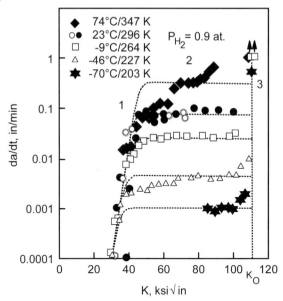

Figure 230: Influence of the temperature and the crack intensity K on the subcritical crack growth da/dt in the material Ti-5Al-2.5 Sn in hydrogen gas (0.9 bar) [216]

The results show that, in contrast to He and N_2, hydrogen considerably increases the crack growth in TiAl6V4 above −73 °C [28]. If the hydrogen is contaminated with oxygen, crack growth is slowed down; refer to point ($H_2 + O_2$) in Figure 229. The reason is the formation of a protective oxide film on the newly formed fracture surface. In general, these findings show that the material TiAl6V4 is not recommended for applications in hydrogen gas above −73 °C.

Another study [216] examined the influence of the temperature from −70 °C to +74 °C on the subcritical crack growth in the material TiAl5Sn2.5 (3.7115) at a constant hydrogen pressure of 0.9 bar. Figure 230 shows the measured ln v-K curves. Whereas the limit value K_0 for initial crack growth depends on the temperature, the crack propagation velocity da/dt in the plateau area II clearly increases with the temperature, being indicative of a transport reaction, i.e. diffusion of hydrogen.

The influence of the temperature (−203 to 77 °C) and the microstructure on the H-induced crack growth in the plateau area II was examined by comparison using the materials TiAl6, TiAl6V4 (3.7165), TiAl5Sn2,5 (3.7115) and TiAl6V6Sn2 (3.7175) [225]. The mechanical load was 50 MPa $m^{1/2}$. The results (Figure 231) clearly show that the crack propagation velocity in the alloys TiAl6 and TiAl6V4 with continuous α-phase is clearly lower than that in the alloys TiAl5Sn2.5 and TiAl6V6Sn2 with continuous β-phase. This is attributed to the higher diffusion rate of hydrogen in the β-phase and, hence, its faster arrival at the area in front of the crack tip. In addition, the maximum susceptibility of both alloys occurs at different temperatures.

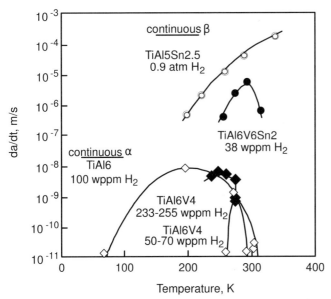

Figure 231: Influence of the temperature on the crack propagation velocity in the materials TiAl6 and TiAl6V4 with continuous α-phase as well as TiAl5Sn2.5 and TiAl6V6Sn2 with continuous β-phase [225]

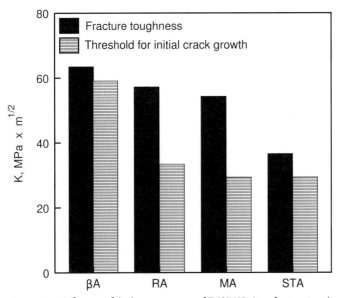

Figure 232: Influence of the heat treatment of TiAl6V6Sn2 on fracture toughness and the threshold value for initial crack growth. β-annealing (βA), recrystallization (RA), as-received state (MA) and solution annealing + aging (STA) [225]

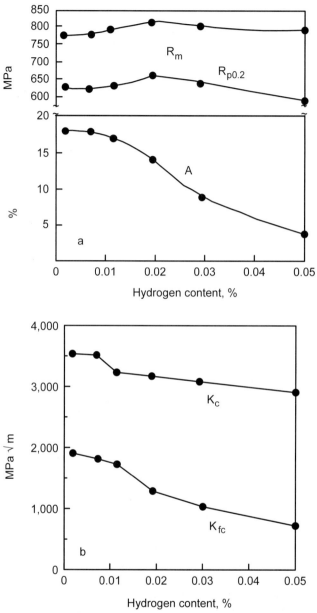

Figure 233: Influence of the hydrogen content in the material TiAl3.2Cr0.84Fe0.4Si0.34B0.01 on the mechanical properties (a) and the fracture toughness under static (K_c) and cyclic load (K_{fc}) at 33 Hz (b) [226]

Since the crack propagation depends on the microstructure of the titanium alloys, also the heat treatment exerts an influence [225]. Examinations of the material TiAl6V6Sn2 with hydrogen concentrations from 10 to 30 ppm revealed that the fracture toughness and the limit value for initial crack growth decrease in the following order of the heat treatments:

annealing in the β-area → recrystallization → as-received state → solution annealing + aging.

The good result of β-annealing is attributed to the formation of a Widmannstätten structure.

The material TiAl3.2Cr0.84Fe0.4Si0.34B0.01 was used to study the influence of hydrogen on the microstructure, the mechanical properties and the crack resistance [226]. Hydrogen of up to 0.05 wt% was absorbed by 3 mm thick sheets in the Sievert's apparatus (refer to Figure 201). From a hydrogen concentration of 0.015–0.2 wt% hydrides could be found with the electron microscope. The influence of the hydrogen content on the strength was determined in tensile tests. Up to a hydrogen concentration of 0.01–0.02 wt% the strength increases due to the dissolved hydrogen and initial hydride formation (Figure 233a). With progressive hydride formation (hydrogen ≥ 0.02%) the elongation values decrease since then embrittlement starts. The dependence of the fracture toughness on the hydrogen content was determined using notched tensile test specimens under static (K_c) and cyclic (K_{fc}) load (33 Hz, R = 0.1) (Figure 232b). A comparison between Figure 233a and Figure 233b shows that not only the elongation values but also the K_c- und K_{fc}-values decrease more or less rapidly as hydride formation starts at about 0.015% H. Moreover, it becomes clear that the influence of hydrogen is stronger under cyclic load compared to static load.

Ultrasonic testing

Various α-, (α+β)- and β-titanium alloys were subjected to ultrasonic tests to study the influence of hydrogen on internal friction [227]. The test method was the pulse-echo method at 10 MHz as well as the resonance method during which vibrations are adjusted to specific frequencies and work in resonance (kHz range). Measurements were performed in the temperature range from –193 to 27 °C in the kHz range and from –193 to 227 °C in the MHz range. First, the specimens were annealed in vacuum (5 × 10^{-7} torr) at 850 °C for 48 h and subsequently charged with hydrogen at 700 °C for 1 h, homogenized for 45 min and cooled in the furnace. Several specimens were cold-formed. Figure 234 uses the example of the material TiAl6V4 with a hydrogen concentration of 1.30 at.% to show the H-induced peak of internal friction and the change of the Young's modulus of elasticity $\Delta E/E_{296}$ at a frequency of 1.5 × 10^5 Hz. Cold forming of the material (here 10%) following the absorption of hydrogen reduces the peak height. Whereas the peak position of the material TiAl6V4 occurs in the kHz range at –129 °C, it is found in a similar form at 10 MHz and –78 °C [227]. The observed peak is attribu-

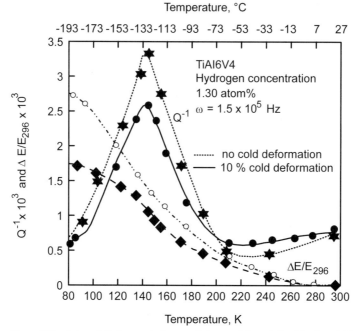

Figure 234: Internal friction and change of the modulus of elasticity in cold-formed and non-deformed TiAl6V4 with 1.3 at.% H, measured with the ultrasonic method at $\omega = 1.5 \times 10^5$ Hz [227]

table to the Snoek effect of the interstitial hydrogen in the β-phase of titanium alloys. The peak height is proportional to the hydrogen content and leads to a relaxation of 3.5×10^{-3} per at.% H. In general, the examination yielded the following findings:

> Ultrasonic measurements for examining the H-induced internal friction of titanium materials are only possible in β-phase-containing alloys.
> A friction peak is not found in any α- or α'-phase alloys.
> The peak height is directly proportional to the hydrogen concentration. The peak occurs with and without cold forming.

The linear relationship between hydrogen concentration and the height of the damping peak facilitates nondestructive determination of the hydrogen content in titanium alloys containing the β-phase. Ultrasonic tests in the MHz range are suited, for instance, for inspecting weld seams potentially contaminated with hydrogen.

Ti-Al-Mn alloys

Titanium alloys with a low aluminum and manganese content (1–2%) belong to the (α + β)-type, but contain only a minor amount of β-phase. Alloys with Al and

Specimen No.	Al	Mn	Fe	C	N	O	H, ppm
TiAl1Mn1							
11/1	0.87	1.01	0.08	0.13	0.02	0.19	6
11/2	0.87	1.01	0.08	0.13	0.02	0.19	55
11/3	0.87	1.01	0.08	0.13	0.02	0.20	150
11/4	0.87	1.01	0.08	0.13	0.02	0.19	420
TiAl2Mn2							
22/1	2.16	2.03	0.02	0.02	0.01	0.12	7
22/2	2.11	1.95	0.02	0.01	0.01	0.13	40
22/3	2.16	2.03	0.02	0.02	0.01	0.12	155
22/4	2.11	1.95	0.02	0.01	0.01	0.16	520

Table 167: Chemical composition of the TiAlMn test materials in wt% and hydrogen content in ppm [228]

Mn contents of 1% and 2%, respectively were used to study the influence of hydrogen on hardening and fracture behavior [228]. The specimens were dehydrogenated in a modified Sievert's apparatus (refer to Figure 201) for 16 h and subsequently charged again with hydrogen up to the desired level at 750 °C, heated to 900 °C for 1 h and cooled in the furnace. The chemical compositions of the test materials and the respective hydrogen contents are summarized in Table 167.

In the dehydrogenated state both alloys exhibit a fine-grained microstructure of the α-phase with β-phase along the grain boundaries. Following hydrogen absorption also the hydride phase can occur along the α/β-phase boundary. The metallographically determined volume fractions of the individual phases are summarized in Table 168. As can be seen, hydrogen concentrations from about 150 ppm exert an influence on the contents of the α- and β-phase since hydrides are formed then. The grain size remains uninfluenced to a major extent.

The influence of the hydrogen content on the mechanical properties of the TiAlMn test materials was determined in the tensile test at a strain rate of 1.6×10^{-4} 1/s. The results are listed in Table 169. Whereas the tensile strength of TiAl1Mn1 remains almost the same and slightly increases in TiAl2Mn2 with

Specimen No.	H, ppm	α vol%	β vol%	Hydride vol%	A.S.T.M grain size α
TiAl1Mn1					
11/1	6	83	17	–	9.8
11/2	55	81.5	18.5	traces	9.8
11/3	150	79	19	2	9.3
11/4	420	75	20	5	9.5
TiAl2Mn2					
22/1	7	79	21	–	10.5
22/2	40	77.5	22.5	–	10.2
22/3	155	77	23	–	10.2
22/4	520	74	21	5	10.3

Table 168: Phase volume fractions and grain sizes of TiAlMn alloys [228]

Specimen No.	TiAl1Mn1				TiAl2Mn2			
	11/1	11/2	11/3	11/4	22/1	22/2	22/3	22/4
Hydrogen content, ppm	6	55	150	420	7	40	155	520
Elastic limit, N/mm²	451	445	401	352	451	377	429	335
Tensile strength, N/mm²	704	710	703	704	704	716	716	727
Fracture elongation, %	26.2	28.0	27.9	13.2	26.0	24.7	23.5	18.7
Reduction of area, %	48.5	49.0	40.5	11.5	41.0	41.0	34.0	22.0

Table 169: Influence of the hydrogen content on the mechanical properties of TiAl1Mn1 and TiAl2Mn2 [228]

increasing hydrogen content, the values of the elastic limit are successively reduced. Fracture elongation and reduction of area as a suitable measure for hydrogen embrittlement clearly decreases from a hydrogen content of 150 ppm attributable to hydride formation according to Table 168.

Ti-Al-Mo and Ti-Mo alloys

The α/β-alloy TiAl5Mo4 and the β-test alloy TiMo30 having a high content of Mo as β-stabilizer were used to study the influence of hydrogen on embrittlement and crack growth [229]. The test materials were annealed in vacuum ($\leq 10^{-5}$ mbar) at 950 °C for 4 h and subsequently charged with hydrogen gas in the Sievert's apparatus (refer to Figure 201) at 800 °C. Following equilibration the specimens were cooled in air (½ h). To produce a Widmannstätten structure in the α/β-alloy, the alloy was heat treated above the β-transition and cooled at different rates to obtain a fine-grained or coarse-grained microstructure. Mechanical loads were applied to notched bend test specimens and flat tensile test specimens. The fatigue behavior of hydrogen-containing specimens was examined at R = 0.05 and a frequency of 25 Hz in air. For comparison several specimens were also cathodically charged with hydrogen. Table 170 shows the chemical composition of the alloy TiAl5Mo4. The influence of the temperature from −53 to 67 °C on the mechanical properties was determined for a hydrogen content of 45 ppm and 1,475 ppm (Table 171). The results show that, as expected, the strength decreases with increasing temperature, however is always clearly higher at higher hydrogen concentrations compared to lower ones, corresponding to the different hardening exponent (Table 171).

Notched bend test specimens were used to study the influence of the hydrogen concentration on the transition from ductile to brittle fracture for the materials TiAl5Mo4 and TiMo30. Figure 235 shows the stress required for the cleavage frac-

Al	Mo	Fe	O	C	Ti
4.9	3.7	0.04	0.083	0.03	balance

Table 170: Chemical composition of the material TiAl5Mo4 in wt% [229]

H, ppm	T, °C	E, GPa	$R_{p0.2}$, MPa	R_m, MPa	n
45	67	115	708	792	0.045
	27	118	784	850	0.049
	−13	120	823	923	0.060
	−53	122	869	947	0.055
1,475	67	107	672	843	0.097
	27	108	793	921	0.080
	−13	109	854	1,002	0.089
	−53	110	922	1,117	0.088

n = hardening exponent

Table 171: Influence of the temperature on the mechanical properties of TiAl5Mo4 at hydrogen contents of 45 ppm and 1,475 ppm [229]

ture depending on the hydrogen content. At lower hydrogen contents the cleavage fracture stress of TiAl5Mo4 is clearly higher compared to TiMo30; with increasing hydrogen concentration (> 1,200 ppm) the values of both materials converge. It should be noted that 1,850 ppm hydrogen can reduce the cleavage fracture stress of TiAl5Mo4 from 3,000 MPa to 1,700 MPa. It is also obvious that cathodic hydrogen charging leads to stronger embrittlement.

The crack propagation velocity da/dN under cyclic load (R = 0.05, 25 Hz) of the alloy TiAl5Mo4 was examined for a low (45 ppm) and a higher hydrogen concentration (1,475 ppm) in the temperature range from −53 to +67 °C [229]. The crack propagation velocity da/dN as a function of the stress intensity factor ΔK under

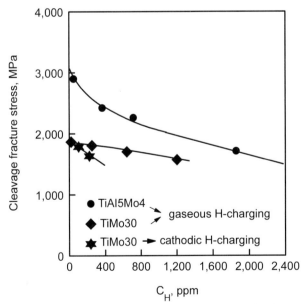

Figure 235: Influence of the hydrogen concentration on the cleavage fracture stress of TiAl5Mo4 and TiMo30 under the condition of gaseous and cathodic hydrogen charging [229]

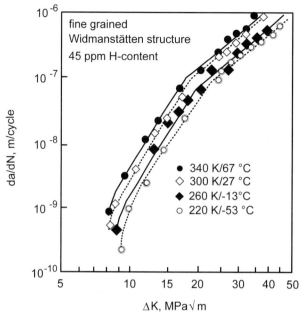

Figure 236: Influence of the temperature on the crack propagation velocity in TiAl5Mo4 at a hydrogen content of 45 ppm [229]

these test conditions is shown in Figure 236 for 45 ppm hydrogen and in Figure 237 for 1,475 ppm hydrogen. According to these results the threshold value K_0 for the subcritical crack growth decreases with increasing temperature at a hydrogen content of 45 ppm and the crack propagation velocity increases. At 1,475 ppm hydrogen the influence of the temperature does not seem to be as clear. The effect of the grain size on K_0 and da/dN (B) is shown in Figure 238. As can be seen, a fine-grained Widmannstätten structure is less favorable. In summary, it is indicated that even 1,500 ppm hydrogen reduce the fatigue limiting value by 30%.

The (α + β)-alloy TiMo2Fe2Cr2 with a hydrogen content of 375 ppm was examined to determine the influence of the temperature (–173 to 127 °C) on hydrogen embrittlement by performing CERT tests with different strain rates [216]. Reduction of area served as the measure for embrittlement (Figure 239). The results show that the strongest embrittlement occurs in the material TiMo2Fe2Cr2 at – 23 °C. In addition, the strain rate plays an important role. Whereas a higher strain rate of 0.42 mm/s hardly exerts any influence, strain rates of ≤ 0.21 mm/s cause a strong reduction of reduction of area at a temperature of about –23 °C. These findings are attributed to a diffusion-controlled process.

In β-titanium alloys hydrogen solubility is very high and can amount to 4,000 ppm (almost 10 at.%). The β-alloy TiMo18 was used to examine the influence of the hydrogen concentration on the mechanical properties [230, 231]. Following chemical polishing, the specimens were charged with hydrogen at 800 °C for 15 min in the Sievert's apparatus (refer to Figure 201). Hydrogen absorption was de-

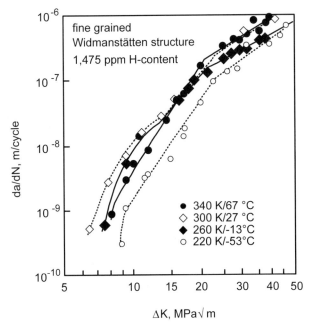

Figure 237: Influence of the temperature on the crack propagation velocity in TiAl5Mo4 at a hydrogen content of 1,475 ppm [229]

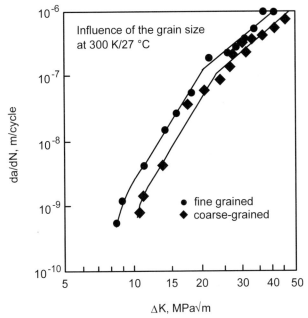

Figure 238: Influence of the grain size of the Widmannstätten structure on the crack propagation velocity in TiAl5Mo4 at 27 °C [229]

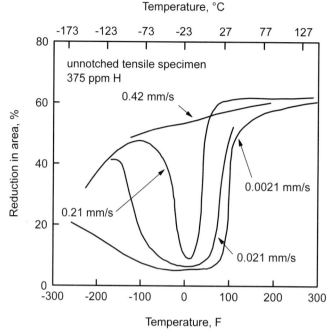

Figure 239: Influence of the temperature and the strain rate on the hydrogen embrittlement of TiMo2Fe2Cr2 (375 ppm hydrogen) [216]

termined by the change in pressure and confirmed by vacuum extraction in individual cases. The mechanical properties were determined in the slow strain rate tensile test (1.3×10^{-4} 1/s). As Figure 240 shows, the elastic limit and the 0.2% yield strength $R_{p0.2}$ linearly decrease with the square root of the hydrogen concentration. Analogously, also the modulus of elasticity reduces (Figure 241). These results are interpreted as the result of the interaction of the hydrogen with the matrix [230].

In contrast to TiMo18, the material TiAl3V8Cr6Mo4Zr4 (beta "C") exhibits an increase in elasticity and the 0.2% yield strengths with the square root of the hydrogen concentration under the same test conditions [231]. This is attributed to the low β-stability of beta "C" and the formation of the ω-phase during hydrogen absorption. Note: It is known that the interaction between absorbed hydrogen and metallic materials can cause both hardening and softening depending on the composition of the alloy (refer to p. 26 in [26]).

The stable β-alloy TiMo30 was used to study the influence of the hydrogen content and the temperature on the strength and crack growth [232]. The specimens were annealed at 1,000 °C for 4 h and cooled at a rate of 10 °C/min. Thereafter, they exhibited a coarse-grained microstructure with a grain size of 85–100 μm. Hydrogen charging took place at 870 °C in an Ar-H_2 atmosphere at a partial hydrogen pressure of 0.2 bar. Figure 243 shows the influence of the hydrogen content (10–1,200 ppm) on the yield strength determined in the compression test for 7 different temperatures (–150 to 90 °C). When the hydrogen concentration increases, the

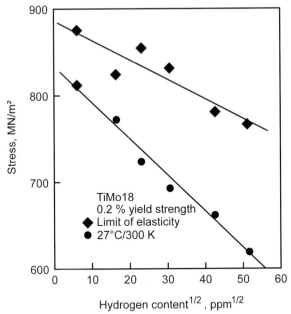

Figure 240: Linear decrease of the elastic limit and 0.2% yield strengths of TiMo18 with the square root of the hydrogen concentration at 27 °C [230]

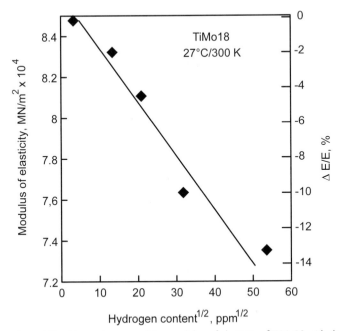

Figure 241: Linear decrease of the modulus of elasticity of TiMo18 with the square root of the hydrogen concentration at 27 °C [230]

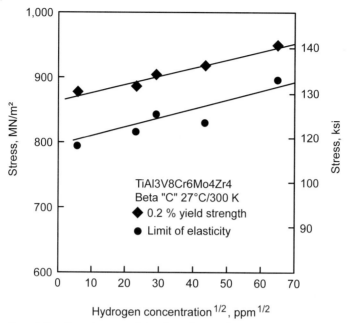

Figure 242: Linear increase of the elastic limit and 0.2% yield strengths of TiAl3V8Cr6Mo4Zr4 with the square root of the hydrogen concentration at 27 °C [231]

strength first increases and slightly decreases thereafter. In addition the strength clearly increases with decreasing temperature. Crack growth was examined using two CT specimens of different thickness (1.78 cm and 0.21 cm) for two different hydrogen concentrations (380 and 540 ppm) (Figure 244) [232]. The threshold value K_0 for initial crack growth is clearly lower for the thicker specimens. In addition, K_0 is reduced with increasing hydrogen concentration.

The technical alloy TiMo11.5Zr6Sn4.5 called Beta III was examined in high-pressure hydrogen and, for comparison, in helium at a gas pressure of 345 bar (34.5 MPa) each [230]. To determine the mechanical properties, round test bars of different heat treatment were subjected to tensile tests in both high-pressure gases. The results are shown in Table 172. Compared to helium, the ductility of the material TiMo11.5Zr6Sn4.5 is strongly reduced in high-pressure hydrogen. Fractographical examinations of the fracture surfaces confirm this finding. Whereas ductile fracture occurred in helium, brittle cleavage fracture occurs in high-pressure hydrogen. Since the results were obtained from short-time tests, additional time-dependent tests, e.g. creep rupture tests, should be performed.

Another study [233] examined the resistance of the metastable Beta III alloy (TiMo11.5Zr6Sn4.5) to hydrogen embrittlement compared to the materials Beta I (TiV13Cr11Al3) and VT 15 (TiMo7Cr11Al3). Hydrogen absorption took place in the Sievert's apparatus (refer to Figure 201) at 850 °C. The results obtained during the slow strain rate tensile test (1×10^{-4} 1/s) regarding the influence of the hydro-

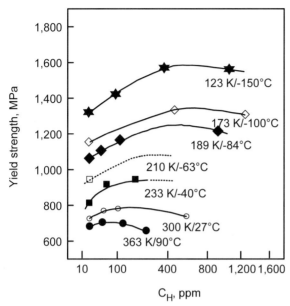

Figure 243: Influence of the hydrogen concentration and the temperature on the yield strength of the material TiMo30 [232]

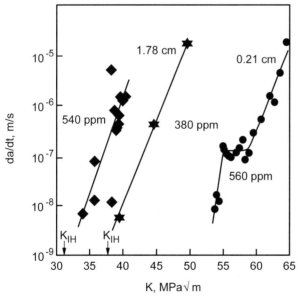

Figure 244: Influence of the hydrogen content on the crack growth da/dt as a function of the stress intensity factor K for thick and thin CT specimens from TiMo30 [232]

Hydrogen

Heat treatment	Gas	Pressure MPa	R_p MPa	R_m MPa	Z %	A %
745 °C/5 min/H_2O	H_2	34.5	544	673	13.1	4.0
	He	34.5	551	785	63.4	22.0
745 °C/5 min/H_2O + 540 °C/8 h	H_2	34.5	989	1,063	6.7	4.0
	He	34.5	987	1,051	16.6	7.2
815 °C/5 min/H_2O	H_2	34.5	555	668	6.3	3.8
	He	34.5	589	764	72.0	31.6
815 °C/5 min/H_2O + 540 °C/8 h	H_2	34.5	989	1,038	7.1	6.6
	He	34.5	949	1,016	13.8	8.8

Table 172: Mechanical properties of TiMo11.5Zr6Sn4.5 in high-pressure hydrogen and helium [230]

gen content on elongation are shown in Figure 245. These findings show that embrittlement of the material Beta III compared to Beta I and VT 15 only starts at high hydrogen contents > 4,000 ppm.

If the material Beta III is first cold-formed and subsequently charged with hydrogen at 590 °C for 8 h, embrittlement starts at clearly lower hydrogen concentrations. Figure 246 shows the influence of cold forming and hydrogen concentration on the relative elongation with reference to the hydrogen-free specimens. The comparatively strong decline of the relative elongation A_H/A_0, in particular already

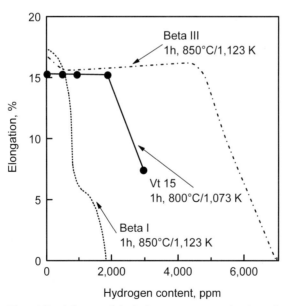

Figure 245: Influence of the hydrogen content on the elongation of the materials VT 15, Beta I and Beta III in the slow strain rate tensile test [233]

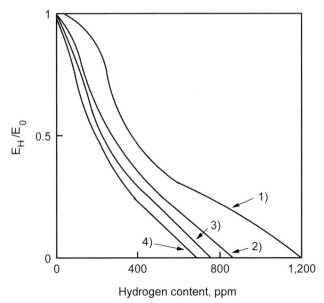

Figure 246: Influence of cold forming and the hydrogen content of Beta III on the relative elongation in the tensile test [233]
1) 10% cold-formed
2) 30% cold-formed
3) 50% cold-formed
4) 70% cold-formed

at low hydrogen contents (cf. Figure 245 and Figure 246), is attributed to the fact that both cold forming and aging at 590 °C support the nucleation of the α-phase.

It should be noted that, despite the higher strength of the β-alloys, the age-hardenable (α + β)-materials are preferably used in many applications.

Ti-V alloys

As with all β-titanium alloys β-Ti-V alloys have high hydrogen solubility, which may assume values up to 12,884 ppm in the material TiV20 [230]. The study [230] examined the influence of hydrogen on the stability of the β-microstructure of TiV20. Since this material has only been a test alloy to date, it will not be further dealt with here.

Titanium aluminides

Due to their combination of properties the new material group of the intermetallic titanium aluminides is an interesting alternative to conventional materials for moving parts in engine and turbine construction as well as in aerospace application [234]. The specific weight of these materials amounts to only half of the specific weight of conventional steels and nickel-based alloys. The strength values meet the requirements on these materials even at temperatures above 700 °C. However, there are certain deficits as to their oxidation resistance at temperatures of 800 °C and higher. Basically, the oxidation resistance of these materials can be improved by the addition of ternary and quaternary elements such that their use at temperatures above 900 °C is possible. Suitable alloying elements include niobium, molybdenum, silicon and boron (at least in certain concentration ranges).

For the material group of the intermetallic titanium aluminides there are four alloying variants according to the Ti-Al phase diagram (Figure 247) [234]:

Ti$_3$Al (α_2)   hcp
TiAl (γ)   fcc
TiAl$_2$
TiAl$_3$.

The single-phase aluminides α_2-Ti$_3$Al (20–36 at.% Al) and γ-TiAl (50.5–57.5 at.% Al) as well as the two-phase γ-aluminides (γ-TiAl + α_2-Ti$_3$Al) are of practical importance.

In a literature guide of 1993 for intermetallic compounds also the most important properties of the titanium aluminides are described with literature references [235]. Since hydrogen also serves as a fuel for powering spacecrafts, the behavior of the titanium materials in compressed hydrogen at high temperatures is very important. As part of the airframe is cooled, hydrogen comes into contact with the material at different temperatures and pressures [236, 237]. Figure 248 shows the

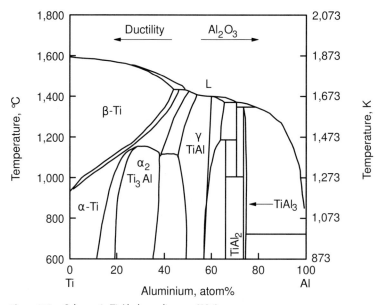

Figure 247: Schematic Ti-Al phase diagram [234]

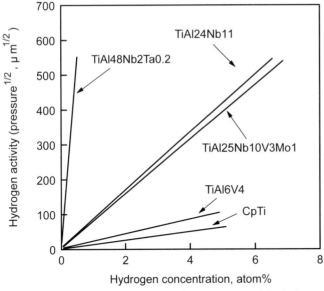

Figure 248: Relationship between the hydrogen activity and the hydrogen solubility for selected titanium materials at 800 °C [236]

relationship between the hydrogen activity and the hydrogen solubility for common titanium materials [236]. As can be seen, the higher the hydrogen activity the lower the hydrogen solubility is. This means that the solubility of hydrogen in titanium aluminides (e.g. TiAl48Nb2Ta0.2, TiAl24Nb11) is clearly lower compared to conventional titanium materials (e.g. TiAl6V4, pure titanium). On the other hand, the hydrogen solubility in the γ-aluminide (TiAl48Nb2Ta0.2) is clearly lower than in the $α_2$-aluminide (TiAl24Nb11). Since the hydrogen solubility is very low at RT, almost the entire hydrogen is present then in the form of hydrides.

According to the results of Thompson [238] the solubility of hydrogen in titanium aluminides increases with increasing temperature and following cooling to room temperature the hydrides are precipitated. Figure 249 uses the example of a γ-aluminide (TiAl50 (at.%)) to show the influence of the temperature on the total hydrogen absorption, reaching 75 at.% (90,000 ppm) at 1,400 °C. Also in the material TiAl48V1 an increasing hydrogen solubility was found with increasing temperature in the flowing hydrogen from 1 bar at a temperature between 400 and 650 °C (Figure 250), [239]. The lower hydrogen absorption at 800 °C was attributed to the formation of a visible oxide film. The water vapor necessary to this end, was generated by the interaction of the hydrogen with the quartz chamber during which SiO and H_2O (g) are formed. Since an increasing tendency for hydrogen absorption at 400 and 500 °C is even found after 8 h (Figure 250) and hence the maximum solubility is not yet reached, oxide formation cannot be excluded and these results are not indicative of a clear temperature dependence of the solubility of hydrogen. In any case, later examinations of the same materials revealed a reduction of the hydrogen solubility with increasing temperature [240].

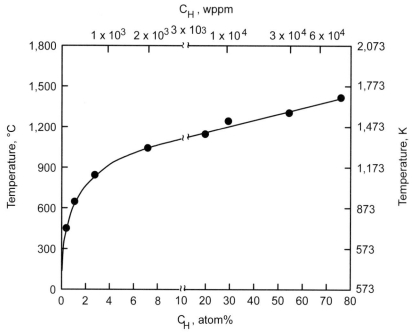

Figure 249: Influence of the temperature on the hydrogen absorption in the γ-aluminide TiAl50 (at.%) – refer to the note in the text [238]

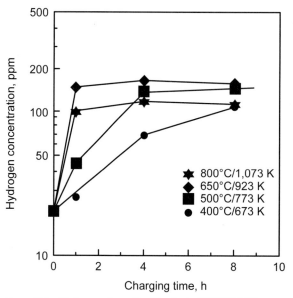

Figure 250: Hydrogen absorption kinetics in TiAl48V1 [239]

	Hydrogen content, ppm (wt%)	
450 °C	600 °C	700 °C
5,200	> 2,000	3,700*
5,000	3,000	2,500

* Furnace cooling in H_2

Table 173: Influence of the temperature on the hydrogen solubility in TiAl25 (at.%) [240]

It is justifiable to question the gravimetrically determined results in Figure 249 in [240]. Since all titanium alloys react exothermally with hydrogen, the capacity for hydrogen absorption necessarily decreases with increasing temperature and, hence, the experimental findings of Thompson in [238] cannot be correct. Most likely the gravimetrically determined weight gain was caused by oxidation with the residual oxygen content or water vapor.

The influence of the temperature on the maximum hydrogen absorption in 0.1 MPa H_2 was examined in various research laboratories using the alloy TiAl25 (at.%). The results are summarized in Table 173 [240]. As can be seen, hydrogen absorption decreases with increasing temperature.

Numerous examinations of the technically interesting γ-based materials TiAl48 (γ-TiAl + $α_2$-Ti$_3$Al) were performed to study the influence of the hydrogen pressure and the temperature on the maximum hydrogen absorption. The results are summarized in Table 174 [240]. As can be seen, hydrogen absorption strongly increases in high-pressure hydrogen (> 10 MPa). However, the values scatter considerably, attributable to the alloy compositions and test conditions which are not exactly identical. In addition, hydride formation occurs in high-pressure hydrogen, which is not the case at lower pressures.

More detailed examinations of hydrogen absorption were performed using a number of titanium aluminides with aluminum contents from 25 to 53 at.% [240]. Hydrogen charging at temperatures from 425 to 800 °C took place in the H_2 flow

H_2 pressure MPa	Temperature °C	H-content ppm
0.1	650	160–215
0.1	800	130–175
0.1	815	235*
0.7	815	120
10	800	1,400–1,850
13.8	650	2,750
13.8	650	6,460
13.8	650	10,800
13.8	815	2,075

* TiAl45V3 + 7 vol% TiB$_2$

Table 174: Influence of the hydrogen pressure and the temperature on the hydrogen absorption in TiAl48 alloys [240]

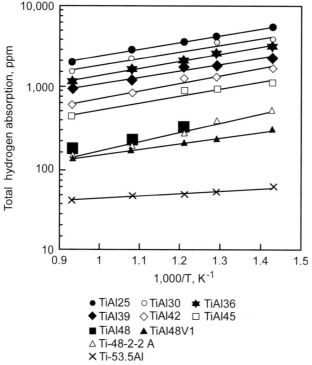

Figure 251: Dependence of the total hydrogen absorption on the inverse temperature for 11 titanium aluminides of different composition [240]

method at a pressure of 0.1 MPa. The hydrogen gas used was ultra-clean and to remove potential traces of oxygen the specimens were wrapped in foils of tantalum or zirconium as a getter material. Subsequently, the hydrogen content was determined by high-temperature vacuum extraction with an accuracy of ± 0.1 ppm. Figure 251 shows the total absorbed content of hydrogen as a function of the inverse temperature for 11 titanium aluminides of different chemical composition. As can be seen, the total hydrogen absorption decreases with increasing temperature (1/T decreasing) and also with an increasing aluminum content of the aluminides. In the examined alloys the absorbed hydrogen amount differs by two orders of magnitude. The widely differing absorption capacity of hydrogen in α_2-Ti$_3$Al and γ-TiAl is illustrated again in Figure 252. It differs by two orders of magnitude.

The summary shows the influence of the aluminum content of the different titanium aluminides α_2-Ti$_3$Al, (γ-TiAl + α_2-Ti3Al) and γ-TiAl on the hydrogen absorption (Figure 253). As can be seen, the solubility of hydrogen is highest in α_2-Ti$_3$Al, clearly lower in (γ-TiAl + α_2-Ti$_3$Al) and comparatively very low in γ-TiAl. The low hydrogen absorption and hydride formation with increasing aluminum content is attributable to the interstitial site occupation by aluminum [240].

The diffusion rate of hydrogen in the different titanium materials depends on the alloy composition [240]. The influence of the material, for instance, at tempera-

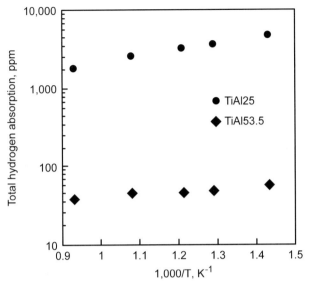

Figure 252: Difference in the hydrogen absorption in the single-phase γ-TiAl (TiAl53.5) compared to the stochiometric α₂-Ti₃Al alloy (TiAl25) [240]

tures around 400 °C (1/T = 1.5 × 1,000/K) is very clear (Figure 254). As can be seen, the diffusion coefficient D in α₂-Ti₃Al is by far the lowest and clearly increases via γ-TiAl and α-Ti up to β-Ti. With increasing temperature the D values converge to one another.

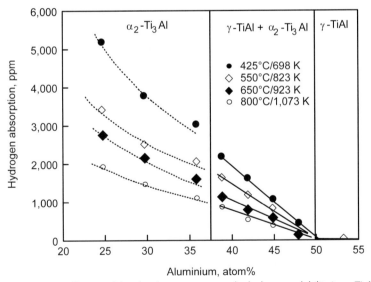

Figure 253: Influence of the aluminum content on the hydrogen solubility in α₂-Ti₃Al, (γ-TiAl + α₂-Ti₃Al) and γ-TiAl aluminides [240]

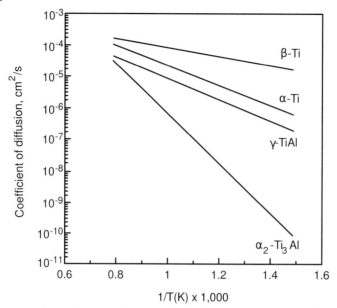

Figure 254: Influence of the temperature on the diffusion coefficient D for hydrogen in different titanium materials [240]

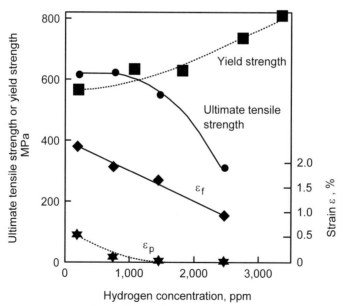

Figure 255: Dependence of the yield strength, ultimate tensile strength and ductility (true strain ε_p and true fracture elongation ε_f) of the α_2-aluminide TiAl24Nb11 (at.%) on the hydrogen concentration in the form of hydrides. Mean values from 2 to 5 tests [238]

Temperature °C	Gas/pressure MPa	R_m MPa	Z %	A %
25	He/34.5	959	2.7	2.8
25	H$_2$/34.5	952	0.4	1.7
204	He/34.5	918	10.1	8.0
204	H$_2$/34.5	835	2.9	2.5

Table 175: Influence of compressed hydrogen on the mechanical properties of TiAl24Nb11 at 25 and 204 °C [236, 237]

Similar to other hydride-forming metals, the hydrides exhibit an influence on the mechanical properties [238, 241]. The dependence of the mechanical properties of the α$_2$-aluminide TiAl24Nb11 on the hydrogen content in the form of hydrides is illustrated in Figure 255. As can be seen, the ductility continuously decreases with increasing hydride content.

In hydrogen gas embrittlement of the α$_2$-aluminides TiAl24Nb11 and TiAl25Nb10V3Mo1 was observed [236, 237]. Table 175 shows relevant results for the material TiAl24Nb11 in 34.5 MPa hydrogen gas at 25 and 204 °C. Comparison tests were performed in helium under the same conditions. Whereas the tensile strength was hardly affected by the influence of hydrogen, the reduction of area Z and the fracture elongation A strongly decrease by embrittlement. The cracks in the hydrogen atmosphere run along the α$_2$/β-phase boundary and in the helium atmosphere within the more ductile β-phase.

In addition, the Ti aluminides TiAl14Nb20V3.2Mo2 and TiAl33Nb5Ta1 as well as high-purity beryllium (< 0.6 BeO) were examined to study the influence of compressed hydrogen (13.8 and 0.1 MPa) on the strength and ductility properties at different temperatures (–130 to 204 °C) [242]. Helium served as the comparison atmosphere. First, the specimen chamber was evacuated (5.3 Pa), then flushed with hydrogen and finally adjusted to the desired hydrogen pressure. The tensile tests with a strain rate of 8×10^{-5} 1/s were performed in-situ in compressed hydrogen. The results are shown in Table 176 to Table 178. They led to the following findings [242]:

- α$_2$-Ti aluminide TiAl14Nb20V3.2Mo2 exhibits strong hydrogen embrittlement at room temperature and 13.8 MPa hydrogen. In contrast, no embrittlement occurs at 204 °C and 0.1 MPa hydrogen.
- Although the results obtained for γ-Ti aluminide TiAl33Nb5Ta1 are not as clear, no damage could be fractographically established at a pressure of 13.8 MPa hydrogen from –130 to 204 °C.
- Even at a pressure of 13.8 MPa hydrogen beryllium remains completely free of embrittlement throughout the entire temperature range from –130 to 204 °C.

Selective heat treatments of the material TiAl24Nb11 were performed to obtain the optimal microstructure for a higher resistance to hydrogen embrittlement [243]. First, the 13 specimens with different heat treatments were subjected to a Vickers hardness test (7,000 g). When incipient cracks occurred, the specimens were dis-

Gas	Pressure MPa	T °C	R_p MPa	R_m MPa	A %	Z %
H_2	13.8	RT	999.6	1,068.6	2.25	1.5
He	13.8	RT	978.9	1,056.9	4.1	4.0
H_2	0.1	204	919.6	1,085.8	9.2	7.6
He	0.1	204	905.2	1,072.0	8.4	8.2

Table 176: Mechanical properties of the α_2-Ti aluminides TiAl14Nb20V3.2Mo2 in in-situ compressed hydrogen at room temperature and 204 °C [242]

Gas	T °C	R_m MPa	A %
H_2	−130	542.6	0.3
He	−130	484.0	0.1
H_2	RT	512.9	0.33
He	RT	537.0	0.85
H_2	104	382.6	0.2
He	104	465.3	0.15
H_2	204	441.9	0.17
He	204	458.5	0.275

Table 177: Mechanical properties of the γ-Ti aluminides TiAl33Nb5Ta1 in in-situ compressed hydrogen at 13.8 MPa [242]

Gas	T °C	R_p MPa	R_m MPa	A %	Z %
H_2	−130	399.2	477.8	1.33	–
He	−130	361.9	454.3	2.5	–
H_2	RT	356.4	486.7	7.0	5.5
He	RT	362.6	466.7	6.3	4.7
H_2	104	273.7	446.0	14.5	10.5
He	104	282.7	449.5	15.5	11.7
H_2	204	255.1	390.2	25.9	22.1
He	204	254.4	388.8	27.6	27.1

Table 178: Mechanical properties of beryllium in in-situ compressed hydrogen 13.8 MPa [242]

carded. Three selected microstructures (globular (α_2+β)-, as well as coarse-grained and fine "wicker-type" lamellar structure with α-plates in a nonparallel arrangement) were tested for their resistance to hydrogen embrittlement. To this end, creep tests were performed first in 13.8 MPa compressed hydrogen for 100 h at 316 °C under a mechanical load of 586 MPa (85–127% R_p at RT) (Table 179).

Since no specimens were broken after a test duration of 100 h, these specimens from the creep tests were deformed until fracture in the slow strain rate tensile test at a strain rate of 5×10^{-6} 1/s at RT. Specimens not charged with hydrogen and charged with hydrogen (13.8 MPa H_2 at 538 °C) having not been subjected to a creep test were used for comparison. The results are summarized in Table 180.

Microstructure	Stress MPa	Stress % R_p	Creep time h	Creep elongation %	Creep rate %/h
Globular	399.9	63	118	–	–
Globular	586.1	92	100	4.30	0.043
Coarse lamellar*	586.1	127	100	3.78	0.038
Coarse lamellar*	586.1	127	100	6.01	0.061
Fine lamellar*	586.1	85	100	1.48	0.015
Fine lamellar*	586.1	85	100	1.26	0.012

* wicker-type

Table 179: Creep tests of the material TiAl24Nb11 in compressed hydrogen (13.8 MPa) at 316 °C [243]

Microstructure	H_2 charging	H_2 ppm	R_p MPa	R_m MPa	A %	Plast. Elongation %
Globular	no	65	639.8	657.1	1.28	0.53
Globular	yes*	340	666.6	666.6	0.96	0.07
Coarse lamellar	no	74	461.9	688.7	3.82	3.07
Coarse lamellar	yes*	270	645.0	757.5	2.19	1.13
Coarse lamellar	yes*	532	768.5	768.5	1.46	0.18
Fine lamellar	yes*	220	704.5	746.6	2.63	1.27
Fine lamellar	yes*	160	730.5	853.6	5.30	4.02
Fine lamellar	no	67	688.4	842.6	3.08	1.88
Fine lamellar	yes**	1,300	535.7	659.0	2.75	1.90
Fine lamellar	yes**	1,300	530.2	690.9	3.25	2.10

* Following prior creep tests at 316 °C under constant load of 586 MPa and 13.8 MPa hydrogen** 100 h in 13.8 MPa hydrogen at 538 °C

Table 180: Slow strain rate tensile tests of the material Ti-24Al-11Nb prior to and after 100 h creep test in compressed hydrogen at elevated temperature [243]

In addition to Table 180, Figure 256 shows the influence of the hydrogen content on the total elongation of the different microstructure of the material TiAl24Nb11 [243]. Whereas the total elongation of the globular and coarse lamellar microstructure clearly decreases with increasing hydrogen content, it remains unchanged in the fine lamellar microstructure at least up to 1,300 ppm hydrogen. As can be seen, the fine lamellar microstructure of the material TiAl24Nb11 is most resistant to hydrogen embrittlement. The different behavior of the individual microstructures regarding hydrogen embrittlement depends on whether the β-phase is continuous as in the coarsely lamellar structure or is not cohesive as in the fine lamellar structure.

The influence of the heat treatment on the behavior of the material TiAl24Nb11 in hydrogen gas was also examined in another study [244]. There, the findings of [243] were confirmed, i.e. that the increase of the strength values by hydrogen absorption depending on the microstructure of the alloys differs. Figure 257 shows the example of the heat treatment on the tensile strength of the material TiAl24Nb11 in vacuum and hydrogen gas at a pressure of 1 bar, measured in the tensile test at different strain rates (5×10^{-2} to 5×10^2 mm/min). Analogously, also the

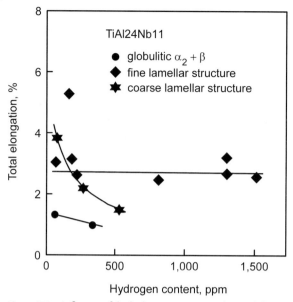

Figure 256: Influence of the hydrogen content on the total elongation of the different microstructure of the material TiAl24Nb11 [243]

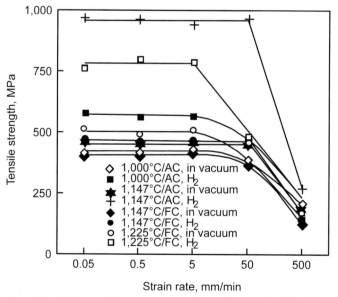

Figure 257: Influence of heat treatment on the tensile strength of the alloy TiAl24Nb11 in hydrogen gas, measured in the tensile test at different strain rates (mean value from 2–6 tests) [244]

nominal fracture stress and the cleavage fracture toughness change. This behavior is attributed to the action of hydrogen as a β-stabilizer. During solution annealing 1,147 °C/air hydrogen as a β-stabilizer can reduce the α-phase and increase the β-phase. As a result, the strength values and the fracture toughness increase. If the solution annealing process is followed by furnace cooling (much slower), the α-phase is hardly reduced and, consequently, minor changes of the strength values occur [244].

The super $α_2$-alloy TiAl14AlNb20V3.2Mo2 and the γ-TiAl material TiAl48Nb2.5-Ta0.3 were examined to study the influence of hydrogen pressure and temperature on the adverse effect on mechanical properties [245]. The strength values determined in air are shown in Table 181. According to the conditions in practice, hydrogen was absorbed both in high-pressure hydrogen (13.8 MPa) and low-pressure hydrogen (0.03 MPa) at 23 and 650 °C or 815 °C. Helium served for comparison. Tensile tests were performed to determine the mechanical properties. The results are summarized in Table 182 and Table 183. Since γ-TiAl turns out to be very

	Super-$α_2$	γ-TiAl
R_p, MPa	893.5	–
R_m, MPa	1,004.6	552.0
A,%	3.7	0.0

Table 181: Strength values of TiAl14AlNb20V3.2Mo2 and TiAl48Nb2.5Ta0.3 in air at room temperature [245]

Atmosphere	Pressure MPa	Temperature °C	R_p MPa	R_m MPa	A %
H$_2$	13.8	23	–	552.0	0.0
H$_2$	0.03	650	400.2	483.0	5.0
H$_2$	13.8	650	345.0	414.0	0.8
H$_2$	13.8	815 + 23*	–	338.1	0.0
He	0.03	650	394.6	489.9	5.2

* 4 h, 815 °C H$_2$ + 23 °C air

Table 182: Mechanical properties of the material TiAl48Nb2.5Ta0.3 following aging tests in hydrogen and helium [245]

Atmosphere	Pressure MPa	Temperature °C	R_p MPa	R_m MPa	A %
H$_2$	0.03	23	865.6	976.0	0.8
H$_2$	13.8	23	870.7	976.0	1.1
H$_2$	0.03	650	554.3	657.9	1.5
H$_2$	0.03	815 + 23*	–	355.3	0.2
He	0.03	650	565.9	665.1	3.3

* 4 h 815 °C H$_2$ + 23 °C air

Table 183: Mechanical properties of the material TiAl14AlNb20V3.2Mo2 following aging tests in hydrogen and helium [245]

brittle in air already at 23 °C (acc. to Table 181: A = 0), no further detriment was found in hydrogen at 23 °C. At 650 °C the hydrogen pressure exerts a major influence. Whereas there is no noticeable change in 0.03 MPa hydrogen compared to helium, fracture elongation is strongly reduced at 13.8 MPa hydrogen. Compared to γ-TiAl the super $α_2$-alloy TiAl14AlNb20V3 is clearly more susceptible to hydrogen embrittlement [245]. Both in low-pressure and high-pressure hydrogen the ductility is strongly reduced at 23 °C compared to air, whereas the influence on the tensile strength is quite low (Table 183). Similar results are obtained in hydrogen gas of low pressure at 650 °C. Hydrogen charging at 815 °C and a subsequent tensile test in air at 25 °C reduce both the tensile strength and the ductility even at a low hydrogen pressure, indicative of strong embrittlement, including hydride formation.

The behavior of two $α_2$-titanium aluminides in hydrogen gas was examined compared to selected cobalt and nickel materials [246]. The test materials and their chemical composition are summarized in Table 184 and the mechanical properties determined by tensile test in hydrogen gas at room temperature are summarized in Table 185. Whereas the tensile strength in hydrogen increases in hydrogen gas compared to air, a decline of the ductility is found in most of the materials. An

Material	Fe	Cr	Ni	Co	Ti	Al	V	Mo	Nb	W
TiAl14Nb21	–	–	–	–	Balance	14	–	–	21	–
TiAl14AlNb20V3.2Mo2	–	–	–	–	balance	14	3.2	2.0	20	–
Nickel 201 (2.4061)	0.1	–	balance	–	–	–	–	–	–	–
HDA-230	1.2	22.0	balance	–	–	0.3	–	–	–	14.0
H1/S88	2.0	21.7	23.0	balance	–	–	–	–	–	14.0

Table 184: Chemical composition of the test materials [246]

Material	Gas	$R_{p0.2}$ MPa	R_m MPa	A %
TiAl14Nb21	air	706	739	3.2
	H_2	661	786	2.6
TiAl14AlNb20V3.2Mo2	air	998	1,223	2
	H_2	1,116	–	–
Nickel 201 (2.4061)	air	203	437	44
	H_2	250	442	29
HDA-230	air	428	771	34
	H_2	454	877	32
H1/S88	air	485	864	32
	H_2	545	1,035	49

Table 185: Mechanical properties of titanium aluminides in hydrogen gas and in air compared to selected cobalt and nickel materials at 25 °C [246]

Material	Gas	T °C	σ MPa	τ h	A %
TiAl14Nb21	13.8 MPa H_2	649	276	2.3	10.7
	air	649	276	3.6	15.9
TiAl14AlNb20V3.2Mo2	13.8 MPa H_2	649	276	19.5	5.1
	13.8 MPa H_2	649	310	10.6	5.0
	air	649	276	55.8	10.4

Table 186: Results of the creep rupture tests of titanium aluminides in air and in hydrogen gas at 649 °C [246]

exception is the cobalt alloy H1/S88, attributable to the strong cold forming of this material.

The creep rupture behavior was examined for Ti aluminides TiAl14Nb21 and TiAl14AlNb20V3.2Mo2 in hydrogen gas and in air at 649 °C [246]. As shown by the results in Table 186, hydrogen embrittlement occurs.

Summarizing the above [246]:

- Absorption of hydrogen at room temperature for 1 h causes an increase of the tensile strength by 10–16% in 4 of the examined 5 materials. An exception is the cobalt alloy.
- Hydrogen absorption in Ti- and Ni-based alloys leads to a reduction of fracture elongation by 6–34%.
- The creep rupture behavior of titanium aluminides (TiAl14Nb21 and TiAl14AlNb20V3.2Mo2) is strongly impaired by hydrogen absorption at 649 °C.

Due to the lower hydrogen solubility and hydride formation in γ-TiAl materials, less hydrogen embrittlement is expected. Table 187 shows the mechanical properties of the material TiAl48Nb2Mn2 determined at a strain rate of 1×10^{-5} 1/s for two different microstructures following hydrogen charging at 500 and 800 °C compared to vacuum [240]. Within a certain range of scatter prominent differences regarding the microstructures A and F as well as aging in vacuum or hydrogen are not found.

Material	Treatment	R_m MPa	R_p MPa	A %	[H] Ppm
A	500 °C, vacuum	394.3	304.7	1.10*	20
	500 °C, H_2	403.3	294.4	1.36	360
A	800 °C, vacuum	430.2	298.5	2.18	20
	800 °C, H_2	404.7	303.3	1.40	180
F	500 °C, vacuum	412.9	312.9	1.20*	20
	500 °C, H_2	404.7	304.7	1.05	375
F	800 °C, vacuum	408.1	305.4	1.43	20
	800 °C, H_2	448.1	342.6	1.60	180

* Broken outside the measurement section

Table 187: Mechanical properties of TiAl48Nb2Mn2 following aging in vacuum and hydrogen and subsequent tensile test at 1×10^{-5} 1/s [240]

				Aging time in h			
			Argon, 815 °C		Hydrogen, 815 °C		
			0 h	82 h	0.5 h	2 h	82 h
Globular microstructure	Oxygen	ppm	1,150	1,100	1,070	1,230	1,500
	Nitrogen	ppm	200	170	180	170	240
	Hydrogen	ppm	39	33	50	90	234
Lamellar microstructure	Oxygen	ppm	125	1,230	2,600	1,390	1,500
	Nitrogen	ppm	180	170	260	200	190
	Hydrogen	ppm	30	90	55	95	239

ppm in wt%

Table 188: Interstitial content of H, N and O in the material TiAl45V3 with 7.5 vol% TiB$_2$ following aging in argon and hydrogen at 815 °C [247]

γ-aluminides (TiAl) are less susceptible to hydrogen embrittlement compared to α$_2$-aluminides (Ti$_3$Al). No hydrogen embrittlement was found in the material TiAl48Nb2.5Ta0.3 even at 13.8 MPa compressed hydrogen in the temperature range from −130 °C to 204 °C. However, if the material is precharged with hydrogen (pressure of 13.8 MPa) at 815 °C (30 min) and subsequently the strength is determined at room temperature, a clear decrease of the mechanical properties can be found.

Table 188 shows the interstitial content of oxygen, nitrogen and hydrogen prior to and after aging in argon and hydrogen at 815 °C for the two microstructural states of the γ-aluminide XD™ (TiAl45V3 with 7.5 vol% TiB$_2$) [236, 247]. Although the hydrogen content increases with increasing aging time, a maximum of 239 wt-ppm is reached after 82 h. Moreover, the fracture toughness of the material TiAl45V3 prior to and after aging in Ar and H$_2$ at 815 °C was determined. As shown by the results in Table 189, although the fracture toughness of the lamellar microstructure is higher than that of the globular microstructure, hydrogen charging at 815 °C remains without almost any effect.

The hot isostatically pressed material TiAl48Cr2 (at.%) was examined under thermocyclic conditions while being simultaneously exposed to hydrogen gas [248]. The alloy had a deformed (α$_2$ + γ)-microstructure. The examinations were

	Fracture toughness in Ar K_I, MPa $\sqrt{m}$			Fracture toughness in H$_2$ K_I, MPa $\sqrt{m}$		
	0 h	2 h	82 h	0 h	0.5 h	2 h	82 h
Globular microstructure	–	–	13.45	–	13.25	–	11.67
	12.21	12.33	11.38	12.21	13.67	12.67	11.67
Lamellar microstructure	–	–	20.70	–	19.86	19.40	18.83
	23.99	23.45	19.91	23.99	21.66	21.18	20.60

Table 189: Influence of the aging time of the material TiAl45V3 with 7.5 vol% TiB$_2$ on the fracture toughness in hydrogen and argon at 815 °C [247]

Temperature change	Helium	Air	Hydrogen
25–900 °C	> 4,100	2,782	3
	> 4,145	2,647	10
	> 4,164	2,106	11
	> 4,626	> 4,000	24
	< 6,467	> 4,000	30
		> 4,000	
25–750 °C	> 4,108		36
	> 4,230		46
			1,828
			> 1,430
			> 3,000

2–6 tests each

Table 190: Cycles to failure under thermocycling conditions of the material TiAl48Cr2 (at.%) in He and H_2 [248]

performed in hydrogen (99.99%) and for comparison in helium (99.995%). Under thermocycling conditions (25–750 °C and 25–900 °C) the 1.3 mm thick flat tensile test specimens notched on one side were rapidly heated with a resistance-type heater and rapidly cooled in a gas stream. Each cycle lasted for 30 s with 10 s heating, 10 s holding and 10 s cooling. Preloading of the specimens was 241 MPa (50% of R_p at RT). The crack growth was observed with a special microscope. Table 190 contains the results for helium and hydrogen [248]. During the test in helium no fracture occurred and the tests were discontinued slightly above 4,000 cycles. In contrast, the specimens withstood only 3 to a maximum of 30 cycles under thermocycling conditions of 25–900 °C in hydrogen gas. At a temperature change in hydrogen gas between 25 and 750 °C the values strongly scattered and were between 36 and > 3,000 cycles. Specimens that withstood more than 50 cycles exhibited an oxide layer visible with the naked eye which, of course, blocks hydrogen absorption. It was assumed that this film was generated by the residual oxygen content or water vapor in the specimen chamber. The extent of hydrogen embrittlement obviously depends on the upper temperature limit and the purity of the hydrogen gas. In the presence of residual contents of oxygen in the hydrogen gas a porous TiO_2 layer is formed above 800 °C, not inhibiting hydrogen absorption, and below 800 °C an Al_2O_3 film is formed, which can block the absorption of hydrogen [248].

In addition, the material TiAl48Cr2 (at.%) was also subjected to thermocycling tests between 25 and 900 in air (water vapor content < 50 ppm) [249]. For comparison with hydrogen the results are also indicated in Table 190. Half of the specimens failed between 2,106 and 2,782 thermal cycles, whereas the other half survived 4,000 cycles. All specimens exhibited a thick, non-protective TiO_2-layer. It is assumed that air accelerates the thermomechanical fatigue through oxidation. The strong scatter of the values is attributed to surface defects (scratches, inclusions).

A 40 Zirconium and zirconium alloys

Similar to titanium, vanadium, niobium and tantalum, also zirconium and its alloys are susceptible to embrittlement by hydrogen. Technical zirconium which always contains hafnium is offered in different grades with the chemical analytical data and mechanical properties listed in Table 191 and Table 192.

Grade ASTM	Weight, %							
	Zr + Hf min.	Hf max.	Fe + Cr max.	H max.	N max.	C max.	O max.	Others
Zr 701 UNS R60701	99.5	4.5	0.05	0.005	0.025	0.05	0.16	
Zr 702 UNS R60702	99.2	4.5	0.20	0.005	0.025	0.05	0.16	–
Zr 704 UNS R60704	97.5	4.5	0.2–0.4	0.005	0.025	0.05	0.18	Sn 1.0–2.0
Zr 705 UNS R60705	99.5	4.5	0.20	0.005	0.025	0.05	0.18	Nb 2.0–3.0
Zr 706 UNS R60706	99.5	4.5	0.20	0.005	0.025	0.05	0.16	Nb 2.0–3.0

Table 191: Chemical compositions of zirconium materials according to ASTM (also VdTÜV-material data sheet 480-06.89 [250] for Zr 702)

Grade ASTM	Min. tensile strength N/mm²	Min. 0.2% yield strength N/mm²	Min. elongation %
Zr 701	340	220	15
Zr 702	379	207	16
Zr 704	413	241	14
Zr 705	552	379	16
Zr 706	510	345	20

Table 192: Mechanical properties of zirconium at room temperature [250]

Zr 702 is the zirconium material most frequently used in chemical engineering. Zirconium materials are also used in the reactor technology, fertilizer production and plastics production, the production of acetic acid, hydrochloric acid, alkalis and treatment of sulfuric acid.

The hafnium-free (< 0.01% Hf) Zircaloy zirconium alloys (Table 193) produced in complex and costly processes are mainly used in reactor construction.

Due to the production process Zirconium contains a certain amount of hydrogen and is able to absorb further hydrogen under operating conditions. If the low hydrogen solubility at room temperature is exceeded in zirconium, the hydrogen, which is no longer soluble, is precipitated as zirconium hydrides. Due to the texture generated during the thermomechanical production of sheets and pipes, these

Zircaloy®-1 UNS R60801	Zr-2.5Sn
Zircaloy®-2 UNS R60802	Zr-1.5Sn-0.5Fe-0.1Cr-0.05Ni-0.01Hf max.
Zircaloy®-4 UNS R60804	Zr-1.5Sn-0.2Fe-0.1Cr-0.01Hf max.
Zr-2.5Nb UNS R60901	Zr-2.5Nb-0.01Hf max.

Table 193: Hafnium-free zirconium alloys for the reactor industry [251]

hydrides tend to align as flat precipitates parallel to the thickness direction of the products. These brittle hydrides with critical stress intensity values K_{IC} of 1 MPa $\sqrt{m}$ to 3 MPa $\sqrt{m}$, however, only impair the fracture toughness of the material if their orientation is perpendicular to the applied stress, i.e. in case of a load in thickness direction. As a result of the increasing hydrogen solubility at higher temperatures these hydrides dissolve at higher operating temperatures and the released hydrogen diffuses to the sites of elevated stress and may accumulate there. During cooling hydrides are formed there again, which may also exhibit an orientation perpendicular to the operating stress. Stresses in these areas may increase up to a level whereby cracks are initiated. Incipient cracks support the orientation of the hydrides since the hydrogen diffuses to the triaxial stress areas at the crack tip. Under cyclic operation this process can repeatedly occur and cause the cracks to gradually grow further.

This delayed growth of the cracks over time is called delayed hydride cracking (DHC). The propagation of such DHC cracks can be schematically described as shown in Figure 258, where the crack velocity V is reflected as a function of the stress intensity factor K_I [252].

The DHC area is characterized by a limiting value K_{IH} below which cracks do not grow as well as an area in which V is virtually independent from K_I. Upon reaching of the fracture toughness K_{IC} the crack growth is unstable.

The growth of these cracks stops again above a critical temperature (T_{crit}) and can be traced by using acoustic emission measuring methods [253]. Table 194 shows the results obtained during fracture-mechanical examinations of cold-deformed DC specimens from a zirconium alloy with 2.5% niobium with different hydrogen contents (about 100 ppm and about 190 ppm). The specimens were loaded to calculated stress intensity factors of about 13 MPa $\sqrt{m}$ to 21 MPa & $\sqrt{m}$.

To date the DHC problem has become known only for zirconium alloys, however not for pure zirconium. Damage has been predominantly found in nuclear reactors, i.e. in the cladding tubes of the fuel elements, and has been the cause of a great many investigations into this problem [252, 254–266].

Zirconium alloys are used for these tubes due to their low absorption cross section for thermal neutrons, the relatively good high temperature strength and the good corrosion resistance in high-temperature water and steam. The atomic hydrogen required for the absorption in the metal or the deuterium is generated under

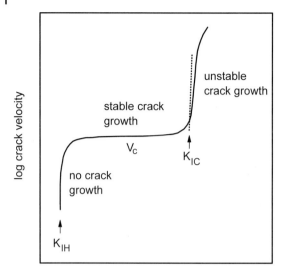

Figure 258: Curve of the crack velocity V as a function of the stress intensity factor K [252]

Test No.	Hydrogen ppm	Load kg	K_I MPa $\sqrt{m}$	$T_{crit.}$ °C
1	103, 116	130	17	296–314
2	187, 195	130	17	282–313
3	118, 146	130	17	328–359
4	118, 146	150	21	325–352
5	118, 146	100	13	280–313

Table 194: Critical crack arrest temperatures of DHC specimens from Zr-2.5%Nb with different hydrogen contents [253]

the operating conditions (pressures of about 11 MPa, temperatures from about 250 °C to 300 °C) in light water reactors or heavy water reactors by the reaction of zirconium with H_2O or D_2O according to:

Equation 22 $\quad Zr + 2 H_2O \rightarrow ZrO_2 + 4 H$

Equation 23 $\quad Zr + 2 D_2O \rightarrow ZrO_2 + 4 D$

Oxidic surface layers and dissolved oxygen reduce or prohibit the absorption of hydrogen, with thin layers being more effective than thick layers. By applying specific measures during the manufacturing process an arrangement of the hydrides parallel to the main stress direction may be obtained by making them less critical.

Although the damage in the cladding tubes of the fuel elements is attributed to hydrogen formed by electrochemical reactions, also examinations of the influence

Sheet	Sn %	Cr %	Fe %	O ppm	H ppm	Zr	R_p MPa
A (2 mm)	1.590	0.110	0.210	1,450	< 5	balance	484
B (1.6 mm)	1.391	0.103	0.193	1,410	< 5	balance	435

Table 195: Chemical compositions and yield strengths of the examined sheets from Zircaloy®-4 [267]

of compressed hydrogen on the zirconium materials were performed in this connection. In [267], for instance, examinations by means of slow strain rate tensile tests regarding the influence of compressed hydrogen on the transition from the ductile to the brittle fracture behavior of the alloy Zircaloy®-4 are described. The two commercial sheets examined had the composition as indicated in Table 195.

The tensile test specimens from sheet A were aged with hydrogen gas in the autoclave at 350 °C and at different pressures for different times and thus charged with hydrogen contents up to 350 ppm. Then the specimens were tested in air. The tensile tests of the specimens from sheet B were performed in hydrogen gas (> 99.9995%) at pressures of 101 kPa, 1,010 kPa and 2,020 kPa. All tensile test specimens were provided with a notch of 7 mm depth and a notch radios of 0.125 mm in the middle of the specimens.

All tests were performed at 25 °C, 100 °C, 200 °C and for the specimens charged with hydrogen also at 300 °C. The strain rate was about 3×10^{-7} 1/cm. The toughness behavior was assessed on the basis of reduction of area and the brittle fracture part on the fracture surface.

Figure 259 shows the results for the specimens charged with hydrogen. At room temperature a transition from ductile to brittle behavior occurs at a hydrogen con-

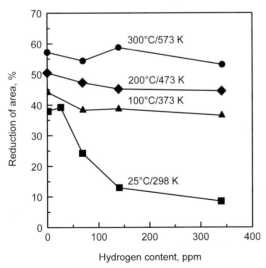

Figure 259: Influence of the hydrogen content on the reduction of area of slow strain rate tensile test specimens from the material Zircaloy®-4 at different test temperatures [267]

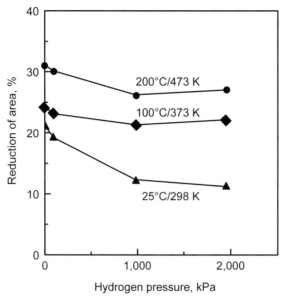

Figure 260: Influence of the hydrogen pressure on the reduction of area of slow strain rate tensile test specimens from the material Zircaloy®-4 at different test temperatures [267]

tent of 30 ppm and 100 ppm. At test temperatures of 100 °C and higher no significant transition to brittle behavior is observed. The authors attribute this effect to the higher toughness of zirconium with increasing temperature.

Figure 260 shows the reduction of area values obtained from the tensile test specimens tested in compressed hydrogen as a function of the test pressure.

Again a clear transition to brittle behavior is found at room temperature and a hydrogen pressure of 101 kPa and at 1,010 kPa. Higher hydrogen pressures do not exhibit any further negative impact. At higher temperatures the embrittling effect of the hydrogen atmosphere is considerably less. Plotting the values of reduction of area compared to the values obtained from a test in argon (Figure 261) makes the ductile-brittle transition at room temperature even clearer.

The hydrogen content of the specimens was determined after the test. As a result, the hydrogen absorption of the tensile test specimens in compressed hydrogen is low and amounted to a value below 30 ppm even at the highest pressure and the highest temperature.

Table 196 indicates the microscopically determined brittle fracture content on the fracture surface of the specimens. In contrast to the values of the reduction of area, the brittle fracture contents increase with increasing hydrogen pressure and increasing temperature. According to the fractographic examinations the clear ductile-brittle transition at room temperature can be attributed to the hydride precipitates and the cracks formed therein. Consequently, the ductility of the hydride precipitates is not improved by the temperature increase and the reduction of area is controlled by the ductility of the matrix of the zirconium at higher temperatures instead of the precipitates of brittle hydrides.

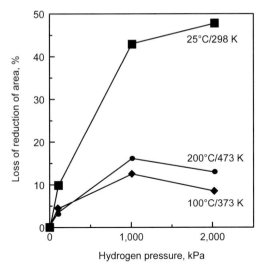

Figure 261: Loss of redeuction of area as a function of hydrogen pressure compared to the test in argon [267]

	101 kPa Ar	101 kPa H$_2$	1,010 kPa H$_2$	2,020 kPa H$_2$
25 °C	0%	0%	5%	8%
100 °C	0%	0%	7%	14%
200 °C	0%	0%	9%	21%

Table 196: Brittle fracture content on the fracture surface of the specimens from Zircaloy®-4 tested in compressed hydrogen [267]

To examine the synergistic effect of internal and external hydrogen, additional studies were performed during which Zircaloy®-4 specimens were charged with hydrogen contents from 90 ppm and 220 ppm by aging in compressed hydrogen (10.4 MPa) at 350 °C for a different duration and then tested in compressed hydrogen at different temperatures [268]. The composition and the mechanical properties of the tested 2.1 mm thick sheet are indicated in Table 197.

Sn %	Cr %	Fe %	O ppm	H ppm	N ppm	C ppm
1.58	0.12	0.21	1,420	< 5	31	163

T °C	R$_p$ MPa	R$_m$ MPa	A %
25	622	678	25
100	558	588	27
200	420	456	29

Table 197: Chemical composition and mechanical values of the tested sheet from Zircaloy®-4 [268]

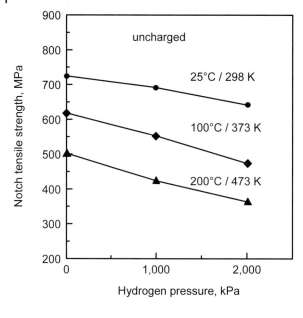

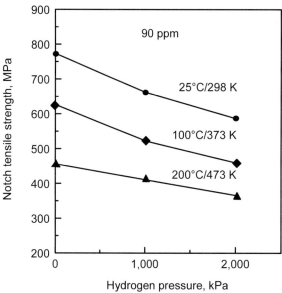

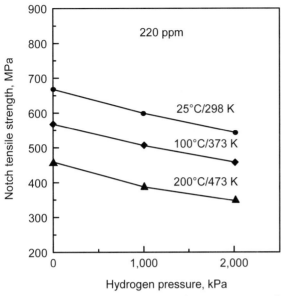

Figure 262: Influence of the external hydrogen pressure on the notch tensile strength of the uncharged specimens and the specimens precharged with hydrogen of 90 ppm and 220 ppm at the three test temperatures [268]

Again, the notched tensile test specimen was tested at a strain rate of 3×10^{-7} cm^{-1} in hydrogen gas at pressures of 1,010 kPa and 2,020 kPa at 25 °C, 100 °C and 200 °C. Figure 262 illustrates the influence of the external hydrogen pressure on the notch tensile strength of the uncharged specimens and the specimens precharged with hydrogen of 90 ppm and 220 ppm at the three test temperatures.

In all specimens the notch tensile strength decreases with increasing hydrogen pressure and increasing temperature. Virtually, this is independent of the fact whether the specimens contain internal hydrogen or not.

Figure 263 shows the values of the reduction of area obtained from the tensile test specimens.

In contrast to the values of the notch tensile strength, the reduction of area values at room temperature show a clear influence of internal hydrogen. At a test pressure of 1,010 kPa hydrogen the specimens not charged with hydrogen also exhibit a transition from ductile to brittle fracture behavior. In specimens with 90 ppm hydrogen this transition is shifted only slightly to lower values, whereas the reduction of area of the specimens charged with 200 ppm hydrogen practically drops to zero. The influence of both the internal and the external hydrogen on the toughness behavior of the specimens at higher test temperatures is only low.

Also the brittle fracture contents microscopically determined on the fracture surface of the specimens (Table 198) confirm the synergystic effects of high internal hydrogen contents and external compressed hydrogen, leading to a 100% brittle fracture at 25 °C and an internal hydrogen content of 200 ppm. This absolutely brittle behavior is attributed to the uniform distribution of hydride precipitates.

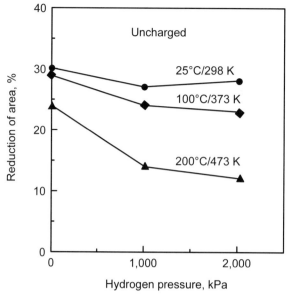

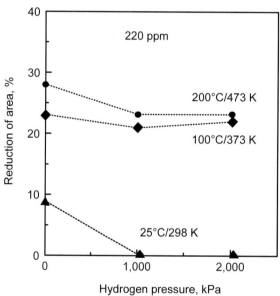

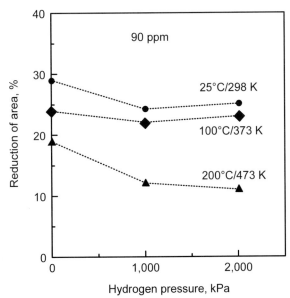

Figure 263: Influence of the external hydrogen pressure on the reduction of area of the uncharged specimens and the specimens precharged with hydrogen of 90 ppm and 220 ppm at the three test temperatures [268]

Temperature, °C		101 kPa Ar	1,010 kPa H_2	2,020 kPa H_2
25	uncharged	0%	4%	8%
	90 ppm H	0%	5%	9%
	220 ppm H	0%	100%	100%
100	uncharged	0%	7%	13%
	90 ppm H	0%	7%	16%
	220 ppm H	0%	7%	16%
200	uncharged	0%	8%	29%
	90 ppm H	0%	10%	20%
	220 ppm H	0%	9%	22%

Table 198: Brittle fracture content on the fracture surface of the uncharged specimens and specimens precharged with 90 ppm and 220 ppm hydrogen tested in compressed hydrogen from Zircaloy®-4 [268]

To examine the influence of oxide layers on the surface as well as the manufacturing conditions and the resulting structure of the material on hydrogen embrittlement, tensile tests were performed at room temperature using specimens from Zircaloy®-4 with the composition indicated in Table 199 following different heat treatments and with and without hydrogen charging [269].

% Sn	% Fe	% Cr	% O	ppm H	Zr
1.50	0.22	0.10	0.13	≤ 5	balance

Table 199: Chemical composition of the examined specimens from Zircaloy®-4 [269]

The 3.1 mm thick sheets were produced by forging at high temperature, quenching, hot rolling and annealing in the α-range as well as several cold-rolling steps. The cold-rolled sheets were tested following stress relief annealing (460 °C, 24 h), on the one hand, and recrystallization annealing (630 °C, 3 h), on the other hand. The tensile test specimens were taken transverse to the rolling direction and charged with hydrogen by aging for different durations (2 h to 120 h) at 400 °C in a gas mixture consisting of 2.5% hydrogen and 97.5% helium (or argon) with a pressure of 0.12 MPa. Although the formation of oxide layers is reduced by this treatment, it is not prevented, in particular not in case of lengthy aging times. The thickness of the oxide layers formed was between 2 µm and 15 µm.

The morphology of the hydrides resulting from this treatment strongly depends on the previous heat treatment. In stress-relief heat treated specimens continuously coherent hydrides precipitate in the rolling direction of the sheets and form a layer-like structure. The morphology of the recrystallized specimens is influenced by the grain size and the orientation of the grain boundaries and smaller and less continuously aligned hydrides in rolling direction are observed.

The tensile tests were performed at a strain rate of 0.5 mm/min in a helium atmosphere. In both conditions of the material hydrogen has a minor effect on the strength values of the tensile test specimens. The ductility values, i. e. elongation and reduction of area, however are clearly reduced above a critical hydrogen content as shown by the summarized results for the reduction of area in Figure 264.

In the specimens with an oxide layer the transition from ductile to brittle fracture characterized by a steep decrease in the reduction of area values occurs at a critical hydrogen content of 560 ppm for the stress-relief heat treated specimens and at 350 ppm hydrogen for the recrystallized specimens. Fractographic examinations reveal numerous incipient cracks in the oxide layer aligned parallel to the fracture surface, which led to the premature fracture of the specimens. If the oxide layer is removed by polishing prior to the test, significantly higher hydrogen contents are necessary to obtain equally low reduction of area values.

Other examinations of different zirconium alloys regarding the influence of the texture of the material, the amount and distribution of hydride precipitates, material properties, internal stresses and existing voids and internal cracks on the crack initiation and crack growth are described in [270–272].

Whereas the highest susceptibility to hydrogen embrittlement was found at room temperature in the slow strain rate tensile tests in compressed hydrogen, the problems caused by DHC in Zircaloy®-2 and Zircaloy®-4 under reactor conditions occurred in the temperature range from about 80 °C to 300 °C. Here, also very slow crack propagation velocities were observed. The maximum values measured were, for instance, 10^{-8} m/s at 260 °C and 6×10^{-8} m/s at 300 °C for Zircaloy®-2. In contrast, during fracture-mechanical tests on Zircaloy®-2 in high-pressure hydrogen crack propagation velocities were measured at 325 °C, which were higher by several orders of magnitude [273].

During these tests CT specimens notched on the side from the cold-rolled and recrystallized sheet were tested in hydrogen with a pressure of 52 bar at 325 °C. The crack length was continuously registered based on the potential drop at the

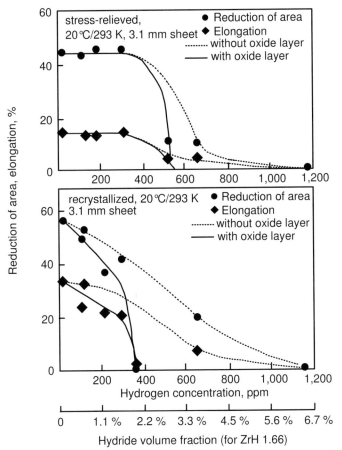

Figure 264: Influence of the hydrogen content on the ductility of Zirkaloy®-4 at room temperature [269]

crack. For comparison tests were performed in air at the same temperature. Figure 265 shows the measured crack extension as a function of time for the specimen in air with an initial load of $K_I = 34$ MPa $\sqrt{m}$ and the specimen in hydrogen with $K_I = 33$ MPa $\sqrt{m}$.

In the beginning the specimen in air exhibited a high crack propagation velocity followed by an area of slower crack propagation in which the crack propagation increases with increasing crack length until the specimen breaks after 5 minutes. The crack path of the specimen in compressed hydrogen is characterized by a very rapid increase of the crack propagation velocity immediately upon application of the load and rupture occurred after 5 seconds. As the results in Figure 266 show, the crack propagation rates strongly depend on the stress intensity at the crack and reach values of up to 10^{-3} m/s.

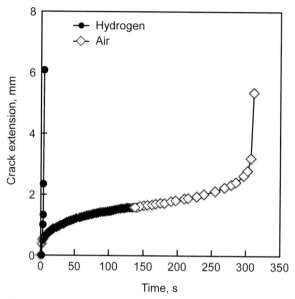

Figure 265: Crack extension of CT specimens from Zircaloy®-2 as a function of time in air and in compressed hydrogen (52 MPa) at 325 °C [273]

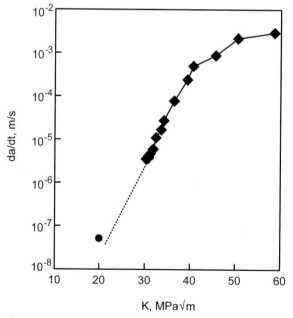

Figure 266: Post crack propagation rate da/dt as a function of the stress intensity factor for CT specimens from Zircaloy®-2 in compressed hydrogen at 325 °C [273]

A 41 Other metals and their alloys

Tungsten

Documents from manufacturers state that tungsten is resistant to hydrogen up to the highest temperatures [274].

Vanadium and vanadium alloys

As with all refractory metals vanadium and vanadium alloys also exhibit excellent mechanical properties at high temperatures and a low susceptibility to radiation damage. Therefore, they are of great interest as materials for the inner wall of fusion reactors. However, potential hydrogen embrittlement is a problem for this application since hydrogen and its isotopes are permanently present in these reactors.

Similar to the other refractory metals niobium and tantalum, also the toughness behavior of vanadium can be impaired by hydrogen. Since examinations regarding the influence of hydrogen on refractory metals most frequently also involved comparison tests on several of these metals and their alloys, the behavior of vanadium and vanadium alloys is partially also dealt with under A 37.

The influence of hydrogen on the toughness behavior of vanadium manifests itself as an increase in the yield stress with decreasing temperature (refer to Figure 169 and Figure 172), on the one hand, and a strong decrease in the reduction of area at temperatures below about –3 °C (refer to Figure 175), on the other hand. The effect described in [194], i.e. the effect of applied stresses on the temperature of formation or dissolution of hydrides, was confirmed in further studies [275]. Here, the influence of an external tensile stress close to the critical yield stress on the dissolution temperature of the hydrides in vanadium alloys was examined by means of transmission electron microscopy and resistance measurements. The dissolution temperature for the hydrides rises under tensile load by 5 °C at hydrogen contents of 0.036 at.% and by 16 °C at hydrogen contents of 0.5 at.%. Since the electrical resistance is directly proportional to the content of dissolved hydrogen (1.12 $\mu\Omega$ cm/at.% H), the hydrogen content was calculated by using resistance readings. To identify the role of the hydrogen during low-temperature hardening, the elastic limit and the electrical resistance of the tensile test specimens were measured at the same time. Figure 267 shows the example of the temperature dependence of the elastic stress limit for hydrogen-free specimens and for specimens charged with 0.2 at.% hydrogen.

Whereas a slow, continuous increase of the strength with decreasing temperature can be observed in the hydrogen-free specimens, the hardness increase of the specimens charged with hydrogen can be divided into sections. At temperatures above about –43 °C the strength depends on the temperature only to a minor extent, followed by a steep increase of the strength up to about –83 °C with decreasing temperature, the increase becoming slower again when the temperature is further reduced. The dissolution temperature of the hydrides in these specimens

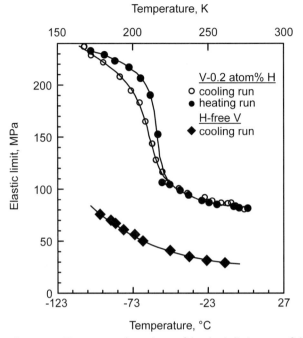

Figure 267: Temperature dependence of the elastic limit stress of charged and uncharged vanadium specimens [275]

was found to be −63 °C, i.e. clearly below the temperature of the steep strength increase. It can be concluded from these results that the hardness increase is attributable to the accumulation of hydrogen at the moving dislocations.

Also previous examinations had referred to the strong decrease in the reduction of area of vanadium in the temperature range below 0 °C as well as the strength increase at low temperatures [276]. These examinations had been performed using high-purity vanadium with the composition indicated in Table 200.

The specimens were charged with hydrogen to a content of 53 ppm at 600 °C. Figure 268 and Figure 269 show the load-elongation curve obtained for the uncharged and the charged specimens at different temperatures.

A significant influence of the hydrogen on the curve path cannot be found. Figure 270 compares the values for the uniform elongation and the reduction of area obtained for the specimens charged with 53 ppm hydrogen and the uncharged specimens.

Al	Ca	Cr	Cu	Fe	Mg	Mn
< 20	<< 30	150	< 25	400	< 15	< 25
Ni	Si	Ti	C	O	N	H
35	< 40	< 25	100	323	74	3

Table 200: Analytical values of the examined vanadium specimens in wt-ppm [276]

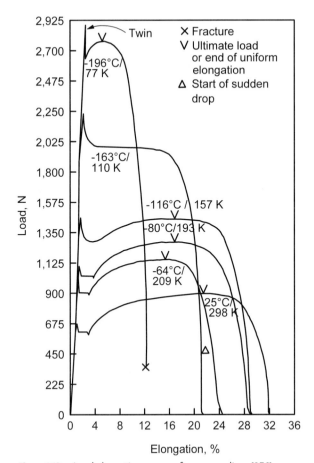

Figure 268: Load-elongation curves of pure vanadium [276]

Even at the lowest temperatures the uncharged specimens did not exhibit any indication of a transition from ductile to brittle fracture. Whereas the reduction of area of the uncharged material down to the temperature of –196 °C remains almost constant, elongation continuously decreases with decreasing temperature.

The absorption of hydrogen further deteriorates the toughness behavior of the specimens below –35 °C. Between –35 °C and –50 °C elongation drops to almost zero. The decrease of the reduction of area in the same temperature range is somewhat less, however reaches similar values. From about –128 °C the values characteristic for toughness increase again.

Regarding the vanadium-niobium alloys it is known that, firstly, the transition temperature from ductile to brittle fracture behavior increases with increasing hydrogen content and, secondly, the susceptibility to hydrogen embrittlement is mitigated by an increased alloy niobium content [277]. In both materials the hydrogen solubility increases by alloying the respective other material, the solubility maximum of 30 at.% being reached in the 50V-50Nb alloy. Despite these relatively high

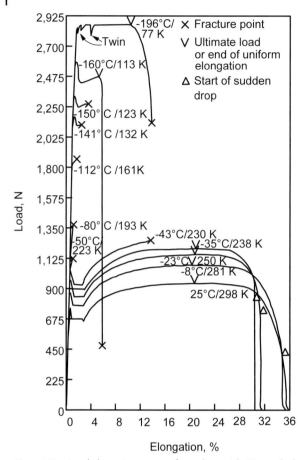

Figure 269: Load-elongation curves of vanadium with 53 ppm hydrogen [276]

hydrogen solubilities several of these alloys are susceptible to hydrogen embrittlement. The studies described in [277] examined the behavior of the alloys listed in Table 201 and the two pure metals following hydrogen charging.

To charge the tensile test specimens with hydrogen, the specimens were first aged in a vacuum chamber at 1,000 °C and 1.3×10^{-5} Pa to activate the surface. Following cooling down to 850 °C hydrogen obtained from the thermal decomposition of uranium hydride was added at the desired pressure. The closed system was kept at 850 °C for 1 h and, thereafter, slowly cooled down. The specimens so charged were analyzed and the gas contents indicated in Table 201 were determined. The hydrogen contents of the uncharged comparison specimens were below 0.03 at.%. The slow strain rate tensile tests of the charged slow strain rate tensile tests of the charged and uncharged specimens were performed at a constant strain rate of 8.3×10^{-5} 1/s.

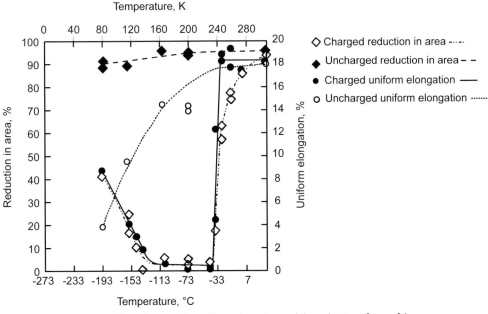

Figure 270: Temperature dependence of the uniform elongation and the reduction of area of the vanadium specimens charged with 53 ppm hydrogen and the uncharged vanadium specimens [276]

	O	N	H	C
100 V	0.021	0.0005	0.29	0.013
90V-10Nb	0.033	0.003	0.22	
75V-25Nb	0.036	0.0066	0.23	
50V-50Nb	0.041	0.013	0.32	
25V-75Nb	0.036	0.016	0.48	
15V-85Nb	0.085	0.038	0.581	
10V-90Nb	0.064	0.029	0.41	
100 Nb	0.073	0.017	0.55	0.015

Table 201: Composition (at.%) of the examined specimens in the vanadium-niobium system [277]

Figure 271 shows the values of the reduction of area as a measure for hydrogen embrittlement for the charged and uncharged tensile test specimens as a function of the test temperature.

The upper set of curves for the uncharged specimens shows that, in general, the materials remain very ductile throughout the temperature range examined. A slight decrease in the ductility is only found at very low temperatures. The bottom

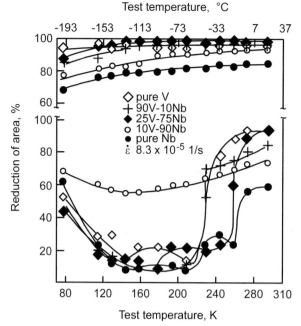

Figure 271: Temperature dependence of the reduction of area for charged (bottom) and uncharged (top) tensile test specimens from vanadium, niobium and several V-Nb alloys [277]

set of curves for the specimens with hydrogen contents of 0.2 at.% to 0.6 at.% demonstrates the drastic drop in ductility at temperatures below room temperature both for the two pure materials and the alloys. Exceptions were only the alloys 10V-90Nb and 15V-85Nb (the latter not shown in the figure) with a much lower ductility drop. Basically, an increase to mean ductility values can be observed in all examined materials.

Another presentation of the test results (Figure 272) shows the values of the reduction of area for three test temperatures depending on the composition of the alloy.

With increasing alloy content the reduction of area of the specimens not charged with hydrogen clearly decreases. Among the charged specimens, the 50V-50Nb alloy exhibits a particularly strong decrease of the reduction of area values at room temperature. The ductility properties of the alloys 10V-90Nb and 15V-85Nb remain on a medium level even at −113 °C, the same also applying to all materials at −195 °C.

According to these results the absorption of hydrogen in the vanadium-niobium alloys with higher vanadium contents leads to strong embrittlement, whereas the alloys with higher niobium contents are less sensitive. Precipitation of hydrides was not detected in the examined materials. Crack initiation starts at the grain boundaries and the crack path is predominantly intercrystalline.

Relevant examinations were also performed for vanadium-titanium alloys indicated in Table 202 [278]. The tensile test specimens from these alloys were charged

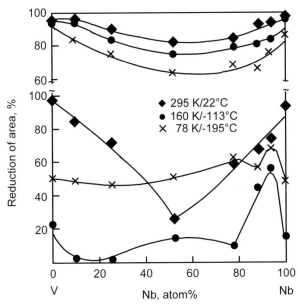

Figure 272: Relationship between the composition of the V- Nb alloys and the ductility of the charged (bottom) and uncharged (top) tensile test specimens [277]

	Ti	O	N	H 1st series	H 2nd series	H 3rd series	H 4th series
V–1Ti	0.7	0.036	0.002	0.33	1.12	2.16	4.66
V-5Ti	4.7	0.021	0.002	0.33	1.12	2.16	4.66
V-10Ti	9.8	0.044	0.002	0.33	1.12	2.16	4.66
V-20Ti	20.2	0.030	0.003	0.33	1.12	2.16	4.66
V-30Ti	29.2	0.041	0.006	0.60	1.12	2.16	4.66

Table 202: Compositions (at.%) of the examined vanadium-titanium alloys [278]

with hydrogen by using the same method as that for the vanadium-niobium alloys. The slow strain rate tensile tests of the charged specimens and the uncharged comparison specimens were again performed at the constant strain rate of 8.3×10^{-5} 1/s.

The embrittlement combined with a ductility loss of the specimens charged with hydrogen, i.e. the transition from ductile to brittle behavior, occurs is a relatively narrow temperature range as can be seen already from the stress-strain curves. Figure 273 shows this using the example of load-elongation curves of the V-1Ti alloy charged with 0.3 at.% hydrogen for various test temperatures.

Above a temperature of −63 °C the alloy shows the typical curve paths with a generally uniform elongation prior to fracture and the specimens show the respective necking. At lower test temperatures of −78 and −93 °C the specimens suddenly

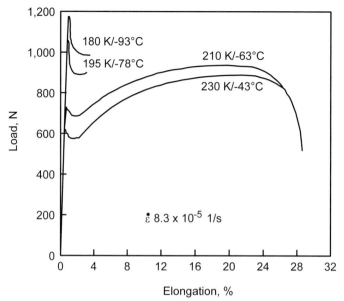

Figure 273: Load-elongation curves of the V-1Ti alloy with 0.3 at.% hydrogen at different test temperatures [278]

break after some few percent of elongation and the reduction of area is correspondingly lower.

The ductility values obtained for the unalloyed vanadium and the vanadium-titanium alloys are shown in Figure 274 as a function of the test temperature for hydrogen-free (< 0.05 at.% H) specimens and specimens charged with 0.3 at.% hydrogen.

The upper curves for the hydrogen-free specimens show that the alloys with 1% titanium and with 5% titanium exhibit practically the same behavior as unalloyed vanadium. In contrast, the curve for the V-20Ti alloy shows significantly lower values for the reduction of area and also a sharp decrease in the values with decreasing temperature. Also the curves for the two alloys V-10Ti and V-30Ti, which are not shown in the picture, have a similar path, the V-10Ti curve running above and the V-30Ti curve running below the V-20Ti curve.

The curves for the specimens charged with 0.3 at.% hydrogen in the bottom part of the figure clearly shows the embrittling effect of hydrogen at lower temperatures. However, the damaging effect of hydrogen is mitigated as the titanium content of the alloys increases. Both the V-1Ti and the V-5Ti alloys show a sharp decrease in the ductility with increasing temperature followed by a temperature range with very brittle behavior and a return to a more ductile behavior at still lower temperature similar to the behavior of the unalloyed vanadium. In contrast, the path of the curve for the V-20Ti alloy charged with hydrogen is generally similar to that of the hydrogen-free alloy, i.e. a titanium content of 20 at.% makes the vanadium-titanium alloys generally unsusceptible to hydrogen embrittlement, at least at hydrogen contents of 0.3 at.%.

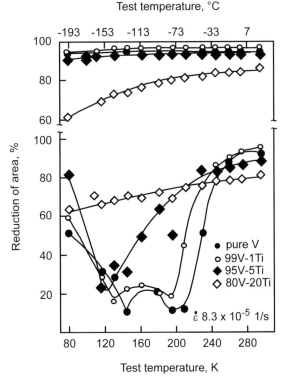

Figure 274: Temperature dependence of ductility (reduction of area) for charged (bottom) and uncharged (top) tensile test specimens from unalloyed vanadium and several vanadium-titanium alloys [278]

As an example for the specimens charged with higher hydrogen contents Figure 275 shows the ductility (reduction of area)-temperature curves for the V-10Ti alloy. As can be seen from the results, sufficiently high hydrogen contents make also this alloy brittle at low temperatures despite its high titanium content. The temperatures at which embrittlement occurs is higher the higher the hydrogen content of the specimens is. At low temperatures hydride precipitates were only found in the unalloyed titanium and the V-10Ti alloy. The higher-alloyed materials did not exhibit any hydride precipitates, not even at the high hydrogen contents and the lowest temperature of −195 °C. In the hydride-precipitating alloys the cracks take an intercrystalline path, whereas they are transcrystalline in the alloys not forming any hydrides.

Relevant examinations to study the influence of hydrogen on the toughness behavior were also performed on vanadium-chromium alloys [279]. The examined materials are summarized in Table 203. The hydrogen contents reached in the specimens charged with the method as described for the vanadium-niobium alloys are also mentioned. The hydrogen content of the uncharged specimens was below 0.02 at.%. The slow strain rate tensile tests of the charged specimens and of the uncharged comparison specimens were again performed at the constant strain rate of 8.3×10^{-5} 1/s.

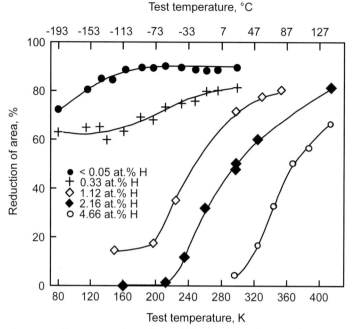

Figure 275: Temperature dependence of the ductility (reduction of area) for tensile test specimens from V-10Ti alloy charged with different hydrogen contents [278]

All specimens with hydrogen contents of 0.07 at.% were metallographically checked for hydride precipitates during cooling from room temperature to −195 °C. With the exception of the V-20Cr alloy, precipitates of hydrides were found in all cases. As the results also indicated in Table 203 show, the temperature for the formation of hydrides decreases with increasing chromium content of the alloy at constant hydrogen content.

	O at.%	N at.%	H at.%	T_H °C
V	0.05	0.010	0.069	−105 ± 7
V-2.5Cr	0.05	0.008	0.065	−123 ± 5
V-5Cr	0.06	0.007	0.075	−131 ± 7
V-10Cr	0.02	0.001	0.071	−149 ± 7
	0.02	0.001	0.180	−123 ± 4
V-20Cr	0.02	0.002	0.064	*)
	0.02	0.001	0.160	*)

T_H = temperature of hydride precipitation
* = no hydride precipitation down to −195 °C

Table 203: Compositions of the examined specimens in vanadium-chromium alloys and temperature of hydride precipitates [279]

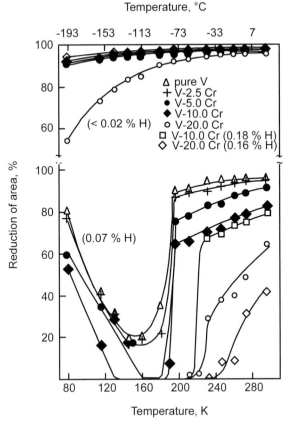

Figure 276: Influence of chromium, hydrogen content and temperature on the reduction of area of vanadium and vanadium-chromium alloys [279]

Figure 276 summarizes the ductility values as a function of the test temperature for all specimens as obtained from the slow strain rate tensile tests. The curves in the upper part of the figure apply to the hydrogen-free specimens and show that chromium contents of up to and including 10 at.% exert only a minor influence on the ductility behavior of vanadium in the temperature range between 22 °C and – 195 °C. However, the vanadium alloy with 20 at.% chromium exhibits a continuous drop in the ductility with a falling tendency in this temperature range.

At hydrogen contents of 0.07 at.% (curves in the bottom picture) all specimens except the V-20Cr alloy exhibit a steep drop in the reduction of area values in a narrow temperature range together with minimum ductility and a return of the ductile behavior at very low temperatures. The V-20Cr alloys becomes brittle with increasing hydrogen content from 0.07 to 0.16 at.% at temperatures of – 43 °C and –63 °C, respectively, and does not exhibit any return to ductile behavior even at low temperatures. Also in the V-10Cr alloy the higher hydrogen content of 0.17 at.% causes an increase in the critical temperature for the transition from ductile to brittle behavior.

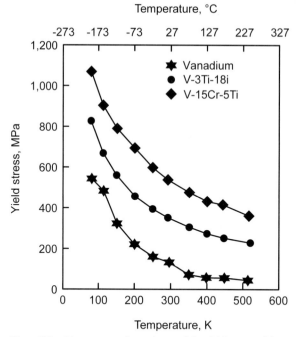

Figure 277: Temperature dependence of the yield stress of the specimens not charged with hydrogen from vanadium and the alloys V-3Ti-1Si and V15Cr-5Ti [280]

In all cases the fractures in the hydrogen-charged specimens are caused by the surface cracks, which take a transcrystalline path in the hydride-forming materials, whereas the V-20Cr alloy not forming hydrides exclusively shows intercrystalline surface cracks.

Examinations to study the influence of the hydrogen content on the mechanical properties of unalloyed vanadium and the alloys V-3Ti-1Si and V-15Cr-5Ti are described in [280]. To this end, slow strain rate tensile tests were performed on specimens from the three mentioned materials at a constant strain rate of 1.67×10^{-3} 1/s in the temperature range from -196 °C to 247 °C. Some of the specimens were charged with hydrogen using Sievert's method described under A 38 for titanium. The hydrogen content of the specimens following charging was 0.3 at.%. The temperature dependence of the yield stress and the elongation of the specimens not charged with hydrogen is shown in Figure 277 and in Figure 278.

Similar to vanadium the two alloys also show a clear increase of the yield strength with decreasing test temperature. The low elongation values for vanadium at −196 °C and −160 °C were attributable to a local necking and the specimens remained ductile down to the lowest temperature.

The temperature dependence of the elongation of the specimens charged with hydrogen is shown in Figure 279.

The specimens charged with 0.3 at.% hydrogen show a steep drop of the elongation and a transition from ductile to brittle fracture behavior below -123 °C for the

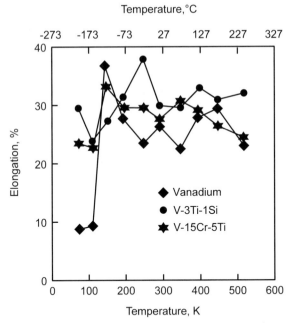

Figure 278: Temperature dependence of the elongation of the specimens not charged with hydrogen from vanadium and the alloys V-3Ti-1Si and V15Cr-5Ti [280]

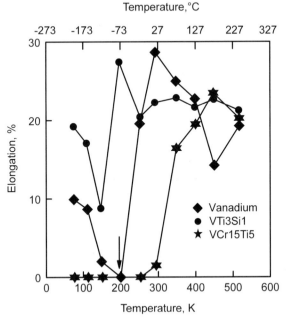

Figure 279: Temperature dependence of the elongation on hydrogen-charged specimens from vanadium and the alloys V-3Ti-1Si and V15Cr-5Ti [280]

alloy V-3Ti-1Si, at -73 °C for unalloyed vanadium and at 27 °C for the alloy V-15Cr-5Ti. In contrast to the alloy V-15Cr-5Ti the elongation values of vanadium and the alloy V-3Ti-1Si are higher again at even lower temperatures.

Uranium

Due to its high density and special nuclear properties uranium is used in various special applications. The demand for materials with medium and high densities of more than 18 g/cm³ has led to the development of various simple binary uranium alloys. By the addition of the elements from the 4th, 5th and 6th subgroups of the Periodic Table to uranium, alloys with the metallurgical diversity of iron alloys are formed which, typically, have densities around 17 g/cm³. Alloys with less than 6 mole% of alloying elements are called low-alloy uranium alloys and are usually high-strength materials with properties obtained by specific heat treatments. Uranium alloys are typically annealed at 800 °C to 900 °C to achieve a homogeneous distribution of the alloying elements in the body-centered cubic lattice (γ-phase). Annealing usually takes place in vacuum to remove the internal hydrogen from the smelt. Subsequently, the material is quenched in water and tempered at temperatures below 650 °C or cooled in the furnace to about 600 °C.

Hydrogen can make these alloys brittle, the hydrogen being absorbed both in the production process and from the environment during operation. With high temperature heat treatments hydrogen from the environment can diffuse into the alloy up to several millimeters. During cooling the hydrogen dissolved at the high temperature is bound as a hydride phase. When deforming the material, this internal hydrogen can impair the deformation capacity and lead to cracks. The same effect is also found for hydrogen absorbed from the environment during loading.

To study the effect of hydrogen on the deformation behavior of uranium alloys, the two alloys U-0.8Ti and U-2.3Nb were charged at different hydrogen pressures and the values of elongation and the reduction of area were determined in tensile tests in comparison with the uncharged specimens [281].

The test material from the alloy U-0.8Ti (from pure titanium 99.97% U + 0.8% Ti) was annealed in vacuum at 800 °C, quenched in water and tempered at 380 °C for 4 h. The round test bars were enclosed in the evacuated quartz tubes and aged in vacuum in the furnace at 800 °C for 2 h to expel the internal hydrogen. Thereafter, the specimens were aged in hydrogen with different pressures between 0.1 mPa and 100 kPa at 800 °C for another 2 h. According to the diffusion coefficient of 8×10^{-9} m²/s at 800 °C hydrogen contents from 0.001 ppm to 22 ppm can be absorbed in this way and remain homogeneously distributed in the material by subsequent quenching in water. The test material from the alloy U-2.3Nb (from pure titanium 99.97% + 2.3% Nb) was annealed in vacuum at 850 °C and thereafter cooled in the furnace to 630 °C. The round test bars were degassed at 0.1 mPa and 850 °C and then charged with hydrogen under a controlled pressure. The tensile test specimens were tested at a tension rate of 0.06 mm/s. The test was per-

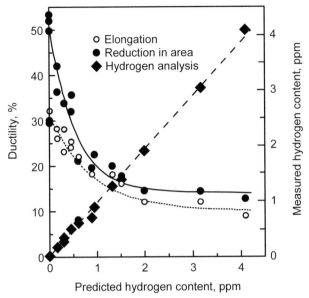

Figure 280: Elongation and reduction of area of the tensile test specimens from the alloy U-0.8Ti as a function of the hydrogen content of the specimens [281]

formed on specimens from the alloy U-0.8Ti in dry air and the specimens from the alloy U-2.3Nb in air at 40% relative moisture.

The values of elongation and reduction of area determined in the specimens are shown in Figure 280 for the alloy U-0.8Ti and in Figure 281 for the alloy U-2.3Nb as a function of the hydrogen content of the specimens.

The ductility characterized by fracture elongation and reduction of area clearly decreases at low hydrogen contents below 1 ppm. Compared to elongation the influence on the reduction of area is stronger. Accordingly, the fracture pattern changes from ductile fracture with the formation of dimples to brittle fracture with an increasing content of cleavage surfaces as shown by the fractographic fracture tests.

Although the strength values of the alloy U-2.3Nb are lower compared to the alloy U-0.8Ti, its ductility is impaired in a very similar manner by the hydrogen absorbed during aging.

An all cases a very good agreement was found between hydrogen contents calculated on the basis of the solubilities, diffusion coefficients and equilibrium conditions as well as mass spectrometrically determined hydrogen contents. For the alloy U-2.3Nb specimens with the lowest hydrogen content were additionally tested to find out whether hydrogen is also absorbed during long-term aging of the specimens in water. As the results in Figure 282 show, this aging does not exert any influence on the elongation and the reduction of area up to a duration of 42 months.

Internal hydrogen also impairs the toughness of the alloy U-5.7Nb with a higher niobium content, however with a different structure and lower strength value. In

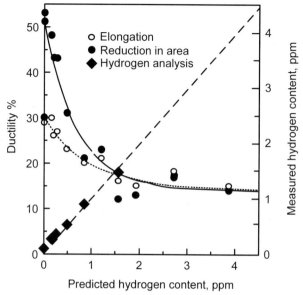

Figure 281: Fracture elongation and reduction of area of the tensile test specimens from the alloy U-2.3Nb as a function of the hydrogen content of the specimens [281]

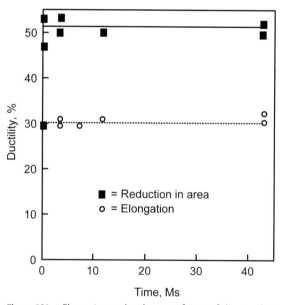

Figure 282: Elongation and reduction of area of the tensile test specimens from the alloy U-2.3Nb following aging in water over a period of several months [281]

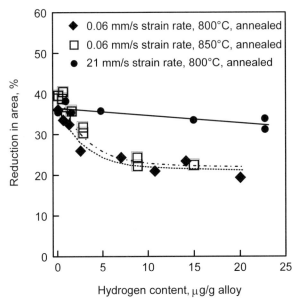

Figure 283: Influence of hydrogen on the reduction of area of tensile test specimens from the alloy U-5.7Nb at different strain rates [282]

this connection, Figure 283 shows the influence of the hydrogen content on the reduction of area of tensile test specimens, which were tested at different strain rates [282]. First, the specimens were solution-annealed in vacuum at 800 °C for 1.5 h, thereafter stored in a hydrogen atmosphere at various pressures at 800 °C for another 1.5 h and finally quenched in water to obtain different hydrogen contents. The maximum hydrogen content of 1.14 ppm is obtained at a hydrogen pressure of 101 kPa (1 bar). The tensile tests were performed both at a relatively high strain rate (21 mm/s) and a very low strain rate (0.06 mm/s).

Under the conditions of a slow strain rate, the reduction of area strongly decreases already at low hydrogen contents of the specimens from an initial value of 50% to assume values of 30% at a hydrogen content of about 10 µg/g, Higher hydrogen contents exert only a minor effect. During high strain rate tensile tests the effect of hydrogen is hardly detectable.

The alloy U-0.8Ti was also tested to determine the influence of very low contents of hydrogen on the toughness values depending on the strain rate of the tensile test specimens and the test temperature [283]. Following annealing in the γ-range (> 800 °C) the alloy was quenched in water to obtain a supersaturated martensitic structure and then tempered in the temperature range from 300 °C to 400 °C to obtain the desired mechanical properties. In this process the martensitic structure is generally preserved and the increase of strength is achieved by fine U_2Ti precipitates. Specimens with hydrogen contents of 0.06 ppm, 0.16 ppm and 1.14 ppm hydrogen were examined. The strain rates during the slow strain rate tensile tests ranged from 10^{-6} 1/s to 10^0 1/s and the test temperatures were in a range from –35 °C to 110 °C. The tests were performed in dry air or in vacuum to exclude hydro-

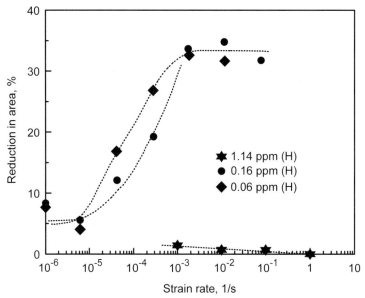

Figure 284: Reduction of area of tensile test specimens from U-Ti0.8 with different hydrogen contents at room temperature as a function of the strain rate [283]

gen absorption during the test. The influence of the hydrogen was evaluated on the basis of the reduction of area and the fractographic fracture pattern. The results of the tensile tests performed at room temperature are shown in Figure 284.

At the highest hydrogen content of 1.14 ppm tested, the reduction of area values are very low even in relatively high strain rate tensile tests. At lower hydrogen contents the decrease in the reduction of area and the transition from ductile to brittle fracture behavior starts at a strain rate of about 10^{-3} 1/s, the higher hydrogen content exerting a stronger negative effect. At very low strain rates even the very low hydrogen content of 0.06 ppm leads to the same low values as the content of 0.16 ppm. This is indicative of the fact that the mechanism of embrittlement is controlled by the diffusion of hydrogen.

Figure 285 shows the influence of the test temperature on the reduction of area of the specimens with 0.16 ppm hydrogen at different strain rates.

Reducing the temperature to 0 °C shifts the critical strain rate of the transition from ductile to brittle fracture behavior from 10^{-3} 1/s at 110 °C or 20 °C to 10^{-2} 1/s. At a test temperature of −35 °C the reduction of area values are very low irrespective of the strain rate. At 110 °C the drop of the reduction of area values starts also at a strain rate of 10^{-3} 1/s, but the drop is slightly lower up to a strain rate of 10^{-5} 1/s compared to 20 °C.

Fracture-mechanical tests of the alloy U-0.75Ti in a hydrogen atmosphere show that there is a direct dependence of the crack propagation velocity on the hydrogen pressure [284]. The composition and the mechanical properties of the alloy obtained after different heat treatments of the rolled sheet metal specimens are indicated in Table 204 and Table 205.

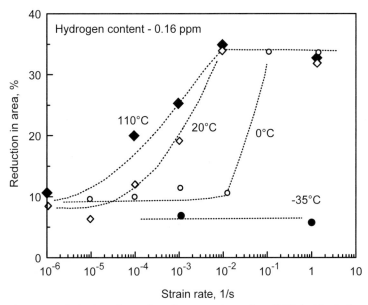

Figure 285: Reduction of area of tensile test specimens from U-Ti0.8 with a hydrogen content of 0.16 ppm at different test temperatures as a function of the strain rate [283]

Ti	Si	Fe	C	N	O	H
%	ppm	ppm	ppm	ppm	ppm	ppm
0.76	17	26	15	60	64	3

Table 204: Composition of the examined alloy U-0.75Ti [284]

Heat treatment	$Re_{0.2}$ MN/m²	Rm MN/m²	Elongation %
800 °C/water	704	1,330	14
800 °C/water 357 °C – 8 h	880	1,440	11
800 °C/water 400 °C – 42 h	1,110	1,600	5

Table 205: Mechanical properties of the examined alloy U-0.75Ti after different heat treatments [284]

The test was performed on DCB specimens in a pressure chamber first evacuated to a pressure of 1.3×10^{-9} bar prior to charging with hydrogen. The dependence of the crack propagation velocity on the hydrogen pressure was determined in the pressure range from 1.3×10^{-3} bar to 0.78 bar at a test temperature of 22 °C.

The results in Figure 286 for the example of sheet specimens show that the crack propagation velocity is proportional to the square root of the hydrogen pressure and increases with increasing strength of the material.

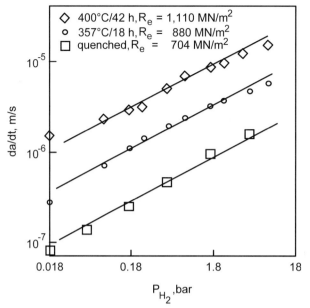

Figure 286: Influence of the hydrogen pressure on the crack propagation velocity of the sheet metal specimens from the alloy U-0.75Ti [284]
◇ K = 12 MN/m$^{3/2}$
□, ○ K = 27 MN/m$^{3/2}$

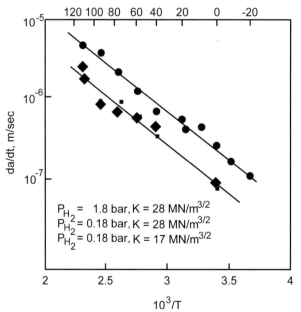

Figure 287: Influence of the temperature on the crack velocity in a hydrogen atmosphere of specimens from U-0.75Ti aged at 357 °C for 18 h [284]

The exponential dependence of the crack velocity in a hydrogen atmosphere on the temperature can be seen in Figure 287.

The studies described in [285] were performed on the uranium alloys listed in Table 206 in connection with the hydrogen content of tensile test specimens, the

	U-10Mo	U-8.5Nb	U-10Nb	U-7.5Nb2.5Zr
% Mo	9.81			
% Nb		8.51	10.33	7.81
% Zr				2.55
C/µg/g	279	59	52	82
N/µg/g	33	35	31	13
O/µg/g	45	66	63	50
Fe/µg/g	118	20	15	70
Ni/µg/g	40	6	4	15
Si/µg/g	96	60	20	60
$Re_{0.2}$/MPa	990	610	1,010	510
Rm/MPa	1,030	910	1,100	880
Hardness/Rc	46	40	49	42

Table 206: Chemical compositions and strength properties of the examined uranium alloys [285]

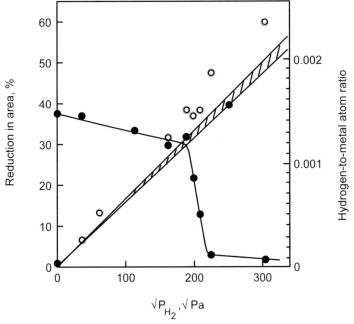

Figure 288: Relationship between reduction of area, hydrogen/metal ratio and hydrogen pressure of the alloy U-10Mo being aged at 850 °C [285]
● reduction of area, ○ hydrogen/metal ratio

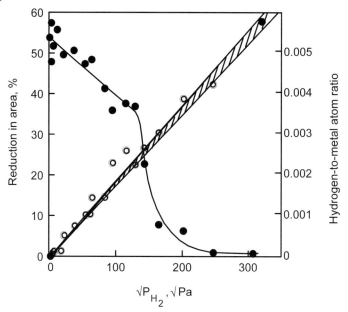

Figure 289: Relationship between reduction of area, hydrogen/metal ratio and hydrogen pressure of the alloy U-8.5Nb being aged at 850 °C [285]
● reduction of area, ○ hydrogen/metal ratio

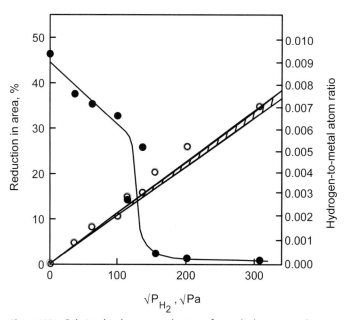

Figure 290: Relationship between reduction of area, hydrogen/metal ratio and hydrogen pressure of the alloy U-10Nb being aged at 850 °C [285]
● reduction of area, ○ hydrogen/metal ratio

hydrogen pressure during aging at 850 °C, reduction of area in the slow strain rate tensile test and the fracture characteristics.

Following melting and casting, the alloys were annealed in the temperature range between 800 °C and 1,000 °C and from this γ-range quenched to obtain a uniform and homogeneous γ-phase. The test was performed on unnotched round test bars with a strain rate of $\varepsilon = 8.3 \times 10^{-5}$ 1/s.

Figure 288 to Figure 291 summarize the results for the individual alloys.

These results show that the toughness of the three alloys U-10Mo, U-8.5Nb and U-10Nb decreases only slightly in the beginning as the hydrogen content increases until the reduction of area value drops to almost zero after a critical hydrogen concentration has been reached. Only the alloy U-7.5Nb2.5Zr did not exhibit this steep drop of the toughness. The authors attribute this phenomenon to the fact that the critical hydrogen content for the steep toughness drop increases with increasing hydrogen solubility in the alloy and that in this alloy the critical hydrogen content was probably not reached under the given test conditions.

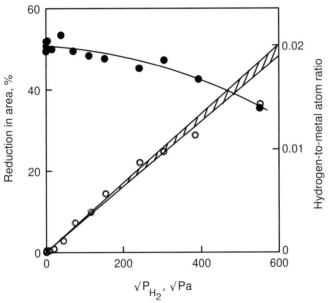

Figure 291: Relationship between reduction of area, hydrogen/metal ratio and hydrogen pressure of the alloy U-7.5Nb2.5Zr being aged at 850 °C [285]
● reduction of area, ○ hydrogen/metal ratio

B
Nonmetallic inorganic materials

B 3 Carbon and graphite

The carbon materials used for the bearing technology without impregnation or with synthetic resin impregnation, polytetrafluoroethylene (PTFE) or metal are resistant to hydrogen at room temperature [286].

The bipolar plates (trade name: e.g. Sigracet® BPP) used in polymer electrolyte membrane fuel cells (PEMFC) consist of graphite and a binding agent. On the gas feed side hydrogen gas flows through the bipolar plates. Depending on the type of the binding agent, the following maximum application temperatures are recommended: 80 °C for bipolar plates with polypropylene as a binding agent, 120 °C with polyvinylidene fluoride and 180 °C with phenolic resin [287]. For better distribution of the supplied hydrogen gas and the reaction products to be discharged a 0.2–0.3 mm thick, high-porosity gas diffusion layer of carbon fibers with low binding agent contents is provided between the bipolar plate and the polymer electrolyte membrane.

Impregnated graphite used for the production of apparatus components is suited for gaseous hydrogen applications of any concentration up to 180 °C [288].

For pure graphite a resistance to hydrogen is indicated up to 1,200 °C; however, in the presence of catalytically acting impurities, material consumption can occur under the influence of hydrogen already at a temperature from 800 °C [289].

B 4 Binders for building materials (e.g. concrete, mortar)

A hydraulically binding cement (trade name: Gorka® 40) consisting of not less than 40% Al_2O_3, not less than 36% CaO, 2–4% SiO_2 and 10–14% Fe_2O_3 and having calcium monoaluminate as the mineralogical main phase is used as a binding agent in mining due to its very good resistance to hydrogen carbon dioxide and methane [290].

B 6 Glass

B 7 Fused silica and silica glass

Hydrogen can be stored [291]:

- in tanks
- cryogenically
- in hydride storage systems ($TiFe_{0.9}Mn_{0.1}H_x$).

A more cost-effective alternative is the storage of hydrogen in glass microspheres described in [291]. To this end, glass microspheres (diameter 25 µm, wall thickness ≈ 2 µm) consisting of 80.9% SiO_2, 2.3% B_2O_3, 8.6% CaO and 8.2% Na_2O (figures in mole%),are used. These glass microspheres are charged with 20–40 MPa compressed hydrogen at 500–700 K. Hydrogen diffuses through the glass walls and fills the microspheres. After cooling to room temperature the hydrogen is stored under pressure in the microspheres and can be released again if needed by heating to about 500 K.

Here, the strong temperature dependence of the diffusion rate of hydrogen in glass is used. The results regarding the influence of the temperature on the glass permeability of hydrogen and helium are available, for instance, for silica glass (Table 207).

°C	He	H_2	°C	He	H_2
150	0.55	–	600	12.3	1.07
200	1.04	0.016	700	16.4	1.89
300	2.36	0.074	800	21.4	3.19
400	4.61	0.28	900	27.2	4.8
500	7.8	0.94	1,000	34.1	7.5

Data from: Hereaeus Quarzschmelze, data on silica glass and fused silica

Table 207: Gas permeability of silica glass for helium and hydrogen at different temperatures in 10^{-10} (normal cm³ × mm)/(s × cm² × mbar) [292]

Figure 292 contains a survey of the influence of the temperature on the diffusion coefficient D for hydrogen and other gases as well as for various ions and molecules in silica and silicate glasses [292].

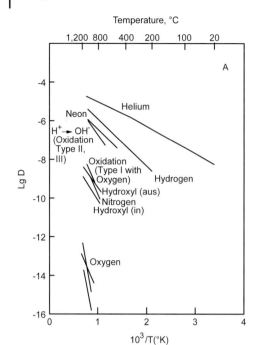

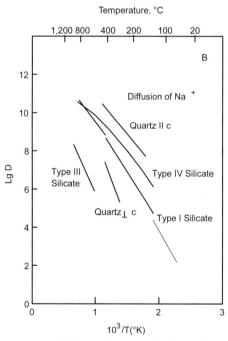

Figure 292: Diffusion of atoms, molecules and ions in silicate glasses (A), diffusion of sodium ions in silica and silicate glasses (B) [292]

B 8 Enamel

To the extent it is fully pore-free, enamel is impermeable to gases, including hydrogen. However, during the enameling process hydrogen is formed, which is absorbed and subsequently stored by the steel and released again during cooling of a component enameled on one side [293]. Figure 293 illustrates this process.

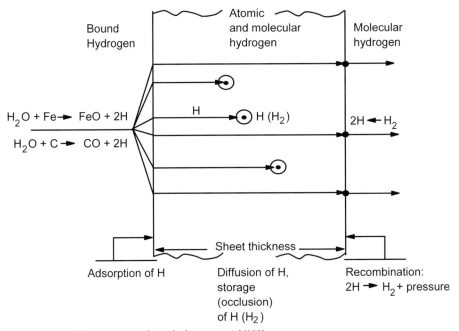

Figure 293: Hydrogen passage through sheet material [293]

According to this concept the important steps are:

- absorption of atomic hydrogen on the sheet surface
- subsequent diffusion of H into the sheet
- accumulation (occlusion) of a portion in atomic or molecular form in the voids in the microstructure, e.g. grain boundaries, lattice vacancies, phase boundaries between ferrite and cementite and in nonmetallic inclusions
- release of non-accumulated H in the sheet with recombination of atomic to molecular H occurring on the sheet surface
- under unfavorable conditions a high pressure gradually builds up at the sheet/enamel phase boundary that enamel spallings in the form of fish scales occur.

Fish scales are crescent shaped flakes of enamel which do not only impair the appearance of enameled objects but also their corrosion protection.

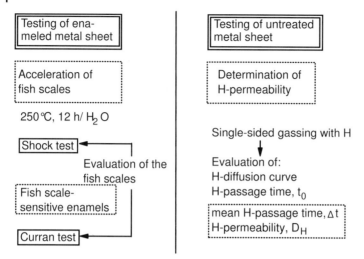

Figure 294: Test for fish scale susceptibility [293]

VOEST-ALPINE perform the following tests for evaluating fish scale susceptibility:

1) Two different fish scale tests of different severity on enameled specimens (shown left in Figure 294):

   a) Shock test
      application of base and cover enamel, aging at 250 °C for 12 h, quenching in water.
   b) Curran test
      The specimens are enameled using an enamel with high fish scale susceptibility (curran powder).

   In both cases an evaluation is made as to whether and with what intensity the fish scales occur. Experience has shown that the curran test is more sensitive than the shock test and, in general, covers the conditions of the practical enameling operation.

2) Test for H permeability
   The principle of these test methods, i.e. tests revealing indirect findings regarding fish scale susceptibility, are shown on the right in the figure. The common feature of most test methods is gassing one side of the sheet with atomic H and evaluating the release of H on the other side. Differences exist in the evaluation criteria. Depending on the method the H-diffusion curve itself, the H-passage time t_0 or a mean H-passage time Δt are used as criteria.

VOEST-Alpine uses the mean passage time Δt to calculate a so-called H diffusion coefficient. The higher this value, the higher is the permeability of the tested material for H and the higher is the tendency to fish scale susceptibility.

Table 208 shows the fish scale susceptibility of various steel grades [293].

Steel grade	Code	Fish scale susceptibility	H permeability cm²/s × 10⁻⁶
Non-killed	U	very low	0.3–1.3
Killed with improved fish scale susceptibility	ASi6, ASi5	very low	0.4–1.5
Vacuum-decarburized, Al-killed, Ti-stab.	Ti vac	very low	0.7–1.6
Vacuum-decarburized	U vac	low	0.7–1.6
Al mould-killed	KB	low	1.0–2.0
Al ladle-killed	A6Pf, A7Pf¹⁾	elevated	1.5–3.0
	A5Pf²⁾	strongly elevated	1.5–4.0

1) globular microstructure
2) elongated microstructure (pancake)

Table 208: Fish scale susceptibility of cold-rolled sheets [293]

[294] describes an instrument for measuring the hydrogen permeability of sheets, measuring continuously the passage time and the volume of hydrogen released from the test sheet. The fish scale susceptibility and the further development of suitable steels to prevent it are described in a number of papers [295–297]).

B 12 Oxide ceramic materials

An aluminum oxide ceramic material with an Al_2O_3 content of 99.7% (trade name: Alsint 99,7®) is resistant to hydrogen up to 1,600 °C [298] and for alumina of undefined purity a temperature up to 1,700° C [289] is mentioned.

Other oxide ceramic materials are resistant to hydrogen up to the following temperatures: quartz (silicon dioxide) and quartz-containing refractory products up to 1,100 °C, beryllium oxide up to 1,700 °C, magnesium oxide up to 1,700 °C and stabilized zirconium dioxide up to 2,150 °C. Above the indicated temperatures the metal oxides are reduced:

Equation 24 $\quad MO + H_2 \rightarrow M + H_2O \; (M = \text{metal})$

Decomposition is accelerated if the reaction products are discharged, the formation of stable lower oxides is possible or oxides with lower decomposition temperatures than the ceramic material are present, e.g. silicon dioxide, in an aluminum oxide ceramic material [289].

Ceramic fibers made from aluminum borosilicate (trade name: Nextel® 312, Nextel® 440) are attacked in a hydrogen-containing atmosphere at 1,200 °C, the SiO_2 being reduced to SiO [299].

B 13 Metal-ceramic materials

Silicon carbide is resistant to hydrogen up to about 1,400 °C, however it quickly reacts with hydrogen at temperatures above 1,400 °C [289]. [300] showed that pure silicon carbide is resistant only up to about 1,300 °C if exposed to hydrogen and that the grain boundaries and the secondary phases in the sintered, reaction-bonded silicon carbide are attacked as low as 1,000 °C.

Other metal-ceramic materials are resistant to hydrogen up to the following temperatures: boron carbide up to 1,200 °C, titanium carbide up to 2,400 °C, tungsten carbide up to 2,200 °C, silicon nitride up to 1,750 °C, boron nitride up to 3,000 °C, titanium nitride up to 1,670 °C and cerium sulfide up to 2,000 °C [289].

C
Organic materials

C Organic materials

Damage types of organic materials

For many polymers chemical decomposition as a result of the reductive effect of gaseous hydrogen is less important at ambient temperature up to the upper application temperatures specified by the decline of mechanical properties. Polymers with C=C bonds can be attacked more easily, in particular if other components of the material exert a catalytic effect. Apart from the polymer as the matrix, organic materials can also contain other components, e.g. fibers, fillers, accelerators, activators, inhibitors, plasticizers, dispersion and flow additives, dyes, pigments, deaerating agents, flame retardants, antioxidants, UV stabilizers, antidegradants or bonding agents. Therefore, it is not only the polymer which should be considered when evaluating the chemical resistance to hydrogen, but also the other components.

Also the absorption of hydrogen and the resulting swelling of the polymers is of less importance since hydrogen solubility in the polymers is low.

In contrast, the permeability of the organic materials plays an important role. The permeation coefficient not depending on the material thickness serves as the measure for the permeability. Table 209 contains the permeation coefficients for hydrogen for various materials. As can be seen, the permeation coefficients for the various polymers are very different without any direct relationship with the chemical structure of the polymer can be found. Usually, the permeation coefficients strongly increase with increasing temperature.

The permeation coefficients for a given temperature indicated for a polymer in the literature may differ widely; reasons are, for instance: different measuring methods; different hydrogen pressures on the gas feed side; inhomogeneities and voids in the specimen having a major impact if the specimen thickness is low, e.g. below 30 µm; different degrees of stretching of foils; variable crystallization degree of partially crystalline polymers; different orientation of the crystallites; different monomer contents in copolymers and block copolymers; different degree of cross-linking of duroplastics and elastomers. In polymers containing fillers the content, size, form and orientation of the particles influence the permeation coefficient.

The codes for plastics are used according to DIN EN ISO 1043-1 [301] and for rubbers according to DIN EN ISO 1629 [302].

Polymer		Tempera-ture °C	Permeation coefficient for hydrogen $10^{-13} \times cm^3$ (STP)/cm s Pa	Comment	References
a) Thermoplastics					
Polyethylene (low density)	PE-LD	20	5.3–6.0		[303]
		23	9.3		[304, 305]
		25	7.4		[306]
		30	7.5		[303]
		30	13.0	pressure diff. 2 bar	[307]
		50	18.5		[303]
Polyethylene (high density)	PE-HD	−40	0.11	pressure diff. ~ 1 bar	[308]
		−20	0.36	pressure diff. ~ 1 bar	[308]
		0	0.95	pressure diff. ~ 1 bar	[308]
		20	2.0	pressure diff. ~ 1 bar	[308]
		20	2.4–2.7		[303]
		20	2.55		[309]
		23	2.8		[304]
		25	2.8	unstretched foil	[305]
		25	2.3	stretched foil	[305]
		30	3.5	unstretched foil	[305]
		30	2.8	stretched foil	[305]
		40	5.3	unstretched foil	[305]
		40	4.4	stretched foil	[305]
		50	6.9–8.1		[303]
		50	7.8	unstretched foil	[305]
		50	6.9	stretched foil	[305]
Polypropylene	PP-H	20	31.0	crystallinity 50%	[306]
		20	5.8–6.9	unstretched foil	[303]
		20	6.38		[309]
		25	8.2		[304]
		25	7.3	unstretched foil	[305]
		25	3.1	stretched foil	[305]
		30	7.5–9.3	unstretched foil	[303]
		30	8.4	unstretched foil	[305]
		30	3.8	stretched foil	[305]
		40	13.0	unstretched foil	[305]
		40	5.7	stretched foil	[305]
		50	20.8–23.1	unstretched foil	[303]
		50	21.6	unstretched foil	[305]
		50	9.2	stretched foil	[305]
Polybutene	PB	25	8.2		[305]
Polyisobutylene	PIB	35	8.2		[310]

STP = Standard temperature and pressure
Shore A = Hardness in Shore A units
RT = room temperature; pressure diff. = pressure difference

Table 209: Permeation coefficient for hydrogen of polymers and fiber-reinforced composites

Table 209: (Continued)

Polymer		Temperature °C	Permeation coefficient for hydrogen $10^{-13} \times cm^3$ (STP)/cm s Pa	Comment	References
Polyvinyl chloride, hard	PVC-U	−60	0.79	pressure diff. ~ 1 bar	[308]
		−20	4.6	pressure diff. ~ 1 bar	[308]
		0	9.3	pressure diff. ~ 1 bar	[308]
		20	16.8	pressure diff. ~ 1 bar	[308]
		20	0.94		[309]
		RT	0.75–1.2	unstretched foil	[303]
		25	1.3		[306]
Polyvinyl chloride, softened with tricresyl phosphate	PVC-P				
0%		27	1.8		[306]
5%		27	1.4		[306]
10.2%		27	1.3		[306]
15.0%		27	1.3		[306]
20.1%		27	1.6		[306]
30.8%		27	2.1		[306]
40.0%		27	2.7		[306]
Polyvinylidene chloride	PVDC	25	0.18–0.41		[304, 305]
Polytetrafluoroethylene	PTFE	25	7.25		[306]
		25	7.4		[306]
Tetrafluoroethylene-hexafluoropropylene copolymer	FEP	25	10.7		[311]
Polyvinylidene fluoride	PVDF	20	0.255		[309]
		23	0.24		[304]
		30	1.8	pressure diff. 2 bar	[307]
		35	0.41		[310]
Polychlorotrifluoroethylene	PCTFE	20	0.705	crystallinity 30%	[306]
		25	0.56		[311]
		40	0.98–1.50		[304, 305]
Polyvinylfluoride	PVF	23	0.26		[304, 305]
		35	0.41		[306, 310]
Polysulfone	PSU	23	8.1		[304, 305]
		23	7.5		[312]
		30	9.1	pressure diff. 2 bar	[307]
		40	8.5	pressure diff. 10 bar	[313]
		40	8.0	pressure diff. 30 bar	[313]
Polyethersulfone	PESU	23	4.4		[312]

STP = Standard temperature and pressure
Shore A = Hardness in Shore A units
RT = room temperature; pressure diff. = pressure difference

Table 209: Permeation coefficient for hydrogen of polymers and fiber-reinforced composites

Table 209: (Continued)

Polymer		Temperature °C	Permeation coefficient for hydrogen $10^{-13} \times cm^3$ (STP)/cm s Pa	Comment	References
Liquid crystal polymer	LCP	40	0.045	0% relative moisture	[314]
		150	2.8	0% relative moisture	[314]
Polyimide	PI	23	1.1		[304, 305]
		40	10.8	pressure diff. 10 bar	[313]
		40	10.3	pressure diff. 30 bar	[313]
Polybenzimidazole	PBI	25	8.2		[304]
Polystyrene	PS	23	18.5		[306]
		25	0.73	stretched foil	[305]
		25	17.0	biaxially oriented	[306]
		30	17.9	pressure diff. 2 bar	[307]
Copolymer of styrene and styrene sulfonic acid, poly Na salt with 15.2 mole% SO_3^-		23	10.13		[306]
with 27.5 mole% SO_3^-		23	5.85		[306]
Copolymer of styrene and styrene sulfonic acid, poly Mg salt with					
15.2 mole% SO_3^-		23	7.20		[306]
with 27.5 mole% SO_3^-		23	5.18		[306]
Graft copolymer of styrene and acrylonitrile on polyacrylester	ASA	20	0.058		[304]
Styrene-butadiene copolymer	SB	30	5.9	pressure diff. 2 bar	[307]
Polyethylene terephthalate	PET	10	0.45	pressure diff. 100 bar	[315]
		23	0.39		[303, 304]
		23	0.43	stretched foil	[305]
		25	0.448		[306]
		30	0.41–0.46		[303]
		30	2.33	pressure diff. 130 bar	[315]
		50	0.75–0.90		[303]
Polycarbonate	PC	−40	0.88	pressure diff. ~ 1 bar	[308]
		−20	2.5	pressure diff. ~ 1 bar	[308]
		0	5.2	pressure diff. ~ 1 bar	[308]
		20	9.9	pressure diff. ~ 1 bar	[308]
		23	6.4		[304, 305]

STP = Standard temperature and pressure
Shore A = Hardness in Shore A units
RT = room temperature; pressure diff. = pressure difference

Table 209: Permeation coefficient for hydrogen of polymers and fiber-reinforced composites

Table 209: (Continued)

Polymer		Temperature °C	Permeation coefficient for hydrogen $10^{-13} \times cm^3$ (STP)/cm s Pa	Comment	References
Polyamide 6	PA6	10	0.18	pressure diff. 130 bar	[315]
		25	0.43		[305]
		30	0.90	pressure diff. 50 bar	[315]
		30	0.60	pressure diff. 100 bar	[315]
		30	1.4	pressure diff. 130 bar	[315]
		30	1.35	pressure diff. 175 bar	[315]
		30	1.0	pressure diff. 190 bar	[315]
		50	2.4	pressure diff. 130 bar	[315]
		75	3.8	pressure diff. 130 bar	[315]
Polyamide 11	PA11	25	1.45		[305]
		30	1.34		[306]
Polymethylmethacrylate	PMMA	30	1.8	pressure diff. 2 bar	[307]
		35	2.78		[306]
Polybenzylmethacrylate		30	8.3	pressure diff. 2 bar	[307]
Polyoxymethylene	POM	20	0.19		[305]
		23	0.19		[304]
Poly(oxy-2,6-dimethyl-1,4-phenylene)	PPE	25	84.6		[306]
Polyvinyl acetate	PVAC	10	2.99		[306]
		30	6.84		[306]
		30	11.3	pressure diff. 2 bar	[307]
Polyvinyl alcohol	PVAL	25	0.00672	relative moisture 0%	[306]
Ethylene vinyl alcohol copolymer	EVOH	30	0.38	pressure diff. 2 bar	[307]
Polyvinyl benzoate		25	6.55		[306]
		70	18.3		[306]
Poly(diphenylacetylene)		35	900	pressure diff. 4.46 bar	[316]
Poly[1-phenyl-2-[p-(trimethylsilyl)phenyl]acetylene]		35	1,950	pressure diff. 4.46 bar	[316]
Poly(1-phenyl-1-propine)		25	225	pressure diff. 4.5 bar	[317]
Poly(1-trimethylsilyl-1-propine)		25	13,500	pressure diff. 4.5 bar	[317]
Poly[1-phenyl-2-[p-(triisopropylsilyl)phenyl]acetylene]		35	750	pressure diff. 3.45 bar	[318]
Cellulose hydrate		25	0.00472	relative moisture 0%	[306]

STP = Standard temperature and pressure
Shore A = Hardness in Shore A units
RT = room temperature; pressure diff. = pressure difference

Table 209: Permeation coefficient for hydrogen of polymers and fiber-reinforced composites

Table 209: (Continued)

Polymer		Temperature °C	Permeation coefficient for hydrogen $10^{-13} \times cm^3$ (STP)/cm s Pa	Comment	References
Cellulose acetate	CA	20	2.63		[306]
		25	4.1		[305]
		35	11.2		[306]
Cellulose nitrate	CN	20	1.5		[306]
Ethyl cellulose	EC	25	40.1		[306]
b) Thermoplastic elastomers					
Thermoplastic polyetherpolyamide elastomers	TPE-A	35	34.5–45	pressure difference about 10 bar	[319]
Thermoplastic polyurethane elastomers	TPE-U	20	4.5	polyester basis, 80 Shore A	[320]
		20	2.0	polyester basis, 95 Shore A	[320]
		20	7.0	polyester basis, 80 Shore A	[320]
		20	4.0	polyester basis, 95 Shore A	[320]
c) Duroplastics					
Epoxy resin bisphenol A-basis/ aromatic polyamine	EP	20	1.4	pressure difference ~ 1 bar	[308]
d) Rubber, elastomers					
Natural rubber	NR				
NR elastomer		20	9.17		[306]
		80	86.1		[306]
Isoprene rubber	IR	25	10.9	trans-polyisoprene, vulcanized	[306]
Butadiene rubber	BR	25	31.6		[306]
1,2-polybutadiene		25	24	pressure diff. 1 bar	[321]
		25	21	pressure diff. 25 bar	[321]
		25	20	pressure diff. 50 bar	[321]
Styrene butadiene rubber	SBR				
SBR elastomer, 52 Shore A		20	6.67		[306]
		80	48.6		[306]

STP = Standard temperature and pressure
Shore A = Hardness in Shore A units
RT = room temperature; pressure diff. = pressure difference

Table 209: Permeation coefficient for hydrogen of polymers and fiber-reinforced composites

Table 209: (Continued)

Polymer		Temperature °C	Permeation coefficient for hydrogen $10^{-13} \times cm^3$ (STP)/cm s Pa	Comment	References
Chloroprene rubber	CR	25	10.2		[306]
		35	20.7		[306]
CR elastomer, 42 Shore A		20	5.00		[306]
		80	61.1		[306]
Isobutene isoprene rubber (Butyl rubber)	IIR	25	5.43		[306]
Chlorobutyl rubber	CIIR	20	1.92		[306]
CIIR elastomer, 68 Shore A		80	26.1		[306]
Acrylonitrile butadiene rubber	NBR				
with 20% acrylonitrile		25	18.9		[306]
with 27% acrylonitrile		25	11.9		[306]
with 32% acrylonitrile		25	8.85		[306]
with 39% acrylonitrile		25	5.35		[306]
NBR elastomer, 50 Shore A		20	4.58		[306]
		80	38.3		[306]
NBR elastomer, 60 Shore A		20	3.75		[306]
		80	47.2		[306]
NBR elastomer, 70 Shore A		20	3.25		[306]
		80	27.2		[306]
Acrylonitrile isoprene rubber	NIR	25	5.58	with 26% acrylonitrile	[306]
Methacrylonitrile isoprene rubber		25	10.2	with 26% methacrylonitrile	[306]
Ethylene propylene diene rubber	EPDM				
EPDM elastomer, 68 Shore A		20	15.6		[306]
		80	86.1		[306]
Chlorosulfonated polyethylene	CSM				
CSM elastomer, 70 Shore A		20	2.94		[306]
		80	20.8		[306]
Acrylate rubber	ACM				
ACM elastomer, 70 Shore A		20	10.6		[306]
		80	80.6		[306]
Fluorinated rubber	FKM				
FKM elastomer, 70 Shore A		20	2.58		[306]
		80	41.7		[306]
Silicone rubber	VMQ				
VMQ elastomer, 50 Shore A		20	225		[306]
		80	583		[306]

STP = Standard temperature and pressure
Shore A = Hardness in Shore A units
RT = room temperature; pressure diff. = pressure difference

Table 209: Permeation coefficient for hydrogen of polymers and fiber-reinforced composites

Table 209: (Continued)

Polymer		Tempera-ture °C	Permeation coefficient for hydrogen $10^{-13} \times cm^3$ (STP)/cm s Pa	Comment	References
Polydimethylsiloxane	MQ	30	281	pressure v 2 bar	[307]
		35	668	pressure diff. 10 bar	[319]
		35	705		[306]
e) Fiber-reinforced composites					
Laminate of epoxy resin bisphenol/aromatic polyamine/ E-glass fiber fabric; glass content: 56 vol%	EP-GF	−40	0.023	pressure diff. ~ 1 bar	[308]
		−20	0.061	pressure diff. ~ 1 bar	[308]
		0	0.14	pressure diff. ~ 1 bar	[308]
		20	0.31	pressure diff. ~ 1 bar	[308]
		25.8	0.40	pressure diff. ~ 1 bar	[308]
Laminate of epoxy resin bisphenol A-basis/aromatic polyamine/carbon fiber fabric; fiber content: 59 vol%	EP-CF	−40	0.027	pressure diff. ~ 1 bar	[308]
		−20	0.072	pressure diff. ~ 1 bar	[308]
		0	0.22	pressure diff. ~ 1 bar	[308]
		20	0.48	pressure diff. ~ 1 bar	[308]
Laminate after 100 cycles RT/ −196 °C		−20	0.086	pressure diff. ~ 1 bar	[308]
		0	0.20	pressure diff. ~ 1 bar	[308]
		20	0.44	pressure diff. ~ 1 bar	[308]
Laminate of epoxy resin/polyamine/carbon fibers; fiber content: 56%	EP-CF	RT	< 0.006	pressure diff. ~ 1 bar	[322]
Laminate after 50 cycles RT/ −196 °C		RT	< 0.006	differential pressure ~ 1 bar	[322]
Laminate polyether ether ketone/ carbon fibers; fiber content: 60%	PEEK-CF	RT	0.0057	differential pressure ~ 1 bar	[322]
Laminate after 50 cycles RT/ −196 °C		RT		not measurable due to material failure (high porosity)	[322]

STP = Standard temperature and pressure
Shore A = Hardness in Shore A units
RT = room temperature; pressure diff. = pressure difference

Table 209: Permeation coefficient for hydrogen of polymers and fiber-reinforced composites

Thermoplastics

Thermoplastic		Tempera-ture °C	Resistance	Remarks/reference to	References
Name	Code				
a) Polyolefins					
Polyethylene, low density	PE-LD	20–60	+		[323, 324]
		60	+		[303]
Polyethylene, medium density	PE-MD	21–60	+		[325]
Polyethylene, high density	PE-HD	20–60	+	pipes, manufacture of chemical equipment	[309, 323, 324, 326328]
		60	+	pipelines	[303, 329]
Polyethylene, high molar mass	PE-HMW	20–60	+	test duration: 60 d	[330]
Polyethylene, very high molar mass	PE-UHMW	20–60	+	test duration: 60 d	[330]
Polyethylene, cross-linked	PE-X	20–60	+		[331]
Polypropylene	PP	20	+	pipes, fittings,	[324, 332]
		20–60	+	manufacture of chemical equipment, pipes	[326, 333]
		60	+		[303
		80	+	pipelines	[329]
		100	–		[303]
	PP-H	20–60	+	manufacture of chemical equipment, pipes	[309, 328]
		100	–	pipelines	[328]
	PP-B	20–60	+	manufacture of chemical equipment	[309]
	PP-R	20–60	+	manufacture of chemical equipment	[309]
Polybutene	PB	20–60	+	pipes, fittings	[324]
Polyisobutylene	PIB	60–100	+		[303]
b) Polyvinyl chloride					
Polyvinyl chloride, without plasticizers	PVC-U	20–60	+	pipes, manufacture of chemical equipment	[309, 324–326, 328, 329, 334]
		60	+		[303
Polyvinyl chloride with plasticizers	PVC-P	20–60	+		[309]

+ = resistant, ⊕ = conditionally resistant, – = not resistant
RT = room temperature

Table 210: Resistance of thermoplastics to hydrogen (gaseous)

Table 210: (Continued)

Thermoplastic		Temperature °C	Resistance	Remarks/ reference to	References
Name	Code				
Polyvinyl chloride, chlorinated	PVC-C	23	+	pipes, fittings	[335]
		20–60	+	pipes, fittings	[324]
		20–80	+	manufacture of chemical equipment, pipes	[309, 328, 329]
Polyvinylidene chloride	PVDC	23	+		[325]
		52	+ to ⊕		[325]
c) Fluorothermoplastics					
Polytetrafluoroethylene	PTFE	120	+	pipe seals	[329]
Tetrafluoroethylene-perfluoropropylvinylether copolymer	PFA	25–75	+	manufacture of chemical equipment	[336]
Tetrafluoroethylene-hexafluoropropylene copolymer	FEP	25–75	+	manufacture of chemical equipment	[336]
Tetrafluoroethylene/perfluoromethyl vinyl ether copolymer	MFA	25–75	+	manufacture of chemical equipment	[336]
Copolymer of tetrafluoroethylene and perfluoro-2-(2-sulfonylethoxy)propyl vinyl ether, Na salt		RT-80	+	membrane in polymer electrolyte membrane fuel cells (PEMFC)	[337]
Ethylene chlorotrifluoroethylene copolymer	ECTFE	20	+	manufacture of chemical equipment	[326]
		25–100	+	manufacture of chemical equipment	[336]
		120	+	pipelines	[329]
		23–149	+		[325, 338]
Ethylene tetrafluoroethylene copolymer	ETFE	≤ 150	+		[325, 339]
Polyvinylidene fluoride	PVDF	80	+	pipelines	[329]
		25–75	+	manufacture of chemical equipment	[336]
		20–100	+	pipelines	[328]
		20–120	+	pipes, manufacture of chemical equipment	[309, 324, 326]
		60–120	+		[303]
		≤ 140	+		[340]

+ = resistant, ⊕ = conditionally resistant, – = not resistant
RT = room temperature

Table 210: Resistance of thermoplastics to hydrogen (gaseous)

Table 210: (Continued)

Thermoplastic		Temperature °C	Resistance	Remarks/reference to	References
Name	Code				
d) High-temperature thermoplastics					
Polyphenylene sulfide	PPS	93	+		[325, 341]
Polyether ether ketone	PEEK	≤ 135	+	seals	[342]
e) Polystyrene and copolymers					
Styrene-acrylonitrile copolymer	SAN	23–52	+	test duration: 28 d	[325]
Acrylonitrile butadiene styrene copolymer	ABS	23–52	+	test duration: 28 days	[325]
		20–60	+	pipes, fittings	[328]
f) Polyester					
Polyethylene terephthalate	PET	RT	+		[303, 343]
Polybutylene terephthalate	PBT	RT	+		[303, 325, 343, 344]
Polycarbonate	PC	RT	+		[303]
g) Polyamides					
Polyamide 6	PA6	RT	+		[303, 325, 343–345]
Polyamide 66	PA66	RT	+		[303, 343–345]
Polyamide 46	PA46	RT	+		[343]
Polyamide 610	PA610	RT	+		[303, 325]
Polyamide 6-3-T	PA6-3-T	RT	+		[303]
Polyamide 11	PA11	RT	+		[303]
		20–90	+		[325, 346]
Polyamide 12	PA12	RT	+		[303]
		20–90	+		[347]
h) Other thermoplastics					
Polyoxymethylene	POM	RT	+		[303, 325, 344]
Polymethylmethacrylate	PMMA	RT	+		[303]

+ = resistant, ⊕ = conditionally resistant, – = not resistant
RT = room temperature

Table 210: Resistance of thermoplastics to hydrogen (gaseous)

Thermoplastic elastomers

For thermoplastic elastomers (TPE) based on polyolefine (TPE-O) (trade name: Alcryn®) limited resistance to hydrogen at 24 °C is indicated [325] and for TPE based on polyester (TPE-E) (trade name: Hytrel®) resistance at 22 °C is indicated [325, 348].

Duroplastics

Reaction resin/duroplastic	Temperature °C	Resistance
Unsaturated polyester resins (UP resins) based on alcoxylated bisphenol A	RT	+
UP resins based on isophthalic acid	RT	+
UP resins based on HET acid (hexachloro endomethylene tetrahydrophthalic acid)	93–121	+
Vinyl ester resins (VE resins)	99–104	+
Diallyl phthalate resins (PDAP)	RT	+ to ⊕
Phenol plastics (phenolic resins)	RT	+

+ = resistant, ⊕ = limited resistance RT = room temperature

Table 211: Resistance of duroplastics to hydrogen (gaseous) [325]

Elastomers

Elastomer		Temperature, °C	Resistance	Remarks/ reference to	References
Rubber basis	Code				
a) Elastomers with R rubbers					
Natural rubber	NR	cold	+ to ⊕		[349]
		cold	⊕		[325]
		RT	+	peristaltic pumps	[350]
		RT	+ to ⊕		[351, 352]
		hot	+ to ⊕		[349]
		hot	⊕		[325]
Styrene butadiene rubber	SBR	cold	+ to ⊕		[349]
		cold	⊕		[325]
		RT	+		[351]
		RT	+ to ⊕		[352]
		hot	+ to ⊕		[349]
		hot	⊕		[325]

+ = resistant, ⊕ = limited resistance, – = not resistant
RT = room temperature

Table 212: Resistance of elastomers to hydrogen (gaseous)

Table 212: (Continued)

Elastomer		Tempera-ture, °C	Resis-tance	Remarks/ reference to	References
Rubber basis	Code				
Chloroprene rubber	CR	cold	+		[325, 349]
		RT	+	peristaltic pumps	[350–352]
		60	+	pipe seals	[329]
		20–80	+	pipe seals	[328]
		hot			[325, 349]
Acrylonitrile butadiene rubber	NBR	cold	+		[325, 349]
		RT	+	peristaltic pumps	[350-351352]
		60	+	pipe seals	[329]
		20–100	+	pipe seals	[328]
		hot			[325, 349]
Hydrogenated nitrile rubber	HNBR	RT	+		[352]
Isobutene isoprene rubber (Butyl rubber)	IIR	cold	+		[325, 349]
		RT	+	peristaltic pumps	[350, 351]
		60	+	pipe seals	[329]
		hot	+		[325, 349]
b) Elastomers with M rubbers					
Ethylene propylene diene rubber	EPDM	cold	+		[325, 349]
		RT	+	peristaltic pumps	[350–351352]
		80	+	pipe seals	[329]
		20–100	+	pipe seals	[328]
		hot	+		[325, 349]
Chlorosulfonated polyethylene	CSM	cold	+		[325, 349]
		RT	+		[351, 352]
		60	+	pipe seals	[329]
		20–80	+	pipe seals	[328]
		hot	+		[325, 349]
Acrylate rubber	ACM	cold	+ to ⊕		[349]
		RT	+ to ⊕		[351, 352]
		hot	+ to ⊕		[349]
Fluorinated rubber	FKM	cold	+		[325, 349]
		RT	+	peristaltic pumps	[350–351352]
		80	+	pipe seals	[329]
		20–100	+	pipe seals	[328]
		hot	+		[325, 349]
Perfluoro rubber	FFKM	RT	+		[351, 352]
		100	+		[325]

+ = resistant, ⊕ = limited resistance, − = not resistant
RT = room temperature

Table 212: Resistance of elastomers to hydrogen (gaseous)

Table 212: (Continued)

Elastomer		Temperature, °C	Resistance	Remarks/reference to	References
Rubber basis	Code				
c) Elastomers with other rubbers					
Silicone rubber	VMQ	cold	⊕ to −		[325, 349]
		RT	+	peristaltic pumps	[350]
		RT	⊕ to −		[325, 351, 352]
		hot	⊕ to −		[349]
Fluorosilicone rubber	FVMQ, FQ	cold	⊕ to −		[325, 349]
		RT	⊕ to −		[325, 351, 352]
		hot	⊕ to −		[349]
Polysulphide rubber	OT	cold	⊕ to −		[325, 349]
		RT	⊕ to −		[351]
		hot	−		[325, 349]
Urethane rubber	AU, EU	cold	+		[325]
		cold	+ to ⊕		[349]
		RT	+	peristaltic pumps	[350–352]
		hot	+		[325]
		hot	+ to ⊕		[349]

+ = resistant, ⊕ = limited resistance, − = not resistant
RT = room temperature

Table 212: Resistance of elastomers to hydrogen (gaseous)

Other resistance data are listed in Section D 2 under O-rings.

D
Materials with special properties

D 1 Coatings and films

Metallic coatings and linings

Compared to the ferritic and martensitic steels, austenitic stainless steels exhibit a hydrogen solubility which is one order of magnitude higher and a diffusion rate which is about two orders of magnitude lower. Metallic coatings of austenitic stainless steels on low-alloy steels can, therefore, lower the partial hydrogen pressure on the base material since only hydrogen diffusing through the coating can exert an effect. Such coatings may be applied by roll cladding, explosive cladding or overlay welding. Coatings of ferritic or martensitic stainless steels do not cause any reduction of the partial hydrogen pressure on the base material or may reduce it only slightly.

The effectiveness of such coatings from austenitic stainless steels is both based on both the layer thickness and the materials used. Several cases are known in which damage in compressed hydrogen at high temperatures occurred in compressed hydrogen resistant 0.5% Mo steels despite cladding with austenitic stainless steel. [109] recommends not to use such cladding or overlay welding for the long-term operation, but use it only if the boundary conditions for the base material according to Figure 64 are exceeded only occasionally and for a short time.

On the other hand, linings with austenitic stainless steels are an effective measure to prevent damage by compressed hydrogen if the hydrogen diffusing through the lining can recombine in the space between the lining and the base material and escape non-pressurized through appropriately arranged degassing holes in the base material.

Organic coatings and linings

Reaction resin/polymer (trade name[1])	Layer thickness mm	Temperature °C	Resistance	References
a) Powder coatings				
Polyethylene, low density (PE-LD), modified	0.3–0.6	21–60	+	[353]
Ethylene chlorotrifluoroethylene copolymer (ECTFE) (E-CTFE Halar®)	0.4–1.0	23–150	+	[354, 355]
Tetrafluoroethylene perfluoro propyl vinyl ether copolymer (PFA) (Edlon® PFA)	max. 1	≤ 130	+	[356]
PFA/Tetrafluoroethylene-hexafluoro-propylene copolymer (FEP) intermediate layer = PFA, cover layer = FEP (Rhenoguard® Jumbo II)	1.0–1.2	≤ 140	+	[357]
PFA/FEP intermediate layer = PFA filled 0.6 mm, cover layer = FEP filled 0.6 mm, top coat = FEP 0.2 mm, (Rhenoguard® Jumbo III)	1.2–1.8	≤ 160	+	[357]
b) Reaction resin coatings				
Phenolic resin, unmodified	0.2	RT	+	[358]
c) Rubber linings				
Soft rubber linings on the basis of	3–5			
natural rubber (NR)		≤ 60	+ to ⊕	[358]
chloroprene rubber (CR)		≤ 60	+ to ⊕	[358]
chlorobutyl rubber (CIIR)		≤ 60	+ to ⊕	[358]
		≤ 82	+ to ⊕	[359]
chlorosulfonated polyethylene (CSM)		≤ 60	+ to ⊕	[358]
Semihard rubber linings on the basis of	3–5			
natural rubber (NR)		≤ 82	+ to ⊕	[358]
synthetic rubber (SBR, IR)		≤ 82	+ to ⊕	[358]
d) Fully adhered thermoplastic webs				
Ethylene chlorotrifluoroethylene copolymer (ECTFE); rear: fabric or soft rubber (EC Duro-Bond® ECTFE Lining)	1.5; 2.3	21–107	+	[359]

\+ = resistant, ⊕ = limited resistance
RT = room temperature
[1] trade names not binding

Table 213: Resistance of coatings and linings to hydrogen (gaseous)

Table 213: (Continued)

Reaction resin/polymer (trade name[1])	Layer thickness mm	Temperature °C	Resistance	References
Ethylene tetrafluoroethylene copolymer (ETFE); rear: fabric (EC Duro-Bond® ETFE Lining)	1.5; 2.3	≤ 110	+	[359]
Polyvinylidene fluoride (PVDF); rear: fabric or soft rubber (EC Duro-Bond® PVDF Lining)	1.5; 2.3; 3.0	21–107	+	[359]
Tetrafluoroethylene perfluoro propyl vinyl ether copolymer (PFA); rear: fiber mesh (EC Duro-Bond® PFA Lining)	1.5; 2.3	≤ 110	+	[359]
Tetrafluoroethylene-hexafluoropropylene copolymer (FEP); rear: fiber mesh (EC Duro-Bond® FEP Lining)	1.5; 2.3	≤ 110	+	[359]
Tetrafluoroethylene/perfluoromethyl vinyl ether copolymer (MFA); rear: fiber mesh (EC Duro-Bond® MFA Lining)	1.5; 2.3	≤ 110	+	[359]
e) Thermoplastic webs mechanically anchored in concrete				
Polyethylene (PE) (Anchor-Lok® PE)	3; 5	20–60	+	[360]
Polypropylene (PP) (Anchor-Lok® PP)	3; 5	20–60	+	[360]
Polyvinyl chloride (PVC-U) (Anchor-Lok® PVC)	3; 5	20–60	+	[360]
Polyvinylidene fluoride (PVDF) (Anchor-Lok® PVDF)	3; 5	20–100	+	[360]

+ = resistant, ⊕ = limited resistance
RT = room temperature
[1] trade names not binding

Table 213: Resistance of coatings and linings to hydrogen (gaseous)

D 2 Gaskets and packings

Organic materials

Flat gaskets

Components of flat gaskets (trade name[1])	Application temperature, °C	Remarks	References
a) Flat gaskets with fibers			
Aramid fibers/graphite/NBR (novapress® multi II)	−40 to 170	at operating pressure of 20 bar	[361]
Aramid fibers/graphite/NBR (minimized content) (novatec® PREMIUM II)	−70 to 170	at operating pressure of 20 bar	[361]
b) Flat gaskets with polytetrafluoroethylene (PTFE)			
PTFE (Gylon® Standard, Gylon® blue, Gylon® white)	≤ 260	maximum operating pressure of 45 bar (thickness 1 and 1.5 mm)	[362]
PTFE/fillers (KLINGER®top-chem 2000)	260	at operating pressure of 20 bar	[363]]
(novaflon® 200, 300)	−200 to 240	at operating pressure of 20 bar	[361]
PFTE multidirectionally expanded (novaflon® 500)	−200 to 240	at operating pressure of 20 bar	[361]
c) Flat gaskets with expanded graphite			
Expanded graphite/one layer of expansion metal (novaphit® Super SSTC)	−200 to 530	at operating pressure of 20 bar	[361]
Expanded graphite/several layers of expansion metal (novaphit® Super HPC)	−200 to 530	at operating pressure of 20 bar	[361]
Expanded graphite, preformed (novaphit® VS)	−200 to 470	at operating pressure of 20 bar	[361]

NBR = acrylonitrile butadiene rubber
[1] Trade names not binding

Table 214: Maximum utilization temperatures of flat gaskets exposed to hydrogen (gaseous)

Sealing bands made of PTFE (trade name: Gore-Tex® series 300 and 600) for metallic vessel flanges are recommended up to 220 °C [364].

Packings

When exposed to hydrogen packings or stuffing box packings from graphite, aramid or polytetrafluoroethylene (PTFE) yarns are suited, which can be impregnated with PTFE or graphite; also packings made from expanded graphite are suited [365, 366].

O-rings

Elastomer basis		Temperature, °C	Resistance	References
Rubber	Code			
Natural rubber	NR	cold	+ to ⊕	[367]
		20	+	[368]
		hot	+ to ⊕	[367]
Styrene butadiene rubber	SBR	cold	+ to ⊕	[367]
		20	+	[368]
		hot	+ to ⊕	[367]
Chloroprene rubber	CR	cold	+	[367]
		20	+	[368, 369]
		hot	+	[367]
Acrylonitrile butadiene rubber	NBR	cold	+	[367]
		20	+	[368, 369]
		hot	+	[367]
Hydrogenated nitrile rubber	HNBR	cold	+	[367]
		20	+	[368, 369]
		hot	+	[367]
Isobutene isoprene rubber (Butyl rubber)	IIR	cold	+	[367]
		20	+	[368]
		hot	+	[367]
Ethylene propylene diene rubber,	EPDM	cold	+	[367]
		20	+	[368, 369]
		hot	+	[367]
Chlorosulfonated polyethylene	CSM	cold	+	[367]
		20	+	[368]
		hot	+	[367]
Acrylate rubber	ACM	cold	+ to ⊕	[367]
		20	+	[368]
		RT	+ to ⊕	[369]
		hot	+ to ⊕	[367]
Fluorinated rubber	FKM	cold	+	[367]
		20	+	[368, 369]
		hot	+	[367]
Perfluoro rubber	FFKM	cold	+	[367]
		20	+	[368, 370]
		hot	+	[367]

+ = resistant, ⊕ = limited resistance
RT = room temperature

Table 215: Resistance of elastomer materials for O-rings and other sealing rings to hydrogen (gaseous)

Table 215: (Continued)

Elastomer basis		Temperature, °C	Resistance	References
Rubber	Code			
Silicone rubber	VMQ	cold	⊕	[367]
		20	+	[368]
		RT	⊕	[369]
		hot	⊕	[367]
Fluorosilicone rubber	FVMQ/FMQ	cold	⊕	[367]
		20	+	[368]
		RT	⊕	[369]
		hot	⊕	[367]
Urethane rubber	AU, EU	cold	+	[367]
		RT	+	[369]
		hot	+	[367]

+ = resistant, ⊕ = limited resistance
RT = room temperature

Table 215: Resistance of elastomer materials for O-rings and other sealing rings to hydrogen (gaseous)

In particular, [367] recommends, EPDM elastomers.

Mechanical seals (end face)

An example of a mechanical seal against hydrogen (gaseous) is listed in Table 216.

Details of mechanical seal	Design/material
Shaft seal assembly	Double seal; Back to back configuration
Auxiliary pipeline	Sealing or quenching liquid of a pressure vessel; circulation by the thermosiphon effect or pump;
Construction type of the mechanical seal	Mechanical seal with O-ring subsidiary seals
sliding ring material	Special chromium molybdenum casting
counter ring material	Carbon graphite, impregnated with synthetic resin
subsidiary seal material	Ethylene propylene diene elastomer (EPDM)
spring material	CrNiMo steel 1.4571

Table 216: Example of mechanical seal for use with hydrogen (gaseous) up to 60 °C [371]

Often piston compressors are used in the production and processing of hydrogen to compress the hydrogen. To a major extent, the reliability of a compressor depends on the resistance of the sealing material under the existing operating conditions. It has turned out that the moisture content and the temperature of the compressed gas exert a considerable influence on the wear behavior of the sealing material typically consisting of reinforced polymer materials. [372] reports the investigation results regarding the wear behavior in hydrogen gas of four commercial, filled PTFE composite materials used as sealing material for compressors. The designations and the physical properties of the tested sealing materials are indicated in Table 217. The tests were performed on rotating specimens under a load of 40 N against surfaces made from cast iron (Rockwell hardness: 37; roughness 0.6 µm to 0.7 µm). The rotational speed was 20 Hz, corresponding to a sliding speed of about 2 m/s. The surface temperature was 150 °C and the test duration was 24 h. The hydrogen pressure was in a range from 1 bar to 4 bar at a flow rate of 1.2 l/min. The moisture content in the gas varied from 1 vppm to 1,000 vppm. The results are summarized together with those obtained during earlier examinations in nitrogen in Figures 295 to 296.

Designation	HY 54	HY 32	HY 50	HY 101	HY 22
Base material	PTFE	PTFE	PTFE, synthetic resin	PTFE, PPS	PTFE
Filling material	carbon, glass	carbon	graphite, MoS$_2$	carbon fiber	carbon
Density, g/cm^3	2.11	2.05	1.80	1.86	2.02
Thermal expansion, 10^{-6}/°C*	40	53	32	n.i.	55
	107	80	32	54	81
Tensile strength, MPa	14	14	n.i.	14	14
Shear strength, MPa	18	14	21	n.i.	15
Bending strength, MPa	21	24	39	n.i.	23

* upper value: radial, bottom value: axial n.i.: not indicated

Table 217: Designations and physical properties of the tested sealing materials [372]

For the material HY 54 (Figure 295 and Figure 296) the examinations in nitrogen revealed a steep increase of the wear rate at a moisture content from 150 vppm to 200 vppm, with the friction coefficient showing a similar tendency. In hydrogen the wear rate and the friction coefficient also depend on the moisture content of the gas. Both values increase with decreasing moisture, however the increase of the wear rate in hydrogen is clearly lower than that in nitrogen. At moisture contents above 200 Vppm the wear rates in both gases are almost on the same level.

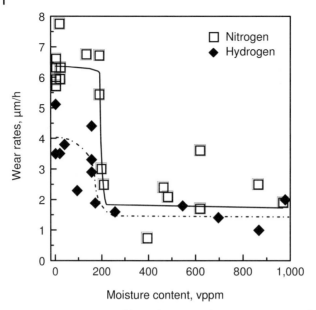

Figure 295: Wear rates of the sealing material HY 54 in nitrogen and in hydrogen as a function of the moisture content of the gases [372]

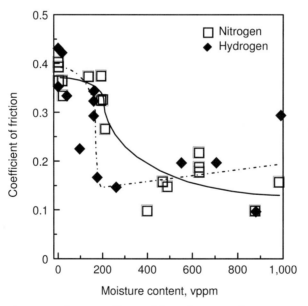

Figure 296: Friction coefficient of the sealing material HY 54 in nitrogen and in hydrogen as a function of the moisture content of the gases [372]

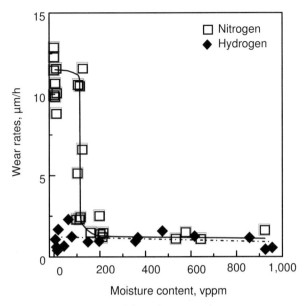

Figure 297: Wear rates of the sealing material HY 32 in nitrogen and in hydrogen as a function of the moisture content of the gases [372]

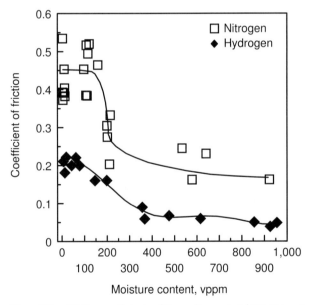

Figure 298: Friction coefficients of the sealing material HY 32 in nitrogen and in hydrogen as a function of the moisture content of the gases [372]

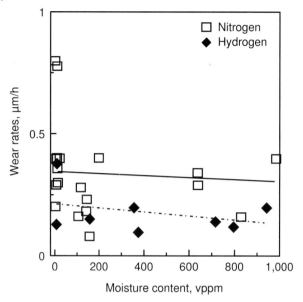

Figure 299: Wear rates of the sealing material HY 50 in nitrogen and in hydrogen as a function of the moisture content of the gases [372]

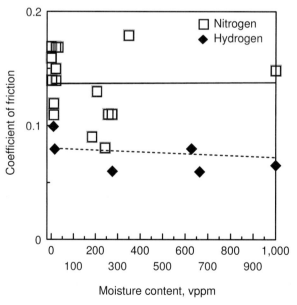

Figure 300: Friction coefficients of the sealing material HY 50 in nitrogen and in hydrogen as a function of the moisture content of the gases [372]

In contrast to the results in nitrogen, no significant influence of moisture on the wear rates was detected for the material HY 32 in hydrogen (Figure 297). Also the friction coefficient in hydrogen gas is influenced only slightly by the moisture content (Figure 298).

As shown in Figure 299 and Figure 300, neither nitrogen nor hydrogen has any influence was found on the wear behavior or the friction coefficient of the material HY 50.

The material HY 101 exhibited very low wear rates in hydrogen, which increased only slightly with increasing moisture content. The friction coefficient was not influenced by the moisture content of the gas.

As these examination results show, it is very important when selecting a sealing material for a compressor to check the wear properties of the material in the relevant gas at the moisture contents to be expected. The results obtained in one gas are not transferable to the behavior in another gas.

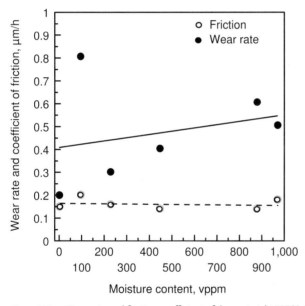

Figure 301: Wear rate and friction coefficient of the material HY 101 in hydrogen [372]

D 3 Composite materials

Glass fiber reinforced plastics (GFRP)

Reaction resin (trade name[1])	Temperature, °C	Resistance	References
VE resin, pipes with resin-rich internal layer 0.5 mm (Fiberdur® VE)	65	+	[373]
VE resin, pipes/component with chemical protection layer 2.5 mm (Fiberdur® CSVE)	65	+	[373]
EP resin/polyamine hot-setting pipes with resin-rich internal layer 0.5 mm (Fiberdur® EP, Wavistrong®)	60	+	[374]
	65	+	[373]
EP resin/polyamine hot-setting pipes with chemical protection layer 2.5 mm (Fiberdur® CSEP)	65	+	[373]
EP resin/polyamine hot-setting centrifugally cast pipes with internal pure resin layer (Fiberdur® Centricast II)	65	+	[373]
EP novolak resin/aromatic polyamine inner lining in valves and cocks from steel	60	+	[375]

+ = resistant
[1] Trade names not binding VE = vinyl ester, EP = epoxy

Table 218: Resistance of GFRP pipes to hydrogen (gaseous)

Several applications of GFRP components containing vinyl ester resins (trade name: Derakane®) after exposure to hydrogen-containing fluids over several years are summarized in Table 219 [376]. The examples show that frequently brominated vinyl ester resins have been used due to their lower flammability.

In the light of the increasing importance of hydrogen as an energy source, the suitability of fiber-reinforced composite materials for storing hydrogen at low temperatures was examined.

In [377] unidirectional, reinforced tensile test specimens (250 mm × 16 mm × 0.5 mm) from two-component epoxy resin on bisphenol A-basis and an aromatic diamine hardening agent (Araldite® LY556/HY917) and E-glass fibers (diameter: 14 µm) were stored in gaseous hydrogen for up to 6 months at 80 °C following preconditioning and tensile tests were performed under quasi static load at room temperature and at −196 °C as well as under dynamic tensile load at −196 °C at intervals of 1 month each. The results shown in Table 220 indicate that the fracture stress and fracture elongation at both temperatures are not significantly influenced by H_2; the fatigue strength at −196 °C even increases up to 30% under the influence of H_2. However, these results do not yet permit conclusions regarding the suitability of the material for components permanently exposed to hydrogen at −196 °C.

In conjunction with the development of a reusable spacecraft [378] reports on the failure of a composite material tank for storing liquid hydrogen. The tank con-

Component (dimensions)	Vinyl ester resin	Fluid	Operating temperature °C
Scrubbing tower (height: 14 m, diameter: 1.58 m)	protective and bearing layer: VE BPA, outer layer: VE BPA-br	H_2, scrubbing liquid	100
Absorption column, (height: 8.87 m, diameter: 1.83 m)	all layers: VE BPA-br; addition of Sb_2O_3	H_2, organic contents	16
Excess-pressure vessel with stack	protective layer: VE Nov + graphite, bearing layer: VE BPA-br	H_2, 99% NaCl solution, 310 g/L, pH value: 13	60
Exhaust air duct, (height: 18.3 m)	all layers: VE BPA-br	H_2 with traces of chlorine and alkali	82
Vessel	all layers: VE BPA-br	H_2, 99%, water, pH-value: 12	50
Electrolytic cell, (length: 3.35 m, height: 3.79 m, width: 0.86 m)	internal layer: fluoro-thermoplastic, main bearing layer: VE BPA, outer layer: VE BPA-br	H_2, sodium chlorate solution, electrolyte liquid	80

VE BPA = vinyl ester resin-based alcoxylated bisphenol A (Derakane® 411-45)
VE Nov = vinyl ester resin-based novolak (Derakane® 470-36)
VE BPA-br = brominated vinyl ester resin-based alcoxylated bisphenol A (Derakane® 510-A40)

Table 219: Applications of GFRP components containing vinyl ester resins and exposure conditions to hydrogen-containing atmospheres [376]

sisted of three layers, i.e. an internal layer consisting of a reaction resin reinforced with carbon fibers, a honeycomb core as a middle layer, and an outer layer. When the tank was filled with liquid hydrogen, microcracks occurred as a result of the different thermal expansion coefficients of the carbon fiber and the polymer matrix, facilitating the diffusion of hydrogen into the middle honeycomb layer. During heating of the emptied tank to −73 °C the microcracks closed again due to the expansion of the polymer matrix and the H_2 gas enclosed in the middle layer built up an overpressure, which led to the detachment of the internal layer from the honeycomb core. Covering the internal layer with a highly hydrogen-impermeable polymer foil was considered to be an appropriate solution to this problem and permeation measurements were performed using several polymer foils – however, these measurements were performed in argon instead of hydrogen.

Regarding the potential use of fiber-reinforced composite materials for components in contact with liquid hydrogen, in [379] laminates with 0°/90° fiber orientation made from glass fibers and unsaturated polyester resin, epoxy resin or polyethylene terephthalate as well as from carbon fibers and epoxy resin or polyether ether ketone were produced. Absorption of hydrogen of the materials was determined using a microbeam balance at −196 °C and at room temperature as well as at a hydrogen pressure of 2 bar. No hydrogen absorption could be detected at −

Specimen preparation	Duration of exposition to H_2 at 80 °C months	Test temperature	Fracture stress MPa	Fracture elongation %
a) Quasi static tensile load				
Non preconditioned	0	RT	1,321	2.8
	0	−196 °C	2,050	4.3
Preconditioned (drying in vacuum at 80 °C/13 d)	0	RT	1,485	3.1
	1	RT	1,460	3.0
	3	RT	1,445	2.9
	6	RT	1,445	2.9
	0	−196 °C	2,190	4.1
	1	−196 °C	2,150	4.0
	3	−196 °C	2,190	4.2
	6	−196 °C	2,110	3.9
b) Dynamic tensile load				
			Fatigue strength after 10^7 load cycles MPa	Increase in %
Preconditioned (drying in vacuum at 80 °C/13 d)	0	−196 °C	484	–
	1	−196 °C	528	9
	2	−196 °C	550	14
	3–6	−196 °C	631	30

RT = room temperature

Table 220: Strength values of unidirectional glass fiber/epoxy resin composite materials after exposure to gaseous hydrogen at 80 °C [377]

196 °C, whereas the absorption of a very low amount of hydrogen only was detected at room temperature only in the polyether ether ketone specimen. The laminates were subjected to tensile tests at −196 °C and concurrently the generated acoustic emissions were measured to determine the extent of microcrack formation based on the acoustic emission intensity. With the exception of the composite material made from glass fibers and unsaturated polyester resin for which a clear acoustic emission was determined at a very low elongation, the composite materials exhibited a low acoustic emission only at an elongation of about 0.3% corresponding to a stress from 100 to 150 MPa. The most favorable behavior was that of the composite material from glass fibers and polyethylene terephthalate, which did not generate any significant acoustic emission even at 0.7% elongation. A stress of 150 MPa corresponds to an internal pressure of 15 bar which can be withstood by a vessel or tank with a diameter of 1 m and a wall thickness of 5 mm until crack formation starts. However, the authors consider it necessary to perform further material tests before the stated fiber-reinforced composite materials can be used in practice, e.g. regarding their thermal shock and permeability.

For the application of rensable rockets a tank was developed for liquid hydrogen with an internal volume of 0.157 m³ consisting of an inner lining of the aluminum alloy EN AW-6061-T3 (EN AW-Al Mg1SiCu) as a gas barrier and an external laminate layer of carbon fibers and a non-specified reaction resin to serve as a strength-providing bearing layer [380]. Since the metallic inner lining acts as a gas barrier, microcrack formation in the polymer matrix as a result of the different thermal expansion coefficients of the carbon fiber and the polymer is tolerable here. To prevent detachment of the inner lining from the laminate layer under thermocycling conditions, sufficient long-term adhesion between the two materials is essential. The hydrogen tank installed in a reusable spacecraft successfully survived several flight tests.

A similar material composition consisting of an aluminum inner lining as a gas barrier and a strength-providing laminate carbon fibers as well as a non-specified reaction resin further was proposed for the manufacture of a hydrogen tank with an internal hydrogen storage alloy, e.g. Mg_2Ni, $TiMn_{1.5}$, $TiFe$, $TiCr_{57.5}V_5$ or $LaNi_5$. Hydrogen is stored, on the one hand, in the storage alloy which occupies only part of the internal volume and, on the other hand, in the remaining volume under high pressure which the tank needs to withstand [381].

E
Material recommendations

Warranty Disclaimer

This book has been compiled from literature data with the greatest possible care and attention. The statements made in this book only provide general descriptions and information.

Even with the correct selection of materials and processing, corrosive attack cannot be excluded in a corrosion system as it may be caused by previously unknown critical conditions and influencing factors or subsequently modified operating conditions.

No guarantee can be given for the chemical stability of the plant or equipment. Therefore, the given information and recommendations do not include any statements, from which warranty claims can be derived with respect to DECHEMA e.V. or its employees or the authors.

The DECHEMA e.V. is liable to the customer, irrespective of the legal grounds, for intentional or grossly negligent damage caused by their legal representatives or vicarious agents.

For the case of slight negligence, liability is limited to the infringement of essential contractual obligations (cardinal obligations). DECHEMA e. V. is not liable in the case of slight negligence for collateral damage or consequential damage as well as for damage that results from interruptions in the operations or delays which may arise from the deployment of this book.

A   Metallic materials

A 1   Silver and silver alloys

Silver and silver alloys may contain considerable amounts of oxygen resulting from the melting process. If such oxygen-containing silver materials come into contact with hydrogen at high temperatures, hydrogen can diffuse into the material and react with the oxygen to form water vapor, leading to blistering or the formation of internal cracks. Therefore, silver materials should not be used in hydrogen at high temperatures.

A 2 Aluminum

A 3 Aluminum alloys

Under normal conditions in dry compressed hydrogen aluminum materials do not absorb hydrogen. The dry molecular hydrogen cannot dissociate at the natural Al_2O_3 scale which is always present, and hence the oxide film is a significant barrier for hydrogen absorption. The adverse affect of aluminum materials by hydrogen can only occur in the presence of moisture. Hydrogen generated by the reaction of moisture with the material can be absorbed on fresh metal surfaces and cause embrittlement. Fresh metal surfaces are generated, for instance, by cyclic mechanical loading.

A 7 Copper

Oxygen-containing copper grades should not be used if they are exposed to hydrogen since, based on the same mechanism as in the silver materials, they are damaged by the reaction of the diffusing hydrogen with oxygen and subsequent blistering. The oxygen-free copper grades are largely resistant to compressed hydrogen.

A 9 Copper-nickel alloys

In copper-nickel alloys susceptibility to hydrogen damage is determined by the nickel content. Embrittlement by hydrogen is not expected in alloys with a copper content of not less than 60%.

A 14 Unalloyed and low-alloy steels/cast steel

The precondition for embrittlement and crack damage to unalloyed and low-alloy steels in compressed hydrogen at ambient temperatures is the formation of diffusible atomic hydrogen. The dissociation of the molecular, adsorbed hydrogen on the metal surface required to facilitate such damage processes necessitates a chemisorption process which can only occur on a clean active surface, e.g. a surface generated by plastic deformation breaking up the oxide film which is permanently present. Plastic deformation in front of the crack tip releases energy in the form of heat of an amount sufficient to facilitate the dissociation of the H_2-molecule. In practice this means that dissociation of hydrogen on steel surfaces is only possible at notches and crack tips, where repeated or permanent plastic deformations occur, e.g. as a result of cyclic loads.

Sites at which Preferential plastic deformation processes take place as a result of the notch effect and a subsequent stress increase under mechanical load, e.g. undercuts in the weld region or surface roughness, are often starting points for hydrogen attack and, therefore, need to be prevented as for a possible.

Moreover, the purity level of the hydrogen plays a role since impurities (e.g. O_2 and H_2O) act as inhibitors (Figure 39, Figure 40).

At temperatures above about 200 °C hydrogen gas contains considerable amounts of atomic hydrogen, which can diffuse into the steel. Depending on the pressure and the temperature hot compressed hydrogen can damage steels in two ways.

- At high temperatures and low hydrogen partial pressures preferential a surface decarburization occurs, not causing crack formation.
- Low temperatures, however above 200 °C and high hydrogen partial pressures enhance the internal decarburization and the formation of methane bubbles and internal cracks, which can eventually lead to fracture. At high temperatures and high pressures, both damage types occur.

Such damage can be prevented to a major extent by alloying elements forming stable carbides in this temperature range. Therefore, the compressed hydrogen resistant steels contain the carbide-forming alloying elements chromium, molybdenum or tungsten which, in contrast to cementite, tie up carbon in the form of much more stable carbides. These compressed hydrogen resistant steels suffer decarburization at temperatures below 450 °C only in the subsurface region. In addition, short-term loading at higher temperatures (1,000 h, 600 °C) leads to a decarburization depth of only about 0.5 mm, which is technically tolerable in most cases. However, grain boundary cracks may occur within the decarburized zone which can widen to form cracks as a result of the internal pressure of the components. This can reduce the fatigue strength-creep elongation.

Today the resistance limits of the various compressed hydrogen resistant steels regarding temperature and hydrogen partial pressure are well known and specified in the so-called Nelson diagram (Figure 64). This diagram serves as the basis for selecting the suitable steels for given operating conditions.

A 17   Ferritic chromium steels with <13% Cr

A 18   Ferritic chromium steels with ≥13% Cr

As with all ferritic steels ferritic chromium stainless steels are susceptible to hydrogen embrittlement due to their bcc lattice structure. In general steels with a high chromium and molybdenum content, e.g. the super ferrite X2CrMoTi29-4, are more susceptible than steels with a lower Cr and Mo content. When welding this material great care must be exercised to prevent hydrogen in the shielding gas and the hydrogen formed by water vapor during the welding process needs to be kept away from the weld region.

A 19.1 Martensitic steels

The comparatively low toughness of martensitic steels is further reduced by the absorption of hydrogen.

A 19.2 Ferritic-austenitic steels/duplex steels

Due to their excellent corrosion resistance ferritic-austenitic steels with 21 to 25% Cr and 4.5 to 6.5% Ni and additions of Mo and N are used in many applications in the chemical industry, in oil and gas production and in seawater applications. Duplex steels with an austenite/ferrite ratio of 50/50 (Mat. No. 1.4462; X2CrNi-MoN22-5-3) exhibit the most favorable behavior, and hence are used most often. In compressed hydrogen both the tensile strength and the fracture elongation are clearly reduced at temperatures below 100 °C. Crack growth increases as the hydrogen pressure increases. Crack initiation and crack growth always occur in the ferrite phase. When the crack has reached the austenite phase boundary it stops first and tries to circumvent the phase boundary.

A 20 Austenitic CrNi steels

In contrast to the ferritic steels, austenitic steels are resistant to hydrogen embrittlement to a major extent. This justifies the widespread use of austenites in hydrogen-containing atmospheres. However, hydrogen resistance only applies to stable austenitic steels. If partial conversion into martensite occurs in unstable austenites, e.g. by forming processes, these steels also become susceptible to hydrogen. The susceptibility to hydrogen embrittlement decreases as the martensite content increases. In unstable austenites the martensite content can amount to more than 50% as a result of cold rolling with a deformation of more than 50%.

A 26 Nickel

Hydrogen can impair the fracture toughness and fracture behavior of nickel and nickel alloys. The extent of embrittlement and the fracture pattern depend on the material, the loading conditions and the temperature. Also the plastic deformation behavior of nickel as well as the behavior under cyclic load can be impaired by hydrogen. Dissolved hydrogen and hydrogen diffusing into the material under plastic strain and simultaneous exposure to compressed hydrogen exert a particularly detrimental effect on the fluidity of the material.

A 27 Nickel-chromium alloys

A 28 Nickel-chromium-iron alloys (without Mo)

A 29 Nickel-chromium-molybdenum alloys

The absorption of hydrogen can exert an adverse effect on the toughness behavior of the nickel-chromium-iron alloys of type NiCrXFe9 with a value X between 10 and 30%, which are widely used in practice. Here, the alloys with a low chromium content are clearly less susceptible compared to those with a higher chromium content. Also the nickel-chromium-molybdenum alloys are less susceptible to hydrogen embrittlement compared to the molybdenum-free nickel-chromium alloys.

Due to its very good mechanical properties, the precipitation-hardening nickel-based super alloy INCONEL® alloy 718 (Alloy 718, UNS N07718, NiCr19Fe19Nb5-Mo3, Mat. No. 2.4668) is not only used at temperatures up to 700 °C, but also as a construction material for pressure vessels in hydrogen plants both at low and higher temperatures. However, at high temperatures this material can absorb hydrogen and, as a result of its high hardness, become susceptible to hydrogen embrittlement even at room temperature.

Fracture-mechanical tests and fatigue tests show that even at room temperature in compressed hydrogen the critical stress intensity values are clearly reduced and crack growth is increased.

The mechanical properties of the nickel-based alloy INCONEL® alloy 625 (NiCr22Mo9Nb, Mat. No. 2.4856) widely used in the chemical industry, in seawater applications, in aviation and in the power plant industry due to its good combination of strength, toughness, weldability and resistance against many aggressive atmospheres, can be affected in compressed hydrogen and the material can become susceptible to delayed fracture.

The susceptibility to hydrogen embrittlement of the precipitation-hardening material INCONEL® alloy 718 is clearly more pronounced than that of the single-phase material INCONEL® alloy 625.

A 30 nickel-copper alloys

The susceptibility of the nickel-copper alloys to hydrogen-induced embrittlement depends on the ratio of both alloying metals. From a nickel content of about 50% the alloys become susceptible to hydrogen embrittlement.

A 37 Tantalum, niobium and their alloys

Although the refractory metals vanadium, niobium and tantalum exhibit excellent corrosion resistance in a great number of media, they are extremely susceptible to embrittlement by hydrogen, which is either dissolved in the metal or can also form hydrides of the respective metal.

A 38 Titanium and titanium alloys

As with all refractory metals titanium materials can become brittle by the absorption of hydrogen and the formation of hydrides. Thermal oxidation treatment in air (15 min at 600 °C) reduces hydrogen absorption. In the presence of moisture in the compressed hydrogen, the formation of a protective oxide surface layer prevents the absorption of hydrogen. Under static or cyclic load the hydrogen-induced crack growth in titanium alloys depends, to a major extent, on the composition of the alloy and the associated microstructure (α-, ($\alpha + \beta$)-, β-alloy). High-strength titanium alloys of the type TiAlV and TiAlSn have proven to be successful in aircraft and spacecraft construction. Titanium aluminides of the type Ti_3Al (20–36 at.%Al) and TiAl (50.5–57.5 at.%Al) as well as the two-phase aluminides (TiAl + Ti_3Al) can be used at temperatures of above 700 °C and, hence, are an alternative to the common materials for spacecrafts.

A 40 Zirconium and zirconium alloys

As with all other refractory metals, also the absorption of hydrogen in zirconium leads to the formation of hydrides and a reduction of the toughness properties, which may cause the occurrence of delayed cracks.

B Nonmetallic inorganic materials

B 3 Carbon and graphite

Graphite impregnated with phenolic resin, which is used in the manufacture of chemical equipment or bipolar plates for fuel cells, is suited for temperatures up to 180 °C. Non-impregnated, high-purity graphite is resistant up to about 1,200 °C, however impurities may lower its resistance to about 800 °C.

B 4 Binders for building materials (e.g. concrete, mortar)

A hydraulically binding cement containing calcium monoaluminate as the mineralogical main phase, which is used, for instance, in mining, is resistant at ambient temperatures.

B 12 Oxide ceramic materials

Oxide ceramic materials are resistant up to high temperatures, i.e. quartz and quartz-containing refractory products up to 1,100 °C, high-purity aluminum oxide up to 1,600 °C, magnesium oxide up to 1,700 °C and zirconium dioxide up to 2,100 °C.

B 13 Metal-ceramic materials

Reaction-bound silicon carbide is resistant up to about 1,000 °C, whereas high-purity silicon carbide is resistant up to 1,300 °C.

C Organic materials/Plastics

Thermoplastics

The thermoplastic materials polyethylene and unplasticized polyvinyl chloride frequently used in the manufacture of chemical equipment are suited for temperatures up to 60 °C and polypropylene for temperatures up to 60–80 °C. The fluorothermoplastics also used in the manufacture of chemical equipment and in the corrosion protection sector are suited for temperatures up to 80–150 °C depending on their structure.
When using these materials, the increasing permeability of the materials with increasing temperature needs to be taken into consideration.

Thermoplastic elastomers

Thermoplastic elastomers based on polyester are resistant at room temperature.

Duroplastics

Duroplastics based on epoxy resins, unsaturated polyester resins, vinyl ester resins or phenol formaldehyde resins are resistant at least at room temperature and can be used up to 60 °C.

Elastomers

Elastomers frequently used for sealing purposes are suited based on chloroprene, acrylonitrile butadiene or hydrated acrylonitrile butadiene rubber up to 60–80 °C and based on ethylene propylene diene, fluoro or perfluoro rubber up to 80–100 °C.

D Materials with special properties

D 1 Coatings and films

Materials with special properties
   Fluorothermoplastic powder coatings can be used up to 130–150 °C, fully adhered fluorothermoplastic webs up to 110 °C and thermoplastic webs mechanically

anchored in concrete made from polyethylene, polypropylene or unplasticized polyvinyl chloride up to 60 °C and from polyvinylidene fluoride up to 100 °C.

Flat gaskets from fibers with rubber binding agents can be used up to about 150 °C, from polytetrafluoroethylene up to 240–260 °C and from graphite up to about 450 °C. Suitable materials for packings are polytetrafluoroethylene and graphite yarns or expanded graphite. For O-ring seals the above elastomers and polytetrafluoroethylene can be used.

Pipes and other components from glass fiber-reinforced epoxy resin or vinyl ester resins can be used up to 60–80 °C.

Bibliography

[1] Römpp Lexikon Chemie – Version 2.0,
Georg Thieme Verlag, Stuttgart/New York, 1999

[2] Häussinger, H.; Lohmüller, R.; Watson, A. M.
Ullmann's Encyclopedia of Industrial Chemistry, 6. Edition, 2002
Electronic Release
Wiley-VCH Verlag, Weinheim

[3] Company publication
Schreckenberg, W.; Palmen, A.; Schubert, M.
Gase-Handbuch, Broschüre 90.1001,
2. Auflage
Messer Griesheim GmbH, Düsseldorf

[4] Wendler-Kalsch, E.
Grundlagen und Mechanismen der Wasserstoff-induzierten Korrosion metallischer Werkstoffe
in: Wasserstoff und Korrosion, 2. Aufl.
Verlag Irene Kuron, Bonn, 2000, p. 7–53

[5] Smialowski, M.
Hydrogen in Steel
Pergamon Press, Oxford, 1962, p. 2

[6] Company publication
Gase – Handbuch, Broschüre 90.1001, 3. Auflage
Messer Griesheim GmbH, Frankfurt

[7] DIN 50922, (10/1985)
Korrosion der Metalle; Untersuchung der Beständigkeit von metallischen Werkstoffen gegen Spannungsrißkorrosion; Allgemeines
Beuth Verlag GmbH, Berlin

[8] DIN EN ISO 7539
Prüfung der Spannungsrißkorrosion
Beuth Verlag GmbH, Berlin
Teil 1: Allgemeine Richtlinien für Prüfverfahren (August 1995)
Teil 2: Vorbereitung und Anwendung von Biegeproben (August 1995)
Teil 4: Vorbereitung und Anwendung von einachsig belasteten Zugproben (August 1995)
Teil 5: Vorbereitung und Anwendung von C-Ring-Proben (August 1995)
Teil 6: Vorbereitung und Anwendung von angerissenen Proben für die Prüfung unter konstanter Kraft oder konstanter Verformung (Juli 2003)
Teil 7: Prüfung mit langsamer Dehngeschwindigkeit (Mai 2005)

[9] ASTM G38 (2001)
Standard Practice for Making and Using C-Ring Stress-Corrosion Test Specimens

[10] ASTM G39 (1999)
Standard Practice for Preparation and Use of Bent-Beam Stress-Corrosion Test Specimens

[11] Company publication
ASTM G49 (1985)
Standard Practice for Preparation and Use of Direct Tension Stress-Corrosion Test Specimens

[12] ASTM G 129-00 (11/2000)
Standard Practice for Slow Strain Rate Testing to Evaluate the Susceptibility of Metallic Materials to Environmental Assisted Cracking

[13] ASTM G 142-98 (4/1998)
Standard Test Method for Determination of Susceptibility of Metals to Embrittlement in Hydrogen Containing Environment at High Pressure, High Temperature, or both

[14] ASTM E 8-03 (7/2003)
Standard Test Methods for Tension Testing of Metallic Materials

[15] ISO 11782-2 (07/1998)
Korrosion von Metallen und Legierungen – Schwingungsrißkorrosion – Teil 1: Lebensdauerprüfung
Beuth Verlag GmbH, Berlin

[16] ISO 11782-2 (07/1998)
Korrosion von Metallen und Legierungen – Schwingungsrißkorrosion – Teil 2: Prüfung des Rißwachstums mittels angerissener Proben
Beuth Verlag GmbH, Berlin

[17] DIN 50905-4, (01/1987)
Korrosion der Metalle; Korrosionsuntersuchungen; Durchführung von chemischen Korrosionsversuchen ohne mechanische Belastung in Flüssigkeiten im Laboratorium
Beuth Verlag GmbH, Berlin

[18] DIN EN ISO 2626 (August 1995)
Kupfer; Wasserstoff-Versprödungsversuch
Beuth Verlag GmbH, Berlin

[19] Fidelle, J. P.; Deloron, J. M.; Roux, C.; Rapin, M.
Influence de traitements de surface et de revêtements sur la fragilisation d'aciers a haute résistance par l'hydrogène sous pression
7ème Congrès International du Traitement de Surface des Métaux, Hannover, 5–7 Mai, 1968

[20] Drodten, P.; König, S.
Bestimmung des Wasserstoffgehaltes im Stahl mittels Heißextraktion
in: Kuron, D.
Wasserstoff und Korrosion, 2. Aufl.
Verlag Irene Kuron, Bonn, 2000, p. 384–398

[21] Klueh, R. L.; Mullins, W. W.
Some observations on hydrogen embrittlement of silver
Transactions of the Metallurgical Society of AIME 242 (1968) 2, p. 237–244

[22] Bond, G. M.; Robertson, I. M.; Birnbaum; H. K.
Effects of hydrogen on deformation and fracture processes in high-purity Aluminium
Acta metall. 36 (1988) 8, p. 2193–2197

[23] Ferreira, P. J.; Robertson, I. M.; Birnbaum, H. K.
Hydrogen effects on the character of dislocations in high-purity aluminum
Acta materialia 47 (1999) 10, p. 2991–2998

[24] Anyalebechi, P. R.
Analysis of the effects of alloying elements on hydrogen solubility in liquid aluminum alloys
SCR. Metall. Matter. 33 (1995) 8, p. 1209–1216

[25] Afanasév, V. K.
Some regularities of plasticity changes in aluminium and aluminium alloys (in Russisch)
IZV. AKAD. NAUK SSSR, MET. (1978) 6, p. 195–199

[26] Wendler-Kalsch, E.
Grundlagen und Mechanismen der Wasserstoff-induzierten Korrosion metallischer Werkstoffe
in: Kuron, D.
Wasserstoff und Korrosion, 2. Aufl.
Irene Kuron, Bonn, 2000, S. 44

[27] Vennet; R. M.; Ansell, G. S.
A study of gaseous hydrogen damage in certain fcc metals
Transactions of of the ASM 62 (1969) Dec., p. 1007–1013

[28] Company publication
Campbell, J. E.
Effects of hydrogen gas on metals at ambient temperature, DMIC Report S-31, April 1970
Defence Metals Information Center, Columbus, Ohio (USA)

[29] Srivatsan, T. S.; Sudarshan, T. S.
Fracture of precipitation strengthened aluminum alloys – role of environment
Engineering Fracture Mechanics 37 (1990) 3, p. 569–589

[30] Kumar, K. S.; John, K. M.; Natarajan, A.; Lakshmanan, T. S.:
Metallurgical aspects of stress assisted failure in high strength aluminum alloy
Prakt. Metallogr. 31 (1994) 11, p. 586–595

[31] Sudarshan, T. S.; Wilson, J. H.; Mabie, H. H.; Louthan, M. R.; Blacksburg
Environmental effects on the torsional fatigue fracture of 2024 aluminum
Aluminium 60 (1984) 5, p. 345–346

[32] Sudarshan, T. S.; Louthan, M. R.
Role of hydrogen and humidity in the torsional fatigue of aluminum alloy 2024-T351
Materials Science and Engineering 73 (1985), p. 131–138

[33] Speidel, M. O.
Hydrogen embrittlement of aluminium alloys?
in: Bernstein, I. M.; Thompson, W.
Hydrogen in Metals
American Society for Metals, Pa., 1973, p. 249–276

[34] Suresh, S.; Palmer, I. G.; Lewis, R. E.
The effect of environment on fatigue crack growth behavior of 2021 aluminum alloy
Fatigue of Engineering Materials and Structures 5 (1982) 2, p. 133–150

[35] Ruiz, J.; Elices, M.
The role of environmental exposure in the fatigue behaviour of an aluminium alloy
Corrosion Science 39 (1997) 12, p. 2117–2141

[36] Christodoulou, L.; Flower, H. M.
In-situ H.V.E.M. observations of hydrogen embrittlement in Al-Zn-Mg alloys
in: Bernstein, I. M.; Thompson, A. W.
Hydrogen Effects in Metals
The Metallurgical Society of AIME, Pittsburgh, Pennsylvania, 1980, p. 500

[37] Gräfen, H.
Zur Frage der Mitwirkung des Wasserstoffs bei der "anodischen" Spannungsrisskorrosion
in: Kuron, D.
Wasserstoff und Korrosion, 2. Aufl.
Irene Kuron, Bonn, 2000, S. 311

[38] Ciaraldi, S. W.; Nelson, J. L.; Yeske, R. A.; Pugh, E. N.
Studies of hydrogen embrittlement and stress-corrosion cracking in an aluminum-zinc-magnesium alloy
in: Bernstein, I. M.; Thompson, A. W.
Hydrogen effects in Metals
The Metallurgical Society of AIME, Pittsburg, Pennsylvania, 1980, p. 437–447

[39] Hilliard, O. H.
Fatigue crack initiation and propagation in aluminum alloys 8090 and 2124 in hydrogenous atmospheres
Dissertation Abstracts International 53, (12) University of Idaho, 1993, p. 1–213

[40] Zhen, L.; Yang, D. Z.; Yu, G. F.; Hou, S. E.
Hydrogen induced fracture in a RSP Al-Li alloy
Scr. Metall. Mater. 31 (1994) 5, p. 595–599

[41] Yukawa, H.; Morinaga, M.; Takahashi, Y.
Interactions between hydrogen and solute atoms in aluminium alloys
Advanced Materials '93, III/A 16A (1994), p. 225–228

[42] Talbot, D. E. J.
Hydrogen in aluminum and its alloys (Dissertation)
Dissertation abstract international 51 (1990) 5, p. 355

[43] Nagao, A.; Kuramoto, S.; Kanno, M.
Detection and visualization of hydrogen in aluminum
Journal of Japan Institute of Light Metals 49 (1999) 2, p. 89–96

[44] Hess, P. D.; Turnbill, G. K.
Effects of hydrogen on properties of aluminium alloys
in: Bernstein, I. M.; Thompson, W.
Hydogen in Metals
American Society for Metals, Pa., 1973, p. 277–287

[45] Das, K.B.; Roberts, E. C.; Bassett, R. G.
An investigation of blistering in 2024-T6 aluminium alloy
in: Bernstein, I. M.; Thompson, A. W.
Hydrogen in Metals
American Society for Metals, Pa., 1973, p. 289–299

[46] Jones, R. H.; Henager, C. H.; Trzaskoma, P. P.; Stoloff, N. S.; Moffat, T. P.; Lichter, B. D.
Environmental effects on advanced materials
Journal of Metals (1988), p. 18–30

[47] Nambu, T.; Fukumori, J.; Morinaga, M.; Matsumoto, Y.; Sakaki, T.
Environmental effects on the ductility of pure chromium
Scripta Metallurgica et Materialia 32 (1995) 3, p. 407–410

[48] Matsumoto, Y.; Fukumori, J.; Morinaga, M.; Furui, M.; Nambu, T.; Sakaki, T.
Alloying effect of 3d transition elements on the ductility of chromium
Scripta Materialia 34 (1996) 11, p. 1685–1689

[49] DIN EN 1976 (05/1998)
Kupfer und Kupferlegierungen, Gegossene Rohformen aus Kupfer
Beuth Verlag GmbH, Berlin

[50] DIN 1708 (01/1998)
Kupfer – Kathoden und Gußformate
Beuth Verlag GmbH, Berlin

[51] Fast, J. D.
Interactions of metals and gases, Vol. 1 Thermodynamics and phase relations
Academic Press, New York, 1965, p. 54

[52] Wyman, L. L.
Copper embrittlement, III
Trans. AIME 111 (1934) 205

[53] Rhines, F. N.; Anderson, W. A.
Hydrogen embrittlement of pure copper and of dilute copper alloys by alternate oxidation and reduction
Trans. AIME 143 (1941) 312

[54] Baukloh, W.; Stromburg, W.
Über die Wasserstoffkrankheit einiger Metalle
Z. Metallkunde 29 (1937), S. 427–430

[55] Matting, A.; Ziegler, R.
Brittleness in copper and copper alloys with particular reference to hydrogen embrittlement
Technical Report
British Engine Boiler and Electrical Insurance Co. Ltd., Manchester (UK), Vol. 7 (1966), p. 59–71

[56] Kauczor, M.
Die Wasserstoffkrankheit des Kupfers
Metall 19 (1965) 11, S. 1186–1187

[57] von Franque, O.; Lindau, E.
Beobachtungen zum Einfluß des Gefüges und der Glühtemperatur auf die Wasserstoffversprödung von Kupfer
Metall 20 (1966) 11, S. 1140–1143

[58] Ransley, J.
Inst. Metals 65 (1939), p. 147

[59] Harper, S.; Calcutt, V. A.; Townsend, D. W.; Eborall, R.
J. Inst. Metals 90 (1962), p. 414, 423

[60] Brokop, H.
Werkstoffschäden durch Wasserstoffversprödung
Damage to materials by the action of hydrogen
Praktische Metallographie 12 (1975) 9, p. 470–475

[61] Belkin, E.; Nagata, P. K.
Hydrogen embrittlement of tough pitch copper by brazing
Welding Research Supplement 54 (1975) 2, p. 54s-62s

[62] Daneliya, I. S.; Rozenberg, V. M.; Solopov, V. I.
Influence of the matrix structure and dispersed oxide particles on the hydrogen embrittlement of copper
Phys. Met. Metall. 44 (1978) 2, p. 80–84

[63] ASTM B 170-99 (07/1999)
Standard Specification for Oxygen-Free Electrolytic Copper – Refinery Shapes
Beuth Verlag GmbH, Berlin

[64] ASTM B 379-99 (07/1999)
Standard Specification for Phosphorized Coppers – Refinery Shapes
Beuth Verlag GmbH, Berlin

[65] ASTM b 577-99 (1998)
Standard Test Methods for Detection of Cuprous Oxide (Hydrogen Embrittlement Susceptibility) in Copper
Beuth Verlag GmbH, Berlin

[66] Goods, S. H.
The evolution of damage in tritium exposed copper
Journal of Materials Research 6 (1991) 2, p. 303–313

[67] Vennett, R. M.; Ansell, G. S.
A study of gaseous hydrogen damage in certain fcc metals
Transactions of the ASM 62 (1969), p. 1007–1013

[68] Caffrey, A. J.; Spaletta, H. W.; Ware, A. G.; Zabriskie, J. M., Hardwick, D. A.; Maltrud, H. R.; Paciotti, M. A.
High pressure deuterium-tritium gas target vessels for muon-catalyzed fusion experiments
Hydrogen effects on material behavior
TMS, The Minerals, Metals & Materials Society, 1990, p. 1047–1055

[69] Parr, R. A.; Johnston, M. H.; Davis, J. H.; Oh, T. K.
Determination of the gaseous hydrogen ductile-brittle transition in copper-nickel alloys
Report No. N90-12715
NASA Technical Memorandum, February 1985

[70] Otsuka, R.; Maruno, T.; Tauji, H.
Correlation between hydrogen embrittlement and metal hydride formation in Ni-Cu and PdAg alloys
Conference : International Congress on Metallic Corrosion, Vol. 2; Toronto, Canada, 3–7 June 1984
National Research Council of Canada, Ottawa, Canada, 1984, 270–277

[71] Erdmann-Jesnitzer, F.; Wessel, A.
Untersuchungen zur Wasserstoffversprödung einiger kubisch-flächenzentrierter Werkstoffe
Archiv Eisenhüttenwesen 52 (1981) 2, S. 77–82

[72] Spitzig, W. A.; Peterson, D. T.; Laabs, F. C.
Effect of hydrogen on the mechanical properties of a deformation processed Cu-20%Nb composite
Journal of materials science 26 (1991) 8, p. 2000–2006

[73] Martens, A.
Untersuchung eiserner Behälter zur Aufbewahrung von Wasserstoffgas
Zeitschrift VDI 40 (1996) 20, S. 717

[74] Hofmann, W.; Rauls, W.
Erzeugung von Fischaugen in Stahl durch äußere Einwirkung von Wasserstoff bei Raumtemperatur
Archiv Eisenhüttenwesen 32 (1961) 3, S. 169–171

[75] Hofmann, W.; Rauls, W.
Die Änderung der Verformungsfähigkeit von Stahl bei der Einwirkung von Druckwasserstoff in der Nähe der Raumtemperatur
Archiv Eisenhüttenwesen 34 (1963), S. 925

[76] Hofmann, W.; Rauls, W.
Ductility of steel under the influence of external high pressure hydrogen
Welding Research Supplement (1965) 5, p. 225s-230s

[77] Hofmann, W.; Voigt, J.
Das Standverhalten niedriglegierter hochfester Stähle an Luft unter Druckwasserstoff im Temperaturbereich von 20 bis 100 °C
Archiv Eisenhüttenwesen 35 (1964) 6, S. 551–559

[78] Wiester, H.-J.; Dahl, W.; Hengstenberg, H.
Über den Einfluß von Wasserstoff auf die Vorgänge beim Zugversuch
Archiv für das Eisenhüttenwesen 34 (1963) 12, S. 915–924

[79] Riecke, E.
Wasserstoff in Eisen und Stahl
Archiv Eisenhüttenwesen 49 (1978), S. 391

[80] Kesten, M.; Gräfen, H.
Druckwasserstoffangriff auf un- und niedriglegierte Stähle unterhalb 200 °C Wasserstoff und Korrosion, 2. Aufl.
Verlag Irene Kuron, Bonn, 2000, S. 101–126

[81] Gräfen, H.; Kuron, D.
Werkstoffverhalten in Wasserstoff
Chemie. Ing.-Tech. 59 (1987) 7, S. 555–563

[82] Gräfen, H.
Wechselwirkung zwischen Gas und Metall unter besonderer Berücksichtigung der mechanischen Belastungsart
Zeitschrift für Werkstofftechnik 9 (1978), S. 391–400

[83] Günther, T.; Gräfen, H.
Wasserstoffversprödung von Feinkornbaustählen in Abhängigkeit von der Legierungszusammensetzung, der Gefügeausbildung und der mechanischen Belastung
Zeitschrift für Werkstofftechnik 10 (1979) S. 373–390

[84] Hanninen, H. E.; Lee, T. C.; Robertson, I. M.; Birnbaum, H. K.
Direct observations of hydrogen effects on fracture of A533B steel
Conference: Corrosion – Deformation Interactions, Fontainebleau, France, 5–7 Oct. 1992
Les Editions de Physique, Les Ulis Cedex A, France, 1993, 377–388

[85] Schmitt, G.; Savakis, S.
Untersuchungen zur Schädigung höherfester niedriglegierter Stähle durch Druckwasserstoff bei statischer und dynamischer Beanspruchung
Werkst. Korros. 42 (1991) 12, S. 605–619

[86] Schmitt, G.; Savakis, S.
Influence of hydrogen purity on HSCC of high strength low alloy steel
Conference: Eurocorr '87 European Corrosion Meeting, Preprints
Karlsruhe, Germany, 6–10 April 1987
DECHEMA, Frankfurt am Main, Germany, 1987, 493–498

[87] Vibrans, G.; Mathes, M.
Zur Plastoermüdung von Stahl unter dem Einfluß von Wasserstoff
Zeitschrift für Werkstofftechnik 10 (1979), S. 57–59

[88] Mathes, M.; Vibrans, G.
Bruchverhalten eines vergüteten und eines kaltgezogenen Stahles im Gleichgewichtszustand mit Druckwasserstoff
Zeitschrift für Werkstofftechnik 14 (1983), S. 284–288

[89] Mathes, K.; Vibrans, G.
Wasserstoffaufnahme von Stahl bis zum Gleichgewicht mit der Gasphase bei Raumtemperatur
Archiv für das Eisenhüttenwesen 53 (1982) 3, S. 119–122

[90] Gräfen, H.; Pöpperling, R.; Schlecker, H.; Schlerkmann, H.; Schwenk, W.
Untersuchungen an Leitungsrohrstählen über eine Korrosionsgefährdung durch wasserstoffhaltige Gase bei hohen Drücken
Werkst. Korros. 39 (1988), S. 517

[91] Gräfen, H.; Pöpperling, R.; Schlecker, H.; Schlerkmann, H.; Schwenk, W.
Zur Frage der Schädigung von Hochdruckleitungen durch Wasserstoff oder wasserstoffhaltige Gasgemische
Gas-Erdgas 130 (1990), S. 16

[92] Hardie, D.; Liu, S.
The effect of stress concentration on hydrogen embrittlement of a low alloy steel
Corrosion Science 38 (1996) 5, p. 721–733

[93] Naumann, F. K.
Einwirkung von Wasserstoff unter hohem Druck auf unlegierten Stahl
Stahl und Eisen 57 (1937) 32, S. 889–899

[94] Spähn, H.
Druckwasserstoffangriff auf unlegierte und niedriglegierte Stähle im Temperaturbereich oberhalb 200 °C
in: Kuron, D.
Wasserstoff und Korrosion, 2. Aufl.
Irene Kuron, Bonn, 2000, S. 129–183

[95] Naumann, K. F.
Der Einfluß von Legierungszusätzen auf die Beständigkeit von Stahl gegen Wasserstoff unter hohem Druck
Stahl und Eisen 58 (1938) 44, S. 1239–1250

[96] Class, I.
Stand der Kenntnisse über die Eigenschaften druckwasserstoffbeständiger Stähle
Stahl und Eisen 80 (1960) 17, S. 1117–1135

[97] Shewmon, P. G.
Hydrogen attack of pressure vessel steels
Materials Science and Technology 1 (1985) 1, p. 2–11

[98] Parthasarathy, T. A.; Lopez, H. H.; Shewmon, P. G.
Hydrogen attack kinetics of 2.25Cr-1Mo steel weld metals
Metallurgical Transactions A 16A (1985) 6, p. 1143–1149

[99] Parthasarathy, T. A.; Shewmon, P. G.
Hydrogen attack behavior of the heat affected zone of a 2.25Cr-1Mo steel weldment
Metallurgical Transactions A 18A (1987) 7, p. 1309–1312

[100] DIN EN 10028-2 (09/2003)
Flacherzeugnisse aus Druckbehälterstählen – Teil 2: Unlegierte und legierte Stähle mit festgelegten Eigenschaften bei erhöhten Temperaturen
Beuth Verlag GmbH, Berlin

[101] Ruoff, S.; Stone, D.; Li, C.-Y.
Hydrogen attack in a 3Cr-1.5Mo steel at elevated temperatures
Materials Science and Engineering 93 (1987) 1–2, p. 217–225

[102] Archakov, Y. I.; Grebeshkova, I. D.
The influence of phase content on steel hydrogen stability at elevated temperatures and pressures
Conference: 10th International Congress on Metallic Corrosion, Vol. 3; Sessions 10–13, Madras, India, 7–11 Nov. 1987
Key Eng. Mater. 3 (1988) p. 2421–2433

[103] Stahl-Eisen-Werkstoffblatt 590-61 (12/1961)
Druckwasserstoffbeständige Stähle
Verlag Stahleisen, Düsseldorf

[104] DIN 17176 (11/1990)
Nahtlose kreisförmige Rohre aus druckwasserstoffbeständigen Stählen
Beuth Verlag GmbH, Berlin

[105] Stahl-Eisen-Werkstoffblatt 595, 2. Ausgabe (08/1976)
und
Stahl-Eisen-Werkstoffblatt 595, 3. Ausgabe (Entwurf Januar 1996)
Stahlguß für Erdöl- und Erdgasanlagen
Verlag Stahleisen, Düsseldorf

[106] DIN EN 10216-2 (08/2002) und (07/2004)
Nahtlose Stahlrohre für Druckbeanspruchungen
Technische Lieferbedingungen
Teil 2: Rohre aus unlegierten und legierten Stählen mit festgelegten Eigenschaften bei erhöhten Temperaturen
Beuth Verlag GmbH, Berlin

[107] VdTÜV – Werkstoffblätter
TÜV-Verlag GmbH, Köln

[108] Company publication
Nelson, G. A.
Hydrogenation Plant Steels, API Proceedings, 1949
American Petroleum Institute (API) Vol. 29m, 163, New York (NY/USA)

[109] API Refining Department
Steels for hydrogen service at elevated temperatures and pressures in petroleum refineries and petrochemical plants
API-Recommended Practice 941, Sixth Edition, 03/2004
American Petroleum Institute, Publishing Services
Washington D.C.

[110] Birring, A. S.; Bartlett, M. L.; Kawano, K.
Ultrasonic detection of hydrogen attack in steels
Corrosion Engineering 45 (1989) 3, p. 259–263

[111] Cantwall, J. E.
High-temperature hydrogen attack
Materials Performance 33 (1994) 7, p. 58–61

[112] Heuser, A.; Wagner, G. H.; Heinke, G.; Cihal, V.
Hydrogen attack on steel for high-pressure hydrogen service as a result of the alteration of carbides by ammonia synthesis gas
Steel research 64 (1993) 8/9, p. 454–460

[113] Jones, R. H.
Application of hydrogen embrittlement models to the crack growth behavior of fusion reactor materials
J. Nucl. Mater. 141–143 (1986), p. 468–475

[114] Jung, P.
A hydrogen problem in fusion material technology
Fusion Technology 33 (1998) 1, p. 63–67

[115] Warren, D.
Hydrogen effects on steel
Mater. Performance 26 (1987) 1, p. 38–48

[116] Huang, J.-H.; Altstetter, C. J.
Internal hydrogen embrittlement of a ferritic stainless steel
Metall. Mater. Trans. A 26A (1995) April, p. 845–849

[117] Perng, T.-P.; Johnson, M.; Altstetter, C.J.
Influence of plastic deformation on hydrogen diffusion and permeation in stainless steels
Acta Metall. 37 (1989) 12, p. 3393–3397

[118] Krysiak, K. F.; Grubb, J. F.; Pollard, B.; Campbell, R. D.
Selection of wrought ferritic stainless steels
ASM Handbook. Vol. 6: Welding, Brazing, and Soldering, Bd. 6
ASM International, 1993, p. 443–455

[119] Perujo, A.; Serra, E.; Alberici, S.; Tominetti, S.; Camposilvan, J.
Hydrogen in the martensitic DIN 1.4914: a review
Journal of alloys and compounds 253–254 (1997), p. 152–155

[120] Perng, T.-P.; Altstetter, C. J.
Cracking kinetics of two-phase stainless steel alloys in hydrogen gas
Metallurgical Transactions A 19A (1988) 1, p. 145–152

[121] Oltra, R.; Bouillot, C. Magnin, T.
Role of hydrogen in stress corrosion cracking of duplex stainless steels
Second International Conference on Corrosion-Deformation Interactions
Institute of Materials, London, 1997, p. 248–253
or:
Oltra, R.; Bouillot, C.; Magnin, T.
Localized hydrogen cracking in the austenitic phase of a duplex stainless steel
Scripta Materialia 35 (1996) 9, p. 1101–1105

[122] Oltra, R., Bouillot, C.
Experimental investigation of the role of hydrogen in stress corrosion cracking of duplex stainless steels
Hydrogen Transport and Cracking in Metals
Institute of Materials, London, 1994, p. 17–26

[123] Huang, J.-H.; Altstetter, C. J.
Cracking of duplex stainless steel due to dissolved hydrogen
Metall. Mater. Trans. A 26 A (1995) 5, p. 1079–1085

[124] Zheng, W.; Hardie, D.
Effect of structural orientation on the susceptibility of commercial duplex stainless steels to hydrogen embrittlement
Corrosion 47 (1991) 10, p. 792–799

[125] Zheng, W.; Hardie, D.
Crack initiation and propagation of 2205 duplex stainless steel in hydrogen gas
in: Conference: Corrosion Control – 7th APCCC, Bd. 1
International Academic Publishers, Beijing, China, 1991, p. 414–418

[126] El-Yazgi, A. A.; Hardie, D.
The embrittlement of a duplex stainless steel by hydrogen in a variety of environments
Corrosion Science 38 (1996) 5, p. 735–744

[127] El-Yazgi, A. A.; Hardie, D.
Effect of heat treatment on susceptibility of duplex stainless steel to embrittlement by hydrogen.
Materials Science and Technology 16 (2000) 5, p. 506–510

[128] Marrow, T. J.; Hippsley, C. A.; King, J. E.
Effect of mean stress on hydrogen assisted fatigue crack propagation in duplex stainless steel
Acta Metallurgia et Materialia 39 (1991) 6, p. 1367–1376

[129] Marrow, T. J.; King, J. E.
The combined effects of aging and hydrogen embrittlement on fatigue crack propagation in a duplex stainless steel
Conference: Fracture of Engineering Materials and Structures, Singapore
Elsevier Science Publishers Ltd., United Kingdom, 1991, p. 747–752

[130] Tsu, I.-F.; Perng, T.-P.
Hydrogen compatibility of femnal alloys
Metallurgical Transactions A, 22A (1991) 1, p. 215–224

[131] Perng, T.-P.; Altstätter, C. J.
Hydrogen effects in austenitic stainless steels
Materials Science and Engineering A, A129 (1990) 1, p. 99–107

[132] Liu, J.; Li, Y. Y.; Hu, Z. Q.
Effect of directional solidification on resistance to hydrogen embrittlement of a stainless steel
Mater. Sci. Eng. A A117 (1989) 1–2, p. 221–226

[133] Ma, L. M.; Liang, G. J.; Fan, C. G.; Li, Y. Y.
Effect of microstructure on hydrogen damage of JBK-75 pecipitate-strengthened austenitic steel
Acta Metallurgica Sinica (Englisch Letters) 10 (1997) 3, p. 206–212

[134] Tan, Y.; Zhou, d. h.; Feng, J.
Influence of internal hydrogen on the hydrogen embrittlement of austenitic stainless steel
Acta Metallurgica Sinica (English Letters) 10 (1997) 3, p. 228–232

[135] Li, Y. Y.; Xing, Z. S.
Effect of hydrogen on the microstructure, mechanical properties and phase transformations in austenitic steels
J. Nucl. Mater. 169 (1989) II, p. 151–157

[136] Liu, J.; Guo, Y.; Xing, Z. Shi, C. H.
Effects of hydrogen and deformation on structure and properties of Fe-Cr-Ni-Mn-N-steels
Acta Metallurgica Sinica, Series A 1 (1988) 3, p. 191–197

[137] Robinson, S. L.; Thomas, G. J.
Accelerated fracture due to tritium and helium in 21-6-9 stainless steel
Metallurgical Transactions A 22A (1991) 4, p. 879–886

[138] Chene, J.; Brass, A. M.
Deuterium and tritium applications to the quantitative study of hydrogen local concentration in metals and related embrittlement
in: Thompson, A. W.
Hydrogen Effects in Materials
Minerals, Metals and Materials Society/AIME, Warrendale, PA, 1996, p. 47–59

[139] Iyer, K. J. L.
The influence of hydrogen on the mechanical properties and structure of a stable 304 stainless steel
Can. Metall. Q 28 (1989) 2, p. 153–158

[140] Yokogawa, K.; Han, G.; He, J.; Fukuyama, S.
Effect of strain-induced martensite on hydrogen environment embrittlement of sensitized austenitic stainless steels at low temperatures
Acta Materialia 46 (1998) 13, p. 4559–4570

[141] Matsuda, F.; Ushio, F.; Nakagawa, M.; Nakata, K.
Hydrogen embrittlement of SUS 316 austenitic stainless steel weldments
Trans. JWRI 14 (1985) 1, p. 63–67

[142] Yang, K.; Xie, Y.; Zhao, X.; Fan, C. G.; Li, Y. Y.
Microstructure and hydrogen embrittlement in Incoloy 907
Scripta Metallurgica et Metallica 25 (1991) 10, p. 2399–2404

[143] Moody, N. R.; Stoltz, R. E.; Perra, M. W.
The relationship between hydrogen-induced thresholds, fracture toughness and fracture modes in In 903
Conference: Corrosion cracking, Salt Lake City, Utah, USA, 2–6 Dec. 1985
American Society for Metals, Metals Park (Ohio/USA), 1986, 43–53

[144] Moody, N. R.; Perra, M. W.; Robinson, S. L.
Hydrogen-induced cracking in an iron-based superalloy
in: Moody, N. R.; Thompson, A. W.
Hydrogen effects on material behavior
TMS, The Minerals, Metals & Materials Society, 1990, p. 625–635

[145] Chen, P. S.; Panda, B.; Bhat, B. N.
NASA-HR-1, a new hydrogen-resistent Fe-Ni-base superalloy
in: Thompson, A.W.; Moody, N. R.
Hydrogen effects in materials
Minerals, Metals and Materials Society, Warrendale, Pa, 1996, p. 1011–1020

[146] Louthan, M. R.; Caskey, G. R.; Donovan, J. A.
Metallographic studies of hydrogen-embrittled nickel
Microstructure Science 3B (1975), p. 823–834

[147] Windle, A. H.; Smith, G. C.
The effect of hydrogen on the deformation and fracture of polycrystalline nickel
Metal Science Journal 4 (1970) 6, p. 136–144

[148] Kandar, M. H.
Embrittlement of nickel by gaseous hydrogen
2nd International Congress "Hydrogen in Metals"
Paris, France, 6–10 June 1977

[149] Morrison, R. A.
Some properties of Ni-Cu foils and their behaviour when heated in hydrogen
Journal Vacuum Science and Technology 12 (1975) 1, p. 598–600

[150] Verpoort, C.; Duquette, D. J.; Stoloff, N. S.; Neu, A.
The influence of plastic deformation on the hydrogen embrittlement of nickel
Materials Science and Engineering 64 (1984) 5, p. 135–145

[151] Matsumoto, T.; Birnbaum, H. K.
Hydrogen embrittlement of nickel
Conference: Hydrogen in Metals, Japan, 26.–29. Nov. 1979
Japan Institute of Metals, Aoba Aramaki, Sendai 980, Japan, 1980, 493–496

[152] Eastman, J.; Matsumoto, T.; Narita, N.; Heubaum, F.; Birnbaum, H. K.
Hydrogen effects in nickel – Embrittlement or enhanced ductility
in: Bernstein, I. M.; Thompson, A. W.
Hydrogen Effects in Metals
The Metallurgical Society of AIME, 1980, p. 397–409

[153] Lynch, S. P.
A fractographic study of hydrogen-assisted cracking and liquid-metal embrittlement in nickel
Journal of Materials Science 21 (1986) 2, p. 692–704

[154] Lynch, S. P.
Hydrogen embrittlement and liquid-metal embrittlement in nickel single crystals
Scripta Metallurgica 13 (1979) 11, p. 1051–1056

[155] Stoltz, R. E.; West, A. J.
Hydrogen assisted fracture in fcc metals and alloys
in: Bernstein, I. M.; Thompson, A. W.
Hydrogen Effects in Metals
The Metallurgical Society of AIME, 1980, p. 541–553

[156] Symons, D. M.
Hydrogen embrittlement of Ni-Cr-Fe alloys
Metallurgical and Materials transactions A 28A (3) 1997, p. 655–663

[157] Symons, D. M.
The effect of hydrogen on the fracture toughness of alloy X-750 at elevated temperatures
Journal of Nuclear Materials 265 (1999) 3, p. 225–231

[158] Symons, D. M.; Thompson, A. W.
The effect of hydrogen on the fracture toughness of Alloy X-750
Metallurgical and Materials Transactions A 28A (1997) 3, p. 817–823

[159] Kekkonen, T.; Hanninen, H.
Effect of heat treatment on the hydrogen embrittlement of Inconel X-750 alloy
Conference: Predictive capabilities in environmental assisted cracking, Miami Beach, Florida, USA, 17–22 Nov. 1985
Publ. American Society of Mechanical Engineers, New York, USA, 1985, 259–272

[160] Sorokina, N. A.; Sergeeva, T. K.; Rusinovich, Y. I.; Rastorgueva, I. A.; Galtsova, V. I.; Shumilov, L. G.
Hydrogen embrittlement resistance of nickel alloys with different alloy contents
Sov. mater. Sci. 21 (1985) 1, p. 25–30

[161] Company publication
Strom für die Weltraumstation, Nickel Magazin, Juni 2002
NDI Nickel Development Institute

[162] Harris jr., J. A.; van Wanderham, M. C.
Various mechanical tests used to determine the susceptibility of metals to high-pressure hydrogen
ASTM STP543, ASTM, (1974), p. 198–220

[163] Chandler; W. T.; Walter, R. J.
Testing to determine the effect of high-pressure hydrogen environments on the mechanical properties of metals
ASTM STP543, ASTM, (1974) p. 170–197

[164] Grove, C. A.; Petzold, L. D.
Corrosion of Nickel-Base Alloys
ASM, Mars, PA, 1985, pp. 165–80.

[165] Company publication
Werkstoffdaten Hochleistungswerkstoffe, Druckschrift N 530 93-08, August 1993
ThyssenKrupp VDM, Werdohl

[166] Fritzemeier, L. G.; Walter, R. J.; Meisels, A. P.; Jewett. R. P.
Hydrogen embrittlement research: A Rocketdyne overview
in: Moody, N. R.; Thompson, A. W.
Hydrogen effects on metal behavior
TMS, The Minerals, Metals & Materials Society, 1989, p. 941–954

[167] Walter, R. J.; Chandler, W. T.
Influence of hydrogen environments on crack growth in Inconel 718
Conference: Environmental degradation of engineering materials, Blacksburg, Va., 10–12 10.1977, Proceedings: 513–522

[168] Hicks, P. D.; Altstetter, C. J.
Hydrogen embrittlement of superalloys
in: Moody, N. R.; Thompson, A. W.
Hydrogen effects on material behavior
TMS, The Minerals, Metals & Materials Society, 1990, p. 613–623

[169] Hicks, P. D.; Altstetter, C. J.
Internal hydrogen effects on tensile properties of iron- and nickel-base alloys
Metall. Trans. A, 21A (1990), 365–372.

[170] Hicks, P. D.; Altstetter, C. J.
Metall. Trans. A, 23A (1992), 237

[171] Fukuyama; S.; Yokogawa, K.; Araki, M.; Koyari, Y.; Aoki, H.; Yamada, Y.
Fatigue crack growth properties of Ni-base alloys in high pressure hydrogen at room temperature
J. Soc. Mater. Sci. Jpn., 38 (1989) p. 539–545

[172] Fukuyama, S.; Yokogawa, K.; Araki, M.; Koyari, Y.; Yamada, Y.
Tensile Properties of Ni-base alloys in high pressure hydrogen at room temperature
J. Soc. Mater. Sci. Jpn., 40 (1991), 736–742

[173] Fukuyama, J.; Yokogawa, K.; Yamada, Y.; Iida, T.
Hydrogen embrittlement of Ni-base superalloys used for H-II rocket engine
J. Iron and Steel Inst. Jpn., 78 (1992), 860–869

[174] He, J.; Fukuyama, S.; Yokogawa, K.; Kimura, A.
Effect of hydrogen on deformation structure of Inconel 718
Mater. Trans., JIM 35 (1994) 10, p. 689–694

[175] Hirose, A.; Arita, Y.; Nakanishi, Y.; Kobayashi, K. F.
Decrease in hydrogen embrittlement sensitivity of INCONEL 718 by laser surface softening
Materials Science and Engineering A219 (1996) 1–2, p. 71–79

[176] Walston, W. S.; Thompson, A. W.; Bernstein, L. M.
The effect of hydrogen on the deformation and fracture behavior of a single crystal nickel-base superalloy
in: Moody, N. R.; Thompson, A. W.
Hydrogen effects on material behavior
TMS, The Minerals, Metals & Materials Society, 1990, p. 581–589

[177] Gayda, J.; Dreshfield, R. L.; Gabb, T. P.
Effect of porosity and gamma-gamma' eutectic content on the low cycle fatigue behaviour of hydrogen-charged PWA 1480
Report No.: N93-31576/0/XAB
GOV. RES. ANNOUNC. INDEX Pp 7 (1991), p. 179–185

[178] Gayda, J.; Gabb, T. P.; Dreshfield, R. L.
The effect of hydrogen on the low cycle fatigue behavior of a single crystal superalloy
in: Moody, N. R.; Thompson, A. W.
Hydrogen effects on material behavior
TMS, The Minerals, Metals & Materials Society, 1990, p. 591–601

[179] DeLuca, D. P.; Cowles, B. A.
Fatigue and fracture of single crystal nickel in high pressure hydrogen
in: Moody, N. R.; Thompson, A. W.
Hydrogen effects on material behavior
TMS, The Minerals, Metals & Materials Society, 1990, p. 603–612

[180] Dollar, M.; Bernstein, I. M.
The effect of hydrogen on deformation substructure, flow and fracture in a nickel-base single crystal superalloy
Acta Metall. 36 (1988) 8, p. 2369–2376

[181] Khan, A. S.; Peterson, D. T.
Solubility and diffusion of hydrogen in In-100 and WASPALOY
in: Moody, N. R.; Thompson, A. W.
Hydrogen Effects on Material Behavior
The Minerals, Metals & Materials Society, 1990, p. 3–10

[182] Walter, R. J.; Frandsen, J. D.; Jewett, R. P.
Fractography of alloys tested in high-pressure hydrogen
in: Moody, N. R.; Thompson, A. W.
Hydrogen Effects on Material Behavior
The Minerals, Metals & Materials Society, 1990, p. 819–827

[183] Takasugi, T.; Nakayama, T.; Hanada, S.
Environmental embrittlement of Ni3(Si, Ti) single crystals
Materials Transactions, JIM 34 (1993) 9, p. 775–785

[184] Wan, X. J.; Zhu, J. H.; Jing, K. L.; Chen, W. J.; Wu, Y.
Environmental embrittlement in A3B-type intermetallic alloys
Journal of materials science & technology (China) 10 (1994) 1, p. 39–53

[185] Takasugi, T.; Hanada, S.
Environmental embrittlement of boron-doped Ni3(Al,Ti) single crystals at room temperature
Journal of Materials Research 8 (1993) 10, p. 2534–2542

[186] Owen, E. A.; Williams, E. S.
Proc. Phys. Soc 65 (1944) 52

[187] Pugachev, V. A.; Nikolaev, E. I.; Busol, F. I.; Nam, B. P.; Shabalin, I. N.
Solubility of hydrogen in palladium-silver alloys
Russian Journal of Physical Chemistry 47 (1973) 1, p. 20–22

[188] Lvov, A. L.; Malysheva, L. A.; Petrovskaya, O. I.
Diffusion of hydrogen through palladium and palladium-silver membranes
Russian Journal of Physical Chemistry 47 (1973) 7, p. 960–963

[189] Lvov, A. L.; Malysheva, L. A.; Petrovskaya, O. I.
Diffusion of hydrogen from converter gas through palladium and palladium-silver membranes
Russian Journal of Physical Chemistry 47 (1973) 7, p. 963–966

[190] Maruno, T.; Tsusi, H.; Otsuka, R.
Hydrogen embrittlement of Pd–Ag alloys
J. JPN. Inst. Met. 47 (1983) 9, p. 768–775

[191] Timofeev, N. I.; Berseneva, F. N.; Gromov, V. I.
Influence of preliminary hydrogen impregnation on the mechanical properties of palladium and its alloys with silver
Sov. Mater. Sci. 17 (1981) 5, p. 417–419

[192] Owen, C. V.; Buck, O.
Low temperature hydrogen effects on the strength and ductility of Nb-Ta alloys
Materials Science and Engineering A108 (1989), p. 117–119

[193] Spitzig, W. A.; Owen, C. V.; Scott, T. E.
Effects of nitrogen on the mechanical behavior of hydrogenated V, Nb, and Ta
Metallurgical Transactions A 17A (1986) 3, p. 527–535

[194] Chen, C. C.; Arsenault, R. J.
Low temperature deformation characteristics of the group Va metal-hydrogen single crystals
Acta Metallurgica 23 (1975) 2, p. 255–267

[195] Owen, C. V.; Spitzig, W. A.
On hydride reorientation in group V metals
Materials Science and Engineering 78 (1986), p. L15-L16

[196] Ravi, K. V.; Gibala, R.
Low temperature strengthening in niobium-hydrogen single crystals
Metallurgical Transactions 2 (1971) 4, p. 1219–1225

[197] Hardie, D.; McIntyre, P.
The low-temperature embrittlement of niobium and vanadium by both dissolved and precipitated hydrogen
Metallurgical Transactions 4 (1973) 5, p. 1247–1254

[198] Owen, C. V.; Cheong, D.-S.; Buck, O.
Grain size and hydrogen concentration effects on the ductility return in a refractory alloy
Metallurgical Transactions A 18A (1987) 5, p. 857–863

[199] Fabriabi, S.; Collins, A.L.W.; Salama, K.
Effects of hydrogen on near-threshold crack propagation in niobium
Metallurgical Transactions A 14A (1983) 4, p. 701–707

[200] Kuron, D.
Einfluß des Wasserstoffs auf Titan, Zirconium, Niob und Tantal
in: Kuron, D.
Wasserstoff und Korrosion
Verlag Irene Kuron, Bonn, 2000, p. 212–256

[201] Wasz, M. L.; Brotzen, F. R.; McLellen, R. B.; Griffin, A. J.
Effect of oxygen and hydrogen on mechanical properties of commercial purity titanium
International Materials Reviews 41 (1996) 1, p. 1–12

[202] Wasz, M. L.; Brotzen, F. R., McLellan, R. B.
The effect of hydrogen on the time-to-rupture behavior of oxygen-strengthened titanium
Scripta Metallurgica et Materialia 28 (1993) 4, p. 483–488

[203] Datta, P. K.; Strafford, K. N.; Dowson, A. L.
Environment/mechanical interaction processes and hydrogen embrittlement of titanium
in: Surianarayana, C.; Prasad, P. M.
Light Metals: Science and Technology
Trans Tech Publications, Switzerland, 1985, p. 203–216

[204] Puttlitz, K. J.; Smith, A. J.
The influence of microstructure on the hydrogen embrittlement of pure and commercially-pure titanium
in: Bernstein, I. M.; Thompson, A. W.
Hydrogen Effects in Metals
The Metallurgical Society of AIME, Pittsburg, Pennsylvania, 1980, p. 427–434

[205] Bargel, H. J.; Schulze, G.
Werkstoffkunde, 6. Aufl.
VDI Verlag, Düsseldorf, 1994, p. 121

[206] Habermann, M.; Kaiser, H.; Kaesche, H.
Wasserstoffeinfluß auf die Kerbschlagzähigkeit und die Rissausbreitung unter konstanter Last von IVa-Metallen technischer Reinheit (Teil I)
Zeitschrift für Metallkunde 84 (1993) 12, p. 832–838

[207] Clarke, C. F.; Hardie, D.; Ikeda, B. M.
The effect of hydrogen content on the fracture of pre-cracked titanium specimens
Corrosion Science 36 (1994) 3, p. 487–509

[208] Clarke, C. F., Hardie, D., Ikeda, B. M.
Hydrogen-induced cracking of commercial purity titanium
Corrosion Science 39 (1997) 9, p. 1545–1559

[209] Shimogori, K.; Satoh, H.; Kamikubo, F.
Investigation of hydrogen absorption-embrittlement of titanium used in actual equipment
in: DGM
Titanium science and technology, Bd. 2
DGM, Oberursel, 1985, p. 1111–1118

[210] Caskey, G. R.
The influence of a surface oxide film on hydriding of titanium
in: Bernstein, J. M.; Thompson, W.
Hydrogen in Metals
American Society for Metals, Pa., 1973, p. 465–473

[211] Shah, K. K.; Johnson, D. L.
Effect of surface pre-oxidation on hydrogen permeation in alpha titanium
in: Bernstein, J. M.; Thompson, W.
Hydrogen in Metals
American Society for Metals, Pa., 1973, p. 475–481

[212] Barta, E. R., Boyer, R. R.; Narayanan, G. H.
Delayed hydrogen embrittlement in commercially pure titanium
Proceedings of the international symposium on testing and failure analysis
ASM International, Metals Park, Ohio, USA, 1988, p. 387–395

[213] Sluzalek, A.
An analysis of titanium weld failure
Proceedings of the international conference on the joining of materials
Helsingor Teknikum, Rasmus Knudsens, Helsingor, Denmark, 1989, p. 72–75

[214] Askeland, D. R.
Materialwissenschaften
Spektrum, Akad. Verl., Heidelberg, Berlin, Oxford, 1996, p. 386

[215] Peters, M.; Hemtenmacher, J.; Kumpfert, J.; Leyens, C.
Structure and properties of titanium and titanium alloys
in: Leyens, C.; Peters, M.
Titanium and Titanium Alloys
Wiley-VCH Verlag, Weinheim, 2003, p. 9, 20–21

[216] Paton, N. E.; Williams, J. C.
Effect of hydrogen on titanium and its alloys
in: Bernstein, I. M.; Thompson, A. W.
Hydrogen in Metals
American Society for Metals, Pa., 1973, p. 409–432

[217] Yeh, M. S.; Huang, J.-H.
Hydrogen-induced subcritical crack growth in Ti-6Al-4V
Materials Science and Engineering A 242 (1998) 1–2, p. 96–107

[218] Nelson, H. G.
Aqueous chloride stress corrosion cracking of titanium – A comparison with environmental hydrogen embrittlement
in: Bernstein, I. M.; Thompson, W.
Hydrogen in Metals
American Society for Metals, Pa., 1973, p. 445–463

[219] Cox, T. B.; Gudas, J. P.
Investigation of the fracture of near-alpha titanium alloys in high pressure hydrogen environments
in: Thompson, A. W.; Bernstein, J. M.
Effect of Hydrogen on Behavior of Materials
The Metallurgical Society of AIME, USA, 1975, p. 287–298

[220] Hardie, D.; Ouyang, S.
Effect of hydrogen and strain rate upon the ductility of mill-annealed Ti6Al4V
Corrosion Science 41 (1999), p. 155–177

[221] Lukas, J. P.
Hydrogen effects on fracture toughness of Ti-6Al-4V determined by a steadily growing stable crack
in: Moody, R.; Thompson, A. W.
Hydrogen Effects on Material Behavior
The Minerals, Metals and Materials Society, 1990, p. 871–880

[222] Meyn, D. A.
Effect of hydrogen on fracture and inert-environment sustained load resistance of α-β titanium alloys
Metallurgical Transactions 5 (1974), p. 2405–2414

[223] Hardie, D.; Ouyang, S.
Effect of microstructure and heat treatment on fracture behaviour of smooth and precracked tensile specimens of Ti6Al4V
Materials Science and Technology 15 (1999) 9, p. 1049–1057

[224] Effect of hydrogen embrittlement on container metals
Light Metals AGE (1969) December, p. 24–26

[225] Somerday, B. P.; Moody, N. R.; Costa, J. E.; Gangloff, R. P.
Environment-induced cracking in structural titanium alloys
Corrosion 98 (1998) Mar., p. 267/1–267/17

[226] Mal'kov, A. V.; Alekseev, V. K.; Mishanova, M. G.
Crack resistance of titanium alloy AT3 under hydrogen embrittlement conditions
Soviet Materials Science 27 (1991) 2, p. 133–135

[227] Guiles, N. L.; Ono, K.
The effect of hydrogen on internal friction of several titanium alloys
in: Thompson, A. W.; Bernstein, J. M.
Effect of Hydrogen on Behavior of Materials
The Metallurgical Society of AIME, USA, 1975, p. 666–675

[228] Haynes, R.; Maddocks, P. J.
The effect of hydrogen on the work-hardening and fracture behaviour of titanium-low-aluminium-low-manganese alloys
Metal Science Journal 3 (1969) Sept., p. 190–195

[229] Gerberich, W. W.; Moody, N. R.; Jensen, C. L.; Hayman, C.; Jatavallabhula, K.
Hydrogen in α/β and all β titanium systems: Analysis of microstructure and temperature interactions on cracking
in: Bernstein, I. M.; Thompson, A. W.
Hydrogen effects in metals
The Metallurgical Society of AIME, Pittsburgh, Pennsylvania, 1980, p. 731–744

[230] Paton, N. E.; Spurling, R. A.; Rhodes, C. G.
Influence of hydrogen on betaphase titanium alloys
in: Bernstein, I. M.; Thompson, A. W.
Hydrogen Effects on Metals
The Metallurgical Society of AIME, Pittsburgh, Pennsylvania, 1980, p. 269–279

[231] Paton, N. E.; Buck, O.
The effect of hydrogen and temperature on the strength and modulus of beta phase Ti alloys
in: Thompson, A. W.; Bernstein, J. M.
Effect of Hydrogen on Behavior of Materials
The Metallurgical Society of AIME, USA, 1975, p. 83–90

[232] Gerberich, W. W.; Jatavallabhula, K.; Peterson, K. A.; Jensen, C. L.
Hydrogen-induced fracture phenomena in a b.c.c. titanium alloy
Conference: Advances in Fracture Research (Fracture 81), Cannes, France, Bd. 2
Pergamon Press Ltd., Oxford, England, 1982, p. 989, 998

[233] Criqui, B.; Fidelle, J. P.; Claus, A.
Effects of internal and external hydrogen on mechanical properties of beta III titanium alloy sheet,
in: Thompson, A. W.; Bernstein, J. M. Effect of Hydrogen on Behavior of Materials
The Metallurgical Society of AIME, USA, 1975, p. 91–101

[234] Brady, M. P.; Pint, B. A.; Tortorelly, P. F.; Wright, I. D.
High-temperature oxidation and corrosion of intermetallics
in: Cahn, R. W.; Haasen, P.; Kramer, E. J.; Schütze, M.
Materials Science and Technology, Corrosion and Environmental Degradation, Bd. 2
Wiley-VCH, Weinheim, New York, 2000, p. 265–279

[235] Company publication
Isserow, S.
Environmental effects on intermetallic compounds: a guide to the literature, Report No.: AD-A269 587/2/XAB, 1993
Us Army Research Laboratory, Watertown (USA), p. 1–20

[236] Eliezer, D.; Froes; F. H.; Manor, E.
Effect of hydrogen on behavior of the intermetallic titanium-aluminides
Sample Quaterly 22 (1991) July, p. 29–35

[237] Eliezer, D.; Froes, F. H.
Environment effects in titanium aluminides alloys
Key Engineering Materials, Switzerland 77–78 (1993), p. 321–328
or:
Eliezer, D.
Environmental effects in aerospace materials
Cercet. Metal. Mater. (Metall. New Mater. Res.) II, (4) (1994), p. 1–10

[238] Thompson, A. W.
Effect of hydrogen in titanium aluminide alloys
Materials Science and Engineering A 153 (1992), p. 578–583
and:
Thompson, A. W.
Environmental effects in titanium aluminide alloys
in: Jones, R. H.; Ricker, R. E.
Environmental Effects on Advanced Materials
The Minerals, Metals & Materials Society, 1991, p. 21–33

[239] Boodey, J. B.; Gao, M.; Wei, R. P.
Hydrogen solubility and hydride formation in a thermally charged gamma-based titanium aluminide
in: Jones, R. H.; Ricker, R. E.
Environmental Effects on Advanced Materials
The Minerals, Metals & Materials Society, 1991, p. 57–65

[240] Boodey, J. B.
Hydrogen interaction with gamma-based titanium aluminides: Hydrogen occlusion and hydrid formation
Dissertation, Lehigh University, USA, 1993, p. 1–157

[241] Chu, W. Y.; Thompson, A. W.; Williams, J. C.
Brittle fracture behavior and influence of hydride in titanium aluminide
in: Moody, N. R.; Thompson, W.
Hydrogen Effects on Material Behavior
The Minerals, Metals & Materials Society, USA, 1990, p. 543–554

[242] Fritzemeier, L. G.; Jacinto, M. A.
Hydrogen environment effects on beryllium and titanium aluminides
in: Moody, N. R.; Thompson, W.
Hydrogen Effects on Material Behavior
The Minerals, Metals & Materials Society, USA, 1990, p. 533–542

[243] Chan, K. S.
Developing hydrogen-tolerant microstructures for an α2-titanium aluminide
Metallurgical Transactions A 23A (1992) 2, p. 497–507

[244] Chu, W. Y.; Thompson, A. W.
Effect of microstructure and hydrogen as a temporary β stabilizer on cleavage fracture behavior in titanium aluminide
in: Moody, N. R. and Thompson, W.
Hydrogen Effects on Material Behavior
The Minerals, Metals & Materials Society, USA, 1990, p. 285–296

[245] Kane, R. D.; Chakachery, E. A.
Evaluation of titanium aluminide and carbon carbon composite materials for hydrogen gas service
in: Jones, R. H.; Ricker, R. E.
Environmental Effects on Advanced Materials
TMS, Minerals-Metals-Materials, Warrendale, Pa. USA, 1991, p. 35–46

[246] Vesely, E. J.; Verma, S. K.; Skrzypchak, M. J.
Effect of hydrogen on room temperature and elevated temperature tensile and creep-rupture properties of selected nickel, cobalt and titanium alloys
in: Moody, N. R.; Thompson, W.
Hydrogen Effects on Material Behavior
The Minerals, Metals and Materials Society, USA, 1990, p. 1057–1066

[247] Christodoulou, L.; Clarke, J. A.
Hydrogen effects on fracture toughness of XD™ titanium aluminides,
in: Moody, N. R.; Thompson, W.
Hydrogen Effects on Material Behavior
The Minerals, Metals and Materials Society, USA, 1990, p. 515–521

[248] Dunfee, W.; Gao, M.; Wei, P.; Wei, W.
Hydrogen enhanced thermal fatigue of γ-titanium aluminide
Scripta Metallurgica et Materialia 33 (1995) 2, p. 245–250

[249] Gao, M.; Dunfee, W.; Wei, R.; Wei, W.
Thermal mechanical fatigue of gamma titanium aluminide in hydrogen and air
in: Soboyejo, W. O.
Fatigue and Fracture of ordered Intermetallic Materials: II, Rosemont USA, 6–10 Oct 1994
The Mineral, Metals & Material Society 1995, p. 3–15

[250] ASTM B551
Standard Specification for Zirconium and Zirkonium Alloy; Strip, Sheet and Pipe
ASTM American Society for Testing and Materials, West Conshohocken (PA/USA)

[251] ASTM B352
Standard Specification for Zirconium and Zirkonium Alloy; Strip, Sheet and Pipe for Nuclear Application
ASTM American Society for Testing and Materials, West Conshohocken (PA/USA)

[252] Cheadle, B. A.; Coleman, C. E.; Ambler, J. F. R.
Prevention of delayed hydride cracking in zirconium alloys
in: Conference: Zirconium in the Nuclear Industry, Strasbourg, France 24–27 June 1985
Publ: ASTM, Philadelphia, Pennsylvania, USA, 1987, p. 224–240

[253] Smith, R. R.; Eadie, R. L.
High temperature limit for delayed hydride cracking
Scripta Metallurgica 22 (1988) 6, p. 833–836

[254] Price, E. G.; Cheadle, B. A.
Fast fracture of a zirconium alloy pressure tube: Cause and implications
Conference: The mechanism of fracture, Salt Lake City, Utah, USA, 2–6 Dec. 1985
American Society for Metals, Metals Park, Ohio, USA, 1986, 511–519

[255] Shalabi, A. F.; Meneley, D. A.
Initiation of delayed hydride cracking in zirconium-2.5 wt.% niobium
Journal of Nuclear Materials 173 (1990) 10, p. 313–320

[256] Puls, M. P.
On the consequences of hydrogen supersaturation effects in Zr alloys to hydrogen ingress and delayed hydride cracking
Journal of Nuclear Materials 165 (1989) 5, p. 128–141

[257] Laursen, T.; Leslie, J. R.; Tapping, R. L.
Deuterium depth distribution in oxidized Zr-2.5 wt.% Nb measured by nuclear reaction analysis
Journal of the Less-Common Metals 172–174 (1991) p. 1306–1312

[258] Huang, F. H.; Mills, W. J.
Delayed hydride cracking behavior for ZIRCALOY-2 tubing
Metallurgical Transactions A 22A (1991) 9, p. 2049–2060

[259] Leger, M.; Donner, A.
The effect of stress on orientation of hydrides in zirconium alloy pressure tube materials
Canadian Metallurgical Quarterly 24 (1985) 3, p. 235–243

[260] Dubey, J. S.; Wadekar, S. L.; Singh, R. N.; Sinha, T. K.; Chakravartty, J. K.
Assessment of hydrogen embrittlement of zircaloy-2 pressure tubes using unloading compliance and load normalization techniques for determining J-R curves
Journal of Nuclear Materials 264 (1999) 20, p. 28

[261] Glavicic, M. G.; Szpunar, J. A.; Lin, Y. P.
The role of oxide texture in the protection of Zr-2.5%Nb alloy pressure tubes against hydrogen ingress
Conference: ICOTOM 12; 12th International Conference on Texture of Materials, Montreal, Quebec, Canada, 9–13 Aug. 1999
Volume 2, p. 1409–1414
National Research Council of Canada, Ottawa (Canada), 1999

[262] Puls, M. P.
Effects of crack tip stress states and hydride-matrix interaction stresses on delayed hydride cracking
Metallurgical Transactions A 21A (1990) 11, p. 2905–2917

[263] Hatano, Y.; Isobe, K.; Hitaka, R.; Sugisaki, M.
Role of intermetallic precipitates in hydrogen uptake of Zircaloy-2
Journal of Nuclear Science and Technology (Japan) 33 (1996) 12, p. 944–949

[264] Kim, Y. J.; Vanderglas, M. L.
Elastic-plastic analysis of hydride blisters in Zircaloy-2 pressure tubes
Journal Pressure Vessel Technology (TRANS. ASME) 110 (1988) 3, p. 276–282

[265] Varias, A. G.; Massih, A. R.
Simulation of hydrogen embrittlement in zirconium alloys under stress and temperature gradients
Journal of Nuclear Materials 279 (2000) 2–3, p. 273–285

[266] Spence, T. C.; Reed, D. S.
Corrosion resistant zirconium castings
Conference: CORROSION 2000, Orlando; FL, USA, 26–31 March 2000
Paper 00505
NACE International, Houston, TX, USA, 2000

[267] Huang, J.-H.; Huang, S.-P.; Ho, C.-S.
The ductile-brittle transition of a zirconium alloy due to hydrogen
Scripta Metallurgica et Materialia 28 (1993) 6, p. 1537–1542

[268] Huang, J.-H.; Yeh, M.-S.
Gaseous hydrogen embrittlement of a hydrided zirconium alloy
Metallurgical and Materials Transactions A 29A (1998) 3, p. 1047–1056

[269] Bai, J. B.
Influence of an oxide layer on the hydride embrittlement in zircaloy-4
Scripta Metallurgica et Materialia 29 (1993) 5, p. 617–622

[270] Levi, M. R.; Sagat, S.
Effect of texture on delayed hydride cracking in Zr-2.5Nb alloy
Conference: Environmental Degradation of Materials and Corrosion Control in Metals, Quebec City, Quebec, Canada, 22–26 Aug. 1999, 323–335
Canadian Institute of Mining, Metallurgy and Petroleum Montreal, Canada, 1999

[271] Puls, M. P.
The influence of hydride size and matrix strength on fracture initiation at hydrides in zirconium alloys
Metallurgical Transactions A 19A (1988) 19A, p. 1507–1522

[272] Puls, M. P.
Determination of fracture initiation in hydride blisters using acoustic emission
Metallurgical Transactions A 19A (1988), p. 2247–2257

[273] Rowe, R. G.
Subcritical cracking of Zircaloy-2 in high-pressure hydrogen gas
Scripta Materialia 38 (1998) 10, p. 1495–1503

[274] Company publication
Wolfram – Tungsten – Tungstene
Metallwerke Plansee GmbH, Reutte (Österreich)

[275] Matsui, H.; Kubota, O.; Koiwa, M.
Effect of stress on hydrogen solubility and the low temperature deformation characteristics in vanadium containing hydrogen
Acta metallurgica 34 (1986) 2, p. 295–302

[276] Sherman, d. h.; Owen, C. V.; Scott, T. E.
The effect of hydrogen on the structure and properties of vanadium
Transactions of the Metallurgical Society of AIME 242 (1968) 9, p. 1775–1784

[277] Owen, C. V.; Cheong, D. S.; Buck, O.; Scott, T. E.
Effects of hydrogen on mechanical properties of vanadium-niobium alloys
Metallurgical Transactions A 15A (1984) 1, p. 147–153

[278] Owen, C. V.; Rowland, T. J.; Buck, O.
Effects of hydrogen on some mechanical properties of vanadium-titanium alloys
Metallurgical Transactions A 16A (1985) 1, p. 59–66

[279] Owen, C. V.; Spitzig, W. A.; Buck, O.
Effects of hydrogen on low temperature hardening and embrittlement of V – Cr – alloys
Metallurgical Transactions A 18A (1987) 9, p. 1593–1601

[280] Yano, S.; Tada, M.; Matsui, H.
Hydrogen embrittlement of MFR candidate vanadium alloys
Journal of Nuclear Materials 179–181 (1991) Mar.-Apr., p. 779–782

[281] Powell, G. L.; Thompson, K. A.
Hydrogen embrittlement of lean uranium alloys
in: Moody, N. R.; Thompson, A. W.
Hydrogen effects on metal behavior
TMS, The Minerals, Metals & Materials Society, 1989, p. 765–774

[282] Powell, G. L.; Northcutt, W. G.
Internal hydrogen embrittlement of uranium 5.7 niobium alloy
Journal of Nuclear Materials 132 (1985) 1, p. 47–51

[283] Odegard, B. C.; Eckelmeyer, K. H.; Dilon, J. J.
The embrittlement of U-0.8Ti by absorbed hydrogen
in: Moody, N. R.; Thompson, A. W.
Hydrogen effects on metal behavior
TMS, The Minerals, Metals & Materials Society, 1989, p. 775–785

[284] Magnani, N. J.
Cracking of U-0.75wt.% Ti in hydrogen and water vapor
in: Thompson, A. W.; Bernstein, I. M.
Effect of hydrogen on behavior of materials
The Metallurgical Society of AIME, 1975, p. 189–199

[285] Powell, G. L.; Köger, J. W.; Bennet, R. K.; Williamson, A. L.; Hemperly, V. C.
Internal hydrogen embrittlement of gamma-stabilized uranium alloys
Corrosion NACE 32 (1976) 11, p. 442–450

[286] Company publication
Chemische Beständigkeit von Kohlenstoff und Graphit, 08 00/05 NÄ, 08/2000
SGL Carbon GmbH, Bonn

[287] SGL Carbon Group
Sigracet®-Komponenten für Brennstoffzellen (Online im Internet)
<http://www.sglcarbon.de>
(Abruf 18.05.2006)

[288] GAB Neumann GmbH, Maulburg
Korrosionsbeständigkeit von imprägniertem Graphit Qualität Diabon (Online im Internet)
<http://www.gab-neumann.de>
(Abruf 16.05.2006)

[289] Lay, L.
Corrosion Resistance of Technical Ceramics, 2. Aufl.
HMSO Publication Centre, London, 1991, p. 53, 111, 113

[290] Gorka Cement, Trzebinia (Polen)
Aluminiumzement Gorkal 40 (Online im Internet)
<http://www.gorka.com.pl>
(Abruf 07.04.2006)

[291] Krikorian, O. H.
Materials problems in production, transport and storage of hydrogen
in: Lochmann, W. J.; Indig, M.;
Materials and corrosion problems in energy systems
National Association of Corrosion Engineers, Houston, Tex., 1980, p. 10-1–10-12

[292] Frischat, G. H.
Glastechnische Ber. 42 (1969), p. 351

[293] Ecker, K.; Papp, G.; Ernsthofer, G.; Giedenbacher, G.
Beitrag zum Emaillierverhalten beruhigter Stähle
Mitteilungen des Vereins Deutscher Emailfachleute 29 (1981) 11, S. 143–156

[294] Buechel, E.; Leontaritis, L.
Gerät zur registrierenden Messung der Wasserstoffdurchlässigkeit von Blechen
VtB-Verfahrenstechnische Berichte 47 (1968) 3, S. 1–4

[295] De Gregorio, P.; Valentini, R.; Solina, A.; Gastaldo, F.; Buonpane, A.
Emaillierstähle und einwandfreie Qualität emaillierter Produkte
Mitt. Ver. Dtsch. Emaillierfachleute 41 (1993) 5, S. 60–68

[296] Maskall, K. A.; Richens, M. J.
The importance of hydrogen in the application of vitreous enamel to ferrous metals
in: Wilcox, J. R.
Hydrogen in steel
Institution of Metallurgists, Bath, 1982, p. 113–118

[297] Ishikawa, Y.; Zen, M.; Nishida, O.
On hydrogen induced fracture of porcelain enamel layers
in: Taplin, D. M. R.
Fracture 77, Bd. 3
University of Waterloo Press, Waterloo, Ontario, Canada, 1977, p. 1007–1014

[298] Company publication
Alsint 99,7® (Al2O3) - Konstruieren mit Oxidkeramik, 07/96/5A, 07/1996
W. Haldenwanger Technische Keramik GmbH & Co. KG, Berlin

[299] Company publication
NextelTM Ceramic Textiles Technical Notebook, 98-0400-5870-7, 08/2001
3M Deutschland GmbH, Neuss

[300] Nickel, K. G.; Gogotsi, Y. G.
Corrosion of Hard Materials
in: Riedel, R.
Handbook of Ceramic Hard Materials
Wiley-VCH Verlag GmbH, Weinheim, 2000, p. 159

[301] DIN EN ISO 1043-1 (06/2002)
Kunststoffe – Kennbuchstaben und Kurzzeichen; Teil 1: Basis-Polymere und ihre besonderen Eigenschaften
Beuth Verlag GmbH, Berlin

[302] DIN ISO 1629 (11/2004)
Kautschuk und Latices – Einteilung, Kurzzeichen
Beuth Verlag GmbH, Berlin

[303] Carlowitz, B.
Kunststoff Tabellen, 4. Aufl.
Carl Hanser Verlag, München Wien, 1995

[304] Oberbach, K.; Baur, E.; Brinkmann, S.; Schmachtenberg, E.
Saechtling Kunststoff Taschenbuch, 29. Aufl.
Carl Hanser Verlag, München Wien, 2004, S. 730–732

[305] Domininghaus, H.
Die Kunststoffe und ihre Eigenschaften, 5. Aufl.
Springer-Verlag, Berlin Heidelberg New York, 1998, S. 231, 1229

[306] Pauly, S.
Permeation and Diffusion Data
in: Brandrup, J.; Immergut, E. H.; Grulke, E.H.
Polymer Handbook, 4. Aufl., Bd. 6, S. 543–569
John Wiley & Sons, New York, 1999

[307] Orme, C. J.; Stone, M. L.; Benson, M. T.; Peterson, E. S.
Testing Polymer Membranes for the Selective Permeability of Hydrogen
Separation Science and Technology 38 (2003) 12/13, p. 3225–3238

[308] Humpenöder, J.
Gaspermeation von Faserverbunden mit Polymermatrices
Dissertation Universität Karlsruhe (TH), Forschungszentrum Karlsruhe GmbH, 1997

[309] Company publication (Disketten)
Trovidur® – Chemische Widerstandsfähigkeit, HTCHEM 1.0, 05/1997
Röchling Trovidur KG, Troisdorf

[310] Mohr, J. M.; Paul, D. R.
Comparison of Gas Permeation in Vinyl and Vinylidene Polymers
Journal of Applied Polymer Science 42 (1991), p. 1711–1720

[311] Company publication
Neoflon® CTFE Molding Powders, EG-71h (0005) AK, 01/1998
Daikin Industries, Ltd., Osaka (Japan)

[312] Company publication
Verhalten von Ultrason® gegen Chemikalien, KTU/AH – 01 d 44781, 10/1998
BASF AG, Ludwigshafen

[313] Kapantaidakis, G. C.; Kaldis, S. P.; Dabou, X. S.; Sakellaropoulos, G. P.
Gas permeation through PSF-PI miscible blend membranes
Journal of Membrane Science 110 (1996), p. 239–247

[314] Company publication
Vectra® Flüssigkristalline Polymere (LCP), B 241 D 10.2001 Europa/RIE/HF/BS, 10/2001
Ticona GmbH, Kelsterbach

[315] Härtel, G.; Rompf, F.
Trennung eines Kohlendioxid/Wasserstoff-Gasgemisches unter hohen Drücken mit polymeren Membranmaterialien
Chemie Ingenieur Technik 69 (1997) 4, S. 506–510

[316] Raharjo, R. D.; Lee, H. J.; Freeman, B. D.; Sakaguchi, T.; Masuda, T.
Pure gas and vapor permeation properties of poly[1-phenyl-2-[p-(trimethylsilyl)phenyl]acetylene] (PTMSDPA) and its desilylated analog, poly[diphenylacetyllene] (PDPA)
Polymer 46 (2005), p. 6136–6324

[317] Morisato, A.; Shen, H. C.; Sankar, S. S.; Freeman, B. D.; Pinnau, I.; Casillas, C. G.
Polymer Characterization and Gas Permeability of Poly(1-trimethylsilyl-1-propine) [PTMSP], Poly(1-phenyl-1-propyne) [PPP], and PTMSP/PPP Blends
Journal of Polymer Science: Part B: Polymer Physics 34 (1996), p. 2209–2222

[318] Nagai, K.; Toy, L. G.; Freeman, B. D.; Teraguchi, M.; Masuda, T.
Influence of Physical Aging and Methanol Conditioning on Gas Permeability and Hydrocarbon Solubility of Poly[1-phenyl-2-[p-(triisopropylsilyl)phenyl]acetylene] (PTPSDPA)
Polymeric Materials Science and Engineering 81 (1999), p. 531–532

[319] Wilks, B.; Rezac, M. E.
Properties of Rubbery Polymers for the Recovery of Hydrogen Sulfide from Gasification Gases
Journal of Applied Polymer Science 85 (2002), p. 2436–2444

[320] Company publication
Thermoplastische Polyurethan-Elastomere. Elastollan®-Materialeigenschaften, Z/M, Fro 163-10-00, 10/2000
Elastogran GmbH, Lemförde

[321] Naito, Y.; Kamiya, Y.; Terada, K.; Mizoguchi, K.; Wang, J.-S.
Pressure Dependence of Gas Permeability in a Rubbery Polymer
Journal of Applied Polymer Science 61 (1996), p. 945–950

[322] Jäckel, M.; Leucke, U.; Jahn, K.; Fietzke, F.; Hegenbarth, E.
Thermal and dielectric properties of epoxy resin at low temperatures after irridiation and H2 permeation of fibre composites at room temperature
Advances in Cryogenic Engineering 40 (1994), p. 1153–1159

[323] Company publication
Chemical Resistance Table Low Density and High Density Polyethylene, ART 254 11.07.2001 Ed. 3, 07/2001
Borealis A/S, Kongens Lyngby (Dänemark)

[324] Technical Report ISO/TR 10358 (06/1993)
Plastics pipes and fittings – Combined chemical-resistance classification table
Beuth Verlag GmbH, Berlin

[325] Chemical Resistance; Volume I – Thermoplastics; Volume II – Thermoplastic Elastomers, Thermosets, and Rubbers, 2. Aufl.
Plastics Design Library, Morris und Norwich, NY/USA, 1994 und 1996

[326] Company publication (CD-ROM)
Chemische Widerstandsfähigkeit, SIM-CHEM 5.0, 05/2003
Simona AG, Kirn

[327] DIN 8075 Beiblatt 1 (02/1984)
Rohre aus Polyethylen hoher Dichte (HDPE). Chemische Widerstandsfähigkeit von Rohren und Rohrleitungsteilen
Beuth Verlag GmbH, Berlin

[328] Company publication
Kunststoff-Rohrleitungssysteme – Chemische Beständigkeit von Kunststoffen und Elastomeren, Fi 1980/1c (2.99), 02/1999
Georg Fischer Rohrleitungssysteme AG, Schaffhausen (Schweiz)

[329] Company publication
Kunststoff-Rohrsysteme – Beständigkeitsliste, 03/2005
Frank GmbH, Mörfelden

[330] Company publication
GUR® (PE-UHMW), GHR® (PE-HMW); Beständigkeit gegen Chemikalien und andere Medien, B 206 BR D-07.2001/042, 2001
Ticona GmbH, Kelsterbach

[331] Company publication
Isoplas Crosslinkable Polyethylene. Guide to chemical resistance, 09/1998
Mikropol Ltd., Stalybridge (England)

[332] Company publication
Chemical Resistance Table Polypropylene, ART 006 11.07.2001 Ed.4, 07/2001
Borealis A/S, Kongens Lyngby (Dänemark)

[333] DIN 8078 Beiblatt 1 (02/1982)
Rohre aus Polypropylen (PP). Chemische Widerstandsfähigkeit von Rohren und Rohrleitungsteilen
Beuth Verlag GmbH, Berlin

[334] DIN 8061 Beiblatt 1 (02/1984)
Rohre aus weichmacherfreiem Polyvinylchlorid. Chemische Widerstandsfähigkeit von Rohren und Rohrleitungsteilen aus PVC-U
Beuth Verlag GmbH, Berlin

[335] Spears Manufacturing Company, Sylmar/Los Angeles (CA/USA)
Spears® LabWaste® CPVC Corrosive Waste Drainage System (Online im Internet)
<http://www.spearsmfg.com>
(Abruf 05.04.2006)

[336] Company publication (CD-ROM)
SymChem – Chemical Resistance Database, Version 1.0, 1998
Symalit AG, Lenzburg (Schweiz)

[337] Institut für Chemische Verfahrenstechnik, Universität Stuttgart
Membran-Brennstoffzellen-Systeme (Online im Internet)
<http://www.icvt.uni-stuttgart.de>
(Abruf 30.05.2006)

[338] Company publication
Chemical Resistance of Halar® Fluoropolymer
Ausimont USA Inc., Morristown (NJ/USA)

[339] Company publication
Tefzel® Fluoropolymer – Chemical Use Temperature Guide, 195062B, 10/1991
Du Pont Polymers, Wilmington (DE/USA)

[340] Company publication
Hylar® Polyvinylidene Fluoride. Chemical Resistance Chart, 0022
Ausimont USA Inc., Morristown (NJ/USA)

[341] Chevron Phillips Chemical Company, LLC, The Woodlands (TX/USA)
Chemical Compatibility Chart (Online im Internet)
<http://www.cpchem.com>
(Abruf 30.03.2006)

[342] Company publication
Beständigkeitsliste für Arlon®
Greene, Tweed u. Co. GmbH, Hofheim

[343] DSM
Resistance of DSM Engineerung Plastics to Chemicals (Online im Internet)
<http://www.dsm.com>
(Abruf 30.03.2006)

[344] Company publication
Verhalten von Ultramid®, Ultraform® und Ultradur® gegen Chemikalien; TI-KTU/AS-28 d, 05/1999
BASF AG, Ludwigshafen

[345] Company publication (CD-ROM)
Bayer Plastics Information – Chemikalienbeständigkeit von Durethan® A und Durethan® B, Version 5.0, 03/2001
Bayer AG, Leverkusen

[346] Company publication
Rilsan PA 11 Beschichtungen – physikalische und chemische Eigenschaften, 2052 D/0394/40, 03/1994
Elf Atochem S. A., Paris (Frankreich)

[347] Company publication
Rilsan® A: Schritt in die Zukunft, Direp 2564 D/09.96/30, 09/1996
Elf Atochem, Puteaux (Frankreich)

[348] Company publication
Hytrel® Thermoplastic Polyester Elastomer. Fluid and chemical resistance guide, L-11964, 06/1999
DuPont de Nemours International S. A., Genf (Schweiz)

[349] Company publication
Beständigkeitsliste für Elastomere, 2003
Artemis Kautschuk- und Kunststofftechnik, Hannover

[350] Company publication
PCM Moineau – Materials Compatibility Guide, 05/2002
PCM Pompes, Vanves (Frankreich)

[351] Du Pont Performance Elastomers
General Chemical Resistance Guide (Online im Internet)
<http://www.dupontelastomers.com>
(Abruf 30.03.2006)

[352] Zrunek Gummiwaren GmbH, Wien (Österreich)
Elastomer Beständigkeiten (Online im Internet)
<http://www.zrunek.at>
(Abruf 30.03.2006)

[353] Company publication
Kunststoffbeschichtungen im Produkte-Segment PE, EVA, PP, 10/1999
Eposint AG, Pfyn (Schweiz)

[354] Company publication
ECTFE Halar® – Thermoplastischer Hightech-Beschichtungs-Kunststoff, 08/2000
Eposint AG, Pfyn (Schweiz)

[355] Company publication
Proco-Kunststoff-Beschichtungen®: Proco-E-CTFE (Halar®), 07/2000, und Beständigkeitsliste, 03/1996
Hüni + CO KG, Friedrichshafen

[356] Company publication
Edlon® PFA Beständigkeitsliste, 01/05.07, 05/1997
Rudolf Gutbrod GmbH, Dettingen/Erms

[357] Company publication
Jumbo – Der Standard für Fluor-Polymer-Beschichtungen und Mitteilung v. 12.04.2001
Rhenotherm Kunststoffbeschichtungs GmbH, Kempen

[358] Coatings and Linings for Immersion Service, 2. Aufl.
NACE International, Houston (TX/USA), 1998

[359] Electro Chemical Engineering & Manufacturing Co., Emmaus (PA/USA)
Fluoropolymer Protection from Corrosive Environments (Online im Internet)
<http://www.electrochemical.net>
(Abruf 30.03.2006)

[360] Atlas Minerals & Chemicals, Inc., Mertztown (PA/USA)
Chemical Construction Materials (Online im Internet)
<http://www.atlasmin.com>
(Abruf 30.03.2006)

[361] Company publication (CD-ROM)
novaDisc – Die Software zur Berechnung dichtungstechnischer Rahmenbedingungen, Version 5.0, 05/2006
Frenzelit-Werke GmbH & Co. KG, Bad Berneck

[362] Company publication
Garlock® Beständigkeitsliste für Flachdichtungen, GP-D7.1.11/02, 11/2002
Garlock GmbH, Neuss

[363] Company publication (CD-ROM)
Klinger Expert® 5.2 – Die leistungsfähige Dichtungsberechnung und Produktdokumentation, 2006
Rich. Klinger Dichtungstechnik GmbH & Co. KG, Gumpoldskirchen (Österreich)

[364] W. L. Gore & Associates, Inc.
Dichtungstechnik (Online im Internet)
<http://www.gore.com/sealants>
(Abruf 30.03.2006)

[365] Company publication
Stopfbuchspackungen, 2003
Freudenberg Process Seals KG, Viernheim

[366] Company publication
Stopfbuchspackungen, P6D/02/5.000/ 08.03/1.2.1, 08/2003
Burgmann Dichtungswerke GmbH & Co. KG, Wolfratshausen

[367] Company publication
Parker Seals – Medien Beständigkeits-Tabelle, 5703 G, 09/2000
Parker Hannifin GmbH, Pleidelsheim

[368] Company publication
Simrit® Katalog Ausgabe 2005, 30 D004 30.00405 IS Re, 2005
Freudenberg Simrit KG, Weinheim

[369] Company publication
O-Ring Werkstoffe, 99D/013/011/04/01, 04/2001
Busak + Shamban GmbH & Co., Stuttgart

[370] Company publication
Isolast® Perfluorelastomer-Dichtungen Beständigkeitsliste, 11/2001
Busak + Shamban GmbH & Co., Stuttgart

[371] Company publication
Burgmann Gleitringdichtungen Konstruktionsmappe 15.3, KM 15D/D5/ 8.000/10.05/1.2.1, 10/2005
Burgmann Industries GmbH & Co. KG, Wolfratshausen

[372] Yu, K.; Bayliss, R.; Hillman, B.
Wear tests of polymer composite compressor seal materials in hydrogen environment
Corrosion 99, San Antonio, TX, USA, 26–30 Apr. 1999, Report No. 052
NACE, Houston, TX, USA, 1999

[373] Company publication
Fiberdur® - Lieferprogramm mit Korrosionstabellen, 02/2003
Fiberdur GmbH & Co. KG, Aldenhoven

[374] Company publication
Wavistrong® Epoxy Pipe Systems – Chemical Resistance List, 02/1998
Future Pipe Industries B.V., Hardenberg (Niederlande)

[375] Advanced Valve Technologies, Broadstairs (Kent/England)
Chemical Resistance Guide (Online im Internet)
<http://www.advalve.com>
(Abruf 17.05.2006)

[376] Company publication
Hydrogen Derakane Case History, Mitteilung E. Kappenstein vom 13.03.2002
Dow Deutschland GmbH & Co. OHG, Rheinmünster

[377] Geiss, G.
Einfluss von Tieftemperatur und Wasserstoff auf das Versagensverhalten von Glasfaser-Verbundwerkstoffen unter statischer und zyklischer Belastung
Dissertation Universität Karlsruhe (TH), Forschungszentrum Karlsruhe GmbH, Karlsruhe, 2001

[378] Grimsley, B. W.; Cano, R. J.; Johnston, N. J.; Loos, A. C.; McMahon, W. M.
Hybrid composites for LH2 fuel tank structure
International SAMPE Technical Conference (2001), 33, p. 1224–1235

[379] Schroder Pedersen, A.
Plastic composite materials for cryogenic storage of hydrogen
Hydrogen Energy Progress XII, Proceedings of the World Hydrogen Energy Conference, 12th, Buenos Aires, June 21–26, 1998 (1998), 3, p. 1883–1892

[380] Higushi, K.; Takeuchi, S.; Sato, E.; Naruo, Y.; Inatani, Y.; Namiki, F.; Tanaka, K.; Watabe, Y.
Development and flight test of metal-lined CFRP cryogenic tank for reusable rocket
Acta Astronautica 57 (2005) 2–8, p. 432–437

[381] Takeichi, N.; Senoh, H.; Yokota, T. et al.
"Hybrid hydrogen storage vessel", a novel high-pressure hydrogen storage vessel combined with hydrogen storage material
International Journal of Hydrogen Energy 28 (2003) 10, p. 1121–1129

Key to materials compositions

Table 1: Chemical compositions of alloys according to German and other standards

German Standard		Materials Compositions	US-Standard
Mat.-No.	DIN-Design	Percent in Weight	SAE/ASTM/UNS
0.6015	EN-JL1020	–	A 48 (25B)
0.6025	EN-JL1040	–	A 48 (40B)
0.6655	GGL-NiCuCr 15 6 2	Fe-max. 3.0C-1.0-2.8Si-0.5-1.5Mn-13.5-17.5Ni-1.0-2.5Cr-5.5-7.5Cu	A 436 Type 1
0.6656	GGL-NiCuCr 15 6 3	Fe-max. 3.0C-1.0-2.8Si-0.5-1.5Mn-13.5-17.5Ni-2.5-3.5Cr-5.5-7.5Cu	A 436 Type 1b
0.6660	GGL-NiCr 20 2	Fe-max. 3.0C-1.0-2.8Si-0.5-1.5Mn-18.0-22.0Ni-1.0-2.5Cr	A 436 Type 2
0.6661	GGL-NiCr 20 3	Fe-max. 3.0C-1.0-2.8Si-0.5-1.5Mn-18.0-22.0Ni-2.5-3.5Cr	A 436 Type 2b
0.6667	GGL-NiSiCr 20 5 3	Fe-≤2.5C-3.5-5.5Si-0.5-1.5Mn-1.5-4.5Cr-18.0-22.0Ni	
0.6676	GGL-NiCr 30 3	Fe-max. 2.5C-1.0-2.8Si-0.5-0.8Mn-2.5-3.5Cr-28.0-32.0Ni	A 436 Type 3
0.6680	GGL-NiSiCr 30 5 5	Fe-≤2.5C-5.0-6.0Si-0.5-1.5Mn-4.5-5.5Cr-29.0-32.0Ni	A 436 Type 4
0.7040	EN-JS1030	–	A 536 (60-40-18)
0.7660	EN-GJSA-XNiCr20-2	Fe-≤3.0C-1.5-3.0Si-0.5-1.5Mn-≤0.080P-1.0-3.5Cr-≤0.50Cu-18.0-22.0Ni	A 439 Type D-2
0.7661	GGG-NiCr 20 3	Fe-≤3.0C-1.5-3.0Si-0.5-1.5Mn-≤0.080P-2.5-3.5Cr-18.0-22.0Ni	A 439 Type D-2B
0.7665	GGG-NiSiCr 20 5 2	Fe-≤3.0C-4.5-5.5Si-0.5-1.5Mn-≤0.080P-1.0-2.5Cr-18.0-22.0Ni	A 439 Type D-2C
0.7670	EN-GJSA-XNi22	Fe-≤3.0C-1.0-3.0Si-1.5-2.5Mn-≤0.080P-≤0.5Cr-≤0.5Cu-21.0-24.0Ni	A 439 Type D-2C

German Standard		Materials Compositions	US-Standard
Mat.-No.	DIN-Design	Percent in Weight	SAE/ASTM/UNS
0.7679	GGG-NiSiCr 30 5 2	Fe-≤2.6C-4.0-6.0Si-0.5-1.5Mn-≤0.08P-29.0-32.0Ni	
0.7680	EN-GJSA-XNiSiCr30-5-5	Fe-≤2.6C-5.0-6.0Si-0.5-1.5Mn-≤0.080P-4.5-5.5Cr-≤0.5Cu-28.0-32.0Ni	A 439 Type D-4
0.7688	EN-GJSA-XNiSiCr35-5-2; GGG-NiSiCr 35 5 2	Fe-≤2.0C-4.0-6.0Si-0.5-1.5Mn-≤0.08P-1.5-2.5Cr-≤0.5Cu-34.0-36.0Ni	A 439 (Type D-5S)
0.9625	EN-GJN-HV550	Fe-≤3.0-3.6C-≤0.080Si-≤0.080Mn-≤0.1P-≤0.1S-1.5-3.0Cr-≤0.5Mo-3.0-5.5Ni	A 532 (IA NiCr-HC)
0.9635	EN-JN3029	Fe-max. 1.8-2.4C-1.0Si-0.5-1.5Mn-0.08P-0.08S-14.0-18.0Cr-3.0Mo-2.0Ni	A 532
0.9640	EN-GJN-HV600(XCr14)	Fe-≤1.8-2.4C-≤1.0Si-≤0.5-1.5Mn-≤0.080P-≤0.080S-14.0-18.0Cr-≤3.0Mo-≤2.0Ni	A 532
0.9650	EN-JN3049	Fe-max. 2.4-3.2C-1.0Si-0.5-1.5Mn-0.08P-0.08S-23.0-28.0Cr-3.0Mo-2.0Ni	A 532
1.0030	St 00	Fe-≤0.30C-≤0.30Si-0.20-0.50Mn-≤0.08P-≤0.05S	
1.0032	St 34-2; (S205GT)	Fe-≤0.15C-≤0.3Si2.0-0.5Mn-0.05P-0.05S-0.007N	1010 (SAE)
1.0035	St 33; S 185		
1.0036	S235JRG1; USt 37-2; G; (S235JRG1+CR)	USt 37-2Fe-≤0.17C-≤1.4Mn-≤0.0045P-≤0.045S-≤0.007N	K02502 (UNS)
1.0037	St 37-2; S235JR	Fe-≤0.17C-≤0.3Si-≤1.4Mn-≤0.045P-≤0.045S-≤0.009N	A 283; SAE 1015
1.0038	RSt 37-2; S235JR	Fe-≤0.17C-≤1.4Mn-≤0.045P-≤0.045S-≤0.009N	UNS K02502
1.0040	USt 42-2	Fe-≤0.25C-≤0.2-0.5Mn-≤0.05P-≤0.05S-≤0.007N	
1.0044	S275JR; St 44-2	Fe-≤0.21C-≤1.50Mn-≤0.045P-≤0.045S-≤0.009N	UNS K03000 A 853 (1020) SAE 1020
1.0050	E295; St 50-2	Fe-≤0.045P-≤0.045S-≤0.009N	
1.0070	E360; St 70-2	Fe-≤0.045P-≤0.045S-≤0.009N	
1.0114	S235J0	Fe-≤0.17C-≤1.40Mn-≤0.040P-≤0.040S-≤0.009N	
1.0116	S235J2G3; St 37-3 N	Fe-max. 0.17C-1.4Mn-0.035P-0.035S	UNS K02001
1.0120	S235JRC	Fe-≤0.17C-≤1.40Mn-≤0.045P-≤0.045S-≤0.009N	
1.0204	UQSt 36	Fe-≤0.14C-≤0.25-0.50Mn-≤0.040P-0.040S	SAE 1008

Key to materials compositions

German Standard		Materials Compositions	US-Standard
Mat.-No.	DIN-Design	Percent in Weight	SAE/ASTM/UNS
1.0208	RSt 35-2; (C10G2)	Fe-0.06-0.12C-≤0.25Si-0.40-0.60Mn-≤0.035P-≤0.035S-≤0.25Cu-≤0.012N	
1.0253	USt 37.0	Fe-≤0.2C-≤0.55Si-≤1.6Mn-≤0.04P-≤0.04S-≤0.007N	
1.0254	P235TR1, St 37.0	Fe-max. 0.16C-0.35Si-1.2Mn-0.025P-0.020S-0.30Cr-0.30Cu-0.08Mo-0.010Nb-0.30Ni-0.04Ti-0.02V	UNS K02501
1.0256	St 44.0	Fe-≤0.21C-≤0.55Si-≤1.60Mn-≤0.040P-≤0.040S-≤0.009N	A 106
1.0301	C10	Fe-≤0.07-0.13C-≤0.4Si-≤0.3-0.6Mn-≤0.045P-≤0.045S	SAE 1010
1.0305	St 35.8	Fe-≤0.17C-≤0.10-0.35Si-≤0.40-0.80Mn-≤0.040P-≤0.040S	UNS K01200
1.0308	E 235	Fe-≤0.17C-≤0.35Si-0.4Mn-≤0.05P-≤0.05S-≤0.007N	SAE 1010
1.0309	DX55D	Fe-≤0.16C-0.17-0.40Si-0.35-0.65Mn-≤0.05P-≤0.050S-≤0.30Cr-≤0.30Ni-≤0.30Cu	UNS K02501
1.0330	St 12; DC01 + ZN	Fe-≤0.12C-≤0.60Mn-≤0.045P-≤0.045S	A 366 (C)
1.0333	USt 13	Fe-≤0.08C-≤0.007N	
1.0336	USt 4	Fe-≤0.09C-≤0.25-0.50Mn-≤0.030P-≤0.030S-≤0.007N	
1.0338	DC04; St 14	Fe-≤0.08C-≤0.40Mn-≤0.03P-≤0.030S	
1.0345	P235GH; H I	Fe-≤0.16C-≤0.35Si-≤0.40-1.20Mn-≤0.030P-≤0.025S-0.02Al-≤0.30Cr-≤0.30Cu-≤0.08Mo-≤0.010Nb-≤0.30Ni-≤0.03Ti-≤0.02V	A 285; A 414
1.0346	H220G1	Fe-≤0.04C-≤0.40Mn-≤0.03P-≤0.02S-≤0.01-0.04Ti	UNS K02202 A 516 (55) (380)
1.0356	TTSt35 N	Fe-≤0.18C-≤0.13-0.45Si-≤0.55-0.98Mn-≤0.035P-≤0.04S	UNS K03000 A 524 (I,II)
1.0375	TH57; T 57	Fe-≤0.1C-Traces Si-0.25-0.45Mn-≤0.04P-≤0.04S-0.007N	
1.0401	C 15	Fe-≤0.12-0.18C-≤0.40Si-≤0.3-0.6Mn-≤0.045P-≤0.045S	SAE 1015
1.0402	C 22	Fe-≤0.17-0.24C-≤0.4Si-≤0.4-0.7Mn-≤0.045P-≤0.045S-≤0.4Cr-≤0.1Mo-≤0.4Ni	SAE 1020
1.0405	St 45.8	Fe-≤0.21C-≤0.10-0.35Si-≤0.40-1.20Mn-≤0.040P-≤0.040S	A 106
1.0408	St 45	Fe-≤0.25C-≤0.035Si-0.40Mn-≤0.050P-≤0.050S	A 108; SAE 1020

German Standard		Materials Compositions	US-Standard
MatNo.	DIN-Design	Percent in Weight	SAE/ASTM/UNS
1.0414	C20D; D 20-2	Fe-≤0.18-0.23C-≤0.30Si-≤0.3-0.6Mn-≤0.035P-≤0.035S-≤0.01Al-≤0.2Cr-≤0.3Cu-≤0.05Mo-≤0.25Ni	UNS G10200; SAE 1020
1.0425	P265GH; H II	Fe-≤0.20C-≤0.4Si-≤0.5-1.4Mn-≤0.030P-≤0.025S-0.02Al-≤0.3Cr-≤0.3Cu-≤0.08Mo-≤0.01Nb-≤0.3Ni-≤0.03Ti-≤0.02V	UNS K01701
1.0426	P280GH	Fe-≤0.08-0.20C-≤0.4Si-≤0.9-1.5Mn-≤0.025P-≤0.015S-≤0.30Cr	A 662 (A)
1.0461	StE 255	Fe-≤0.18C-≤0.40Si-≤0.50-1.30Mn-≤0.035P-≤0.030S-≥0.02Al-≤0.30Cr-≤0.20Cu-≤0.08Mo-≤0.02N-≤0.03Nb-≤0.30Ni	UNS K02202 A 516 (55) (380)
1.0473	P355GH; 19 Mn 6	Fe-≤0.1-0.22C-≤0.6Si-≤1.0-1.7Mn-≤0.03P-≤0.025S-≤0.30Cr-≤0.3Cu-≤0.08Mo-≤0.3Ni	A 299
1.0481	17 Mn 4; P 295 GH	Fe≤0.08-0.20C-≤0.4Si-≤0.90-1.50Mn-≤0.030P-≤0.025S-0.02Al-≤0.30Cr-≤0.30Cu-≤0.08Mo-≤0.010Nb-≤0.3Ni-≤0.03Ti-≤0.02V	A 414, 515
1.0482	19 Mn 5	Fe-≤0.17-0.22C-≤0.30-0.60Si-≤1.00-1.30Mn-≤0.045P-≤0.045S-≤0.30Cr	UNS K12437; A 537
1.0490	S275N	Fe-≤0.18C-≤0.40Si-0.50-1.40Mn-≤0.035P-≤0.030S-≥0.0200Al-≤0.30Cr-≤0.35Cu-≤0.10Mo-≤0.015N-≤0.050Nb-≤0.30Ni-≤0.03Ti-≤0.05V	UNS K03000
1.0501	C 35	Fe-≤0.32-0.39C-≤0.4Si-≤0.5-0.8Mn-≤0.045P-≤0.045S-≤0.4Cr-≤0.1Mo-≤0.4Ni	SAE 1035
1.0503	C 45	Fe-≤0.42-0.50C-≤0.4Si≤0.5-0.8Mn-≤0.045P-≤0.045S-≤0.4Cr-≤0.1Mo-≤0.4Ni	SAE 1045
1.0505	StE 315	Fe-≤0.18C-≤0.45Si-≤0.70-1.50Mn-≤0.035P-≤0.030S-≤0.30Cr-0.020Al-≤0.20Cu-≤0.020N-≤0.03Nb-≤0.08Mo-0.30Ni	A 573
1.0528	C 30	Fe-≤0.27-0.34C-≤0.4Si-≤0.5-0.8Mn-≤0.045P-0.045S-≤0.4Cr-≤0.1Mo≤0.4Ni	SAE 1030
1.0540	C50	Fe-0.47-0.55C-≤0.40Si-0.60-0.90Mn-≤0.045P-≤0.045S-≤0.40Cr-≤0.10Mo-≤0.40Ni	A 689 (1050) ASTM A 866 (1050) ASTM
1.0545	S355N	Fe-≤0.20C-≤0.50Si-≤0.90-1.65Mn-≤0.035P-0.030S-0.02Al-≤0.30Cr-≤0.35Cu-≤0.10Mo-≤0.015N-≤0.050Nb-≤0.50Ni-≤0.03Ti-≤0.12V	UNS K12709
1.0553	S355J0	Fe-≤0.20C-≤0.55Si-≤1.60Mn-≤0.04P-≤0.04S-≤0.009N	
1.0562	StE 355; P 355 N	Fe-≤0.20C-≤0.50Si-≤0.90-1.70Mn-≤0.03P-≤0.025S-0.02Al-≤0.30Cu-≤0.30Cr-≤0.08Mo-≤0.02N-≤0.05Nb-≤0.50Ni-0.03Ti	A 633 (C)

Key to materials compositions

German Standard		Materials Compositions	US-Standard
Mat.-No.	DIN-Design	Percent in Weight	SAE/ASTM/UNS
1.0564	N-80	Fe-≤0.030P-≤0.030S	–
1.0570	St 52-3 N; S 355 J2G3	Fe-≤0.20C-≤0.55Si-≤1.60Mn-≤0.035P-≤0.035S	SAE 1024
1.0580	E 355	Fe-≤0.22C-≤0.55Si-≤1.60Mn-≤0.025P-≤0.025S	A 513 (1024) (ASTM)
1.0589	GL-E 36	Fe-≤0.18C-≤0.50Si-0.90-1.60Mn-≤0.040P-≤0.040S-≥0.0200Al-≤0.20Cr-≤0.35Cu-≤0.08Mo-0.020-0.050Nb-≤0.40Ni-0.05-0.10V	UNS K11852
1.0601	C60	Fe-0.57-0.65C-≤0.40Si-0.60-0.90Mn-≤0.045P-≤0.045S-≤0.40Cr-≤0.10Mo-≤0.40Ni	A 830 (1060) ASTM A 713 (1060) ASTM
1.0605	C 75	Fe-≤0.7-0.8C-≤0.15-0.35Si-≤0.6-0.8Mn-≤0.045P-≤0.045S	SAE 1074
1.0616	C86D	Fe-≤0.83-0.88C-≤0.10-0.30Si-≤0.50-0.80Mn-≤0.035P-≤0.035S-≤0.01Al-≤0.15Cr-≤0.25Cu-≤0.05Mo-≤0.20Ni	SAE 1086
1.0619	GP240GH	Fe-≤0.18-0.23C-≤0.60Si-≤0.50-1.20Mn-≤0.03P-≤0.02S	A 216
1.0664	St 160/180	Fe-≤0.80C-≤0.20Si-≤0.70Mn-≤0.04P-≤0.04S	
1.0670	P-105	Fe-≤0.70C-≤0.03-0.30Si-1.0Mn-≤0.04P-≤0.04S-≤0.007N	–
1.0721	10S20	Fe-0.07-0.13C-≤0.40Si-0.70-1.10Mn-≤0.060P-0.150-0.250S	SAE 1109
1.0854	M125-35P	consult producer	
1.0912	46Mn7	Fe-≤0.42-0.50C-≤0.15-0.35Si-≤1.6-1.9Mn-≤0.05P-≤0.05S-≤0.007N	SAE 1345
1.1013	RFe 100	Fe-≤0.05C-≤0.10Si-≤0.20-0.35Mn-≤0.03P-≤0.035S-≤0.04-0.10Al	
1.1104	EStE 285; P275NL2	Fe-≤0.16C-≤0.4Si-≤0.5-1.5Mn-≤0.025P-≤0.015S-0.02Al-≤0.30Cr-≤0.30Cu-≤0.02N-≤0.5Ni	P275NL2
1.1106	P355NL2; EStE 355	Fe-≤0.18C-≤0.5Si-≤0.9-1.7Mn-≤0.025P-≤0.015S-≤0.3Cr-≤0.3Cu-≤0.3Mo-0.5Ni-≤0.02N	A 707
1.1121	C10E; Ck 10	Fe-≤0.07-0.13C-≤0.40Si-0.30-0.60Mn-≤0.035P-≤0.035S	SAE 1010
1.1127	36Mn6	Fe-≤0.34-0.42C-≤0.15-0.35Si-≤1.4-1.65Mn-≤0.035P-≤0.035S	
1.1136	G24Mn4	Fe-0.20-0.28C-0.30-0.60Si-0.90-1.20Mn-≤0.035P-≤0.035S	

German Standard		Materials Compositions	US-Standard
MatNo.	DIN-Design	Percent in Weight	SAE/ASTM/UNS
1.1151	Ck 22; C22E	Fe-≤0.17-0.24C-≤0.4Si-≤0.4-0.7Mn-≤0.035P-≤0.035S-≤0.4Cr-≤0.1Mo-≤0.4Ni	SAE 1023
1.1166	34Mn5	Fe-0.30-0.37C-0.15-0.30Si-1.20-1.50Mn-≤0.035P-≤0.035S	G15360 (UNS) A 711 (1536) (ASTM) 1536 (SAE)
1.1176	G36Mn5	Fe-≤0.32-0.40C-≤0.15-0.35Si-≤1.20-1.50Mn-≤0.035P-≤0.035S	UNS H10380; A 830; SAE 1038
1.1186	C40E; Ck 40	Fe-≤0.37-0.44C-≤0.4Si-≤0.5-0.8Mn-≤0.035P-≤0.035S-≤0.4Cr-≤0.1Mo-≤0.4Ni	SAE 1040
1.1191	C45E; Ck 45	Fe-≤0.42-0.50C-≤0.4Si-≤0.50-0.80Mn-≤0.035P-≤0.035S-≤0.40Cr-≤0.10Mo-≤0.4Ni	SAE 1045
1.1520	C70U	Fe-≤0.65-0.74C-≤0.10-0.30Si-≤0.10-0.35Mn-≤0.030P-≤0.030S	
1.1525	C80U; C80W1	Fe-≤0.75-0.85C-≤0.10-0.25Si-≤0.10-0.25Mn-≤0.020P-≤0.020S	SAE W 108
1.1545	C105U; C105W1	Fe-≤1.0-1.1C-≤0.10-0.25Si-≤0.10-0.25Mn-≤0.020P-≤0.020S	SAE W 110
1.1730	C45U; C45W	Fe-≤0.42-0.50C-≤0.15-0.40Si-≤0.60-0.80Mn-≤0.03P-≤0.03S	A 830 (1045); SAE 1045
1.2210	115CrV3	Fe-1.10-1.25C-0.15-0.30Si-0.20-0.40Mn-≤0.030P-≤0.030S-0.50-0.80Cr-0.07-0.12V	L 2 SAE A 681 (L2) ASTM
1.2311	40CrMnMo7	Fe-0.35-0.45C-0.20-0.40Si-1.30-1.60Mn-≤0.035P-≤0.035S-1.80-2.10Cr-0.15-0.25Mo	
1.2343	X37CrMoV5-1; X38CrMoV5-1	Fe-0.33-0.41C-0.80-1.20Si-0.25-0.50Mn-≤0.030P-≤0.020S-4.80-5.50Cr-1.10-1.50Mo-0.30-0.50V	H 11 SAE A 681 (H 11) ASTM T 20811 UNS
1.2344	X40CrMoV5-1	Fe-0.35-0.42C-0.80-1.20Si-0.25-0.50Mn-≤0.030P-≤0.020S-4.80-5.50Cr-1.20-1.50Mo-0.85-1.15V	H 13 SAE T 20813 UNS A 681 (H13) ASTM
1.2365	32CrMoV12-28 X32CrMoV33	Fe-≤0.28-0.35C-≤0.10-0.40Si-≤0.15-0.45Mn-≤0.030P-≤0.030S-≤2.70-3.20Cr-≤2.60-3.00Mo-≤0.40-0.70V	SAE H 10
1.2550	60WCrV8; 60WCrV7	Fe-0.55-0.65C-0.70-1.00Si-0.15-0.45Mn-≤0.030P-≤0.030S-0.90-1.20Cr-0.10-0.20V-1.70-2.20W	
1.2567	30WCrV17-2	Fe-0.25-0.35C-0.15-0.30Si-0.20-0.40Mn-≤0.035P-≤0.035S-2.20-2.50Cr-0.50-0.70V-4.00-4.50W	
1.2787	X23CrNi17	Fe-≤0.10-0.25C-≤1.00Si-≤1.00Mn-≤0.035P-≤0.035S≤15.5-18.0Cr-≤1.0-2.5Ni	

Key to materials compositions

German Standard		Materials Compositions	US-Standard
Mat.-No.	DIN-Design	Percent in Weight	SAE/ASTM/UNS
1.2823	70Si7	Fe-≤0.65-0.75C-≤1.50-1.80Si-≤0.60-0.80Mn-≤0.03P-≤0.03S	
1.2842	90MnCrV8	Fe-≤0.85-0.95C-≤0.10-0.40Si-≤1.80-2.20Mn-≤0.030P-≤0.030S-≤0.20-0.50Cr-≤0.05-0.20V	UNS T31502; SAE O2; A 681 (O2); SAE O2
1.3247	HS2-9-1-8 S2-10-1-8	Fe-1.05-1.15C-≤0.70Si-≤0.40Mn-≤0.030P-≤0.030S-7.50-8.50Co-3.50-4.50Cr-9.00-10.00Mo-0.90-1.30V-1.20-1.90W	UNS T11342
1.3355	HS 18-0-1	Fe-≤0.70-0.78C-≤0.45Si-≤0.4Mn-≤0.030P-≤0.030S-≤3.8-4.5Cr-≤1.0-1.2V-≤17.5-18.5W	A 600
1.3505	100Cr6	Fe-≤0.93-1.05C-≤0.15-0.35Si-≤0.25-0.45Mn-≤0.025P-≤0.015S-≤0.05Al-1.35-1.60Cr-≤0.30Cu-≤0.10Mo	SAE 52100; A 29
1.3551	80MoCrV42-16	Fe-≤0.77-0.85C-≤0.40Si-≤0.15-0.35Mn-≤0.025P-≤0.015S-≤3.90-4.30Cr-≤0.30Cu-≤4.00-4.50Mo-≤0.90-1.10V-≤0.25W	SAE M50; A 600 (M50)
1.3728	AlNiCo 9/5	Fe-≤11.0-13.0Al-5.0Co-2.0-4.0Cu-≤57.0Fe-21.0-28.0Ni-≤1.0Ti	
1.3813	X40MnCrN19	Fe-≤0.30-0.50C-≤0.80Si-≤17.0-19.0Mn-≤0.10P-≤0.030S-≤3.0-5.0Cr-≤0.08-0.12N	
1.3817	X40MnCr18	Fe-0.30-0.50C-≤1.00Si-17.00-19.00Mn-≤0.060P-≤0.030S-≤3.00-5.00Cr-≤0.100N-≤1.00Ni	
1.3914	X2CrNiMnMoNNb21-15-7-3	Fe-≤0.03C-≤0.75Si-6.00-8.00Mn-≤0.025P-≤0.010S-20.00-22.00Cr-3.00-3.50Mo-0.350-0.500N-0.100-0.250Nb-14.00-16.00Ni	
1.3940	GX2CrNiN18-13	Fe-≤0.03C-≤1.00Si-≤2.00Mn-≤0.035P-≤0.020S-16.50-18.50Cr-0.100-0.200N-12.00-14.00Ni	
1.3951	G-X 4 CrNiMoN 22 15	Fe-≤0.05C-0.80-1.10Si-0.50-1.00Mn-≤0.030P-≤0.030S-22.00-23.50Cr-1.10-1.30Mo-0.150-0.250N-17.00-18.00Ni	
1.3952	X2CrNiMoN18-14-3	Fe-≤0.03C-≤1.00Si-≤2.00Mn-≤0.045P-≤0.015S-16.50-18.50Cr-2.50-3.00Mo-0.150-0.250N-13.00-15.00Ni	
1.3955	GX12CrNi18-11	Fe-≤0.15C-≤1.00Si-≤2.00Mn-≤0.045P-≤0.030S-16.50-18.50Cr-≤0.75Mo-10.00-12.00Ni	
1.3964	X2CrNiMnMoNNb21-16-5-3	Fe-≤0.03C-≤1.00Si-4.00-6.00Mn-≤0.025P-≤0.010S-20.00-21.50Cr-3.00-3.50Mo-0.200-0.350N-≤0.250Nb-15.00-17.00Ni	NITRONIC 50

German Standard		Materials Compositions	US-Standard
MatNo.	DIN-Design	Percent in Weight	SAE/ASTM/UNS
1.3974	X2CrNiMnMoNNb23-17-6-3	Fe-≤0.03C-≤1.00Si-≤4.50-6.50Mn-≤0.025P-≤0.01S-≤21.00-24.50Cr-≤2.80-3.40Mo-≤0.30-0.50N-≤0.10-0.30Nb-≤15.50-18.00Ni	
1.3981	NiCo 29 18; (X3NiCo29-18)	Fe-≤0.050C-≤0.30Si-≤0.50Mn-≤17.0-18.0Co-≤28.0-30.0Ni	UNS K94610
1.4000	X6Cr13	Fe-≤0.08C-≤1.0Si-≤1.0Mn-≤0.04P-≤0.015S-≤12.0-14.0Cr	SAE 403, 410S
1.4001	X 7 Cr 14	Fe-≤0.08C-≤1.00Si-≤1.00Mn-≤0.045P-≤0.030S-≤13.00-15.00Cr	429 (SAE) A240 (410S) (ASTM) 410S (SAE)
1.4002	X6CrAl13	Fe-≤0.08C-≤1.0Si-≤1.0Mn-≤0.04P-≤0.015S-≤0.10-0.3Al-≤12-14Cr	SAE 405
1.4003	X2CrNi12	Fe-≤0.03C-≤1.0Si-≤1.5Mn-≤0.04P-≤0.015S-≤10.5-12.5Cr-≤0.03N-≤0.3-1.0Ni	UNS S40977
1.4005	X12CrS13	Fe-≤0.08-0.15C-≤1.0Si-≤1.5Mn-≤0.04P-≤0.15-0.35S-≤12.0-14.0Cr-≤0.6Mo	SAE 416
1.4006	X12Cr13	Fe-≤0.08-0.15C-≤1.00Si-≤1.5Mn-≤0.04P-≤0.015S-≤11.0-13.5Cr-≤0.75Ni	SAE 410
1.4008	GX7CrNiMo12-1	Fe-≤0.10C-1.0Si-1.0Mn-≤0.035P-≤0.025S-12.0-13.5Cr-0.20-0.50Mo-1.0-2.0Ni	UNS J91150
1.4015	X8Cr18	Fe-≤0.10C-≤1.50Si-≤1.50Mn-≤0.030P-≤0.030S-≤16.50-18.50Cr	S43080 (UNS)
1.4016	X6Cr17	Fe-≤0.08C-≤1.0Si-≤1.0Mn-≤0.04P-≤0.015S-≤16.0-18.0Cr	SAE 430
1.4021	X20Cr13	Fe-≤0.16-0.25C-≤1.0Si-≤1.5Mn-≤0.04P-≤0.015S-≤12.0-14.0Cr	SAE 420
1.4024	X15Cr13	Fe-≤0.12-0.17C-≤1.0Si-≤1.0Mn-≤0.045P-≤0.03S-≤12.0-14.0Cr	SAE 420
1.4028	X30Cr13	Fe-≤0.26-0.35C-≤1.0Si-≤1.5Mn-≤0.04P-≤0.015S-≤12.0-14.0Cr	A 743; UNS J91153
1.4031	X39Cr13	Fe-≤0.36-0.42C-≤1.00Si-≤1.00Mn-≤0.04P-≤0.015S-≤12.5-14.5Cr	UNS S42080
1.4034	X46Cr13	Fe-≤0.43-0.5C-≤1.0Si-≤1.0Mn-≤0.04P-≤0.015S-≤12.5-14.5Cr	
1.4057	X17CrNi16-2	Fe-≤0.12-0.22C-≤1.0Si-≤1.5Mn-≤0.04P-≤0.015S-≤15.0-17.0Cr-≤1.5-2.5Ni	SAE 431
1.4085	GX70Cr29	Fe-≤0.50-0.90C-2.0Si-1.0Mn-≤0.045P-≤0.030S-27.0-30.0Cr	
1.4104	X14CrMoS17	Fe-≤0.10-0.17C-≤1.0Si-≤1.5Mn-≤0.04P-≤0.15-0.35S-≤15.5-17.5Cr-≤0.2-0.6Mo	SAE 430 F

Key to materials compositions

German Standard		Materials Compositions	US-Standard
Mat.-No.	DIN-Design	Percent in Weight	SAE/ASTM/UNS
1.4105	X6CrMoS17	Fe-≤0.08C-≤1.50Si-≤1.50Mn-≤0.040P-≤0.15-0.35S-≤16.00-18.00Cr-≤0.20-0.60Mo	
1.4110	X55CrMo14	Fe-≤0.48-0.60C-≤1.0Si-≤1.0Mn-≤0.04P-≤0.015S-≤13.0-15.0Cr-≤0.5-0.8Mo-≤0.15V	
1.4112	X90CrMoV18	Fe-≤0.85-0.95C-≤1.0Si-≤1.0Mn-≤0.04P-≤0.015S-≤17.0-19.0Cr-≤0.9-1.3Mo-≤0.07-0.12V	SAE 440 B
1.4113	X6CrMo17-1	Fe-≤0.80C-≤1.0Si-≤1.00Mn-≤0.04P-≤0.03S-≤16.0-18.0Cr-≤0.90-1.40Mo	SAE 434
1.4116	X50CrMoV15	Fe-≤0.45-0.55C-1.0Si-1.0Mn-≤0.040P-≤0.015S-14.0-15.0Cr-0.5-0.8Mo-0.10-0.20V	
1.4117	X38CrMoV15	Fe-≤0.35-0.40C-1.0Si-1.0Mn-≤0.045P-≤0.030S-14.0-15.0Cr-0.4-0.6Mo-0.10-0.15V	
1.4120	X20CrMo13	Fe-≤0.17-0.22C-≤1.0Si-≤1.0Mn-≤0.04P-≤0.015S-12-14Cr-≤0.9-1.3Mo-≤1.0Ni	
1.4122	X39CrMo17-1	Fe-≤0.33-0.45C-≤1.0Si-≤1.5Mn-≤0.04P-≤0.015S-≤15.5-17.5Cr-≤0.8-1.3Mo-≤1.0Ni	
1.4125	X105CrMo17	Fe-≤0.95-1.20C-1.0Si-1.0Mn-≤0.040P-≤0.015S-16.0-18.0Cr-0.4-0.8Mo	SAE 617
1.4126	X 110 CrMo 13	Fe-≤1.05-1.15C-≤1.0Si-≤1.0Mn-≤0.040P-≤0.030S- ≤17.0-18.0Cr-≤0.8-1.0Mo	
1.4131	X 1 CrMo 26 1	Fe-≤0.010C-≤0.40Si-≤0.40Mn-≤0.020P-≤0.020S-≤0.015N-≤25.0-27.5Cr-≤0.75-1.50Mo-≤0.50Ni	
1.4133		Fe-≤0.01C-1.0Si-1.0Mn-≤0.030P-≤0.030S-26.0-30.0Cr-1.7-2.3Mo-0.010N	
1.4136	GX70CrMo29-2	Fe-≤0.50-0.90C-2.0Si-1.0Mn-≤0.045P-≤0.030S-27.0-30.0Cr-2.0-2.5Mo	
1.4138	GX120CrMo29-2	Fe-≤0.90-1.30C-2.0Si-1.0Mn-≤0.045P-≤0.030S-27.0-30.0Cr-2.0-2.5Mo	
1.4153	X80CrVMo13-2	Fe-≤0.76-0.86C-≤1.00Si-≤1.00Mn-≤0.045P-≤0.030S-≤12.00-14.00Cr-≤0.40-0.60Mo-≤1.50-2.10V	
1.4300	X 12 CrNi 18 8	Fe-≤0.12C-≤1.0Si-≤2.0Mn-≤0.045P-≤0.030S-≤17.0-19.0Cr-≤8.0-10.0Ni	
1.4301	X5CrNi18-10	Fe-≤0.07C-≤1.0Si-≤2.0Mn-≤0.045P-≤0.015S≤17-19.5Cr-≤0.11N-≤8.0-10.5Ni	SAE 304
1.4302	X5CrNi19-9	Fe-≤0.05C-≤1.40Si-≤1.90Mn-≤0.025P-≤0.015S-≤18.2-19.8Cr-≤8.70-10.30Ni	UNS S30888

German Standard		Materials Compositions	US-Standard
MatNo.	DIN-Design	Percent in Weight	SAE/ASTM/UNS
1.4303	X4CrNi18-12	Fe-≤0.06C-≤1.0Si-≤2.0Mn-≤0.045P-≤0.015S-≤17-19Cr-≤0.11N-≤11.0-13.0Ni	SAE 305/308
1.4304	X5CrNi18-12E	Fe-≤0.12C-≤1.0Si-≤2.0Mn-≤0.045P-≤0.030S-≤17-19Cr-≤8.0-10.0Ni	
1.4305	X8CrNiS18-9; X10CrNiS189	Fe-≤0.10C-≤1.0Si-≤2.0Mn-≤0.045P-≤0.15-0.35S-≤17-19Cr-≤1.0Cu-≤0.11N-≤8.0-10.0Ni	SAE 303
1.4306	X2CrNi19-11	Fe-≤0.03C-≤1.0Si-≤2.0Mn-≤0.045P-≤0.015S-≤18-20Cr-≤0.11N-≤10.0-12.0Ni	SAE 304 L
1.4307	X2CrNi18-9	Fe-≤0.03C-≤1.0Si-≤2.0Mn-≤0.045P-≤0.015S-≤17.5-19.5Cr-≤0.11N-≤8.0-10.0Ni	
1.4308	GX5CrNi19-10	Fe-≤0.07C-≤1.5Si-≤1.5Mn-≤0.04P-≤0.030S-≤18.0-20.0Cr-≤8.0-11.0Ni	SAE 304 H
1.4310	X10CrNi18-8	Fe-≤0.05-0.15C-≤2.0Si-≤2.0Mn-≤0.045P-≤0.015S-≤16.0-19.0Cr-≤0.8Mo-≤0.11N-≤6.0-9.5Ni	SAE 301
1.4311	X2CrNiN18-10	Fe-≤0.03C-≤1.0Si-≤2.0Mn-≤0.045P-≤0.015S-≤17.0-19.5Cr-≤8.5-11.5Ni-≤0.12-0.22N	SAE 304 LN
1.4312	GX10CrNi18-8	Fe-≤0.12C-≤2.0Si-≤1.5Mn-≤0.045P-≤0.03S-≤17.0-19.5Cr-≤8.0-10.0Ni	A 743
1.4313	X3CrNiMo13-4	Fe-≤0.05C-≤0.7Si-≤1.5Mn-≤0.04P-≤0.015S-≤12.0-14.0Cr-≤0.3-0.7Mo-≤0.02N-≤3.5-4.5Ni	UNS J91540
1.4315	X5CrNiN19-9	Fe-≤0.06C-≤1.0Si-≤2.0Mn-≤0.045P-≤0.03S-≤18.0-20.0Cr-≤0.12-0.22N-≤8.0-11.0Ni	
1.4317	GX4CrNi13-4	Fe-≤0.06C-≤1.00Si-≤1.00Mn-≤0.035P-≤0.025S-≤12.00-13.50Cr-≤0.70Mo-≤3.50-5.00Ni	UNS J91540 A 743 (CA-6NM)
1.4318	X2CrNiN18-7	Fe-≤0.03C-≤1.0Si-≤2.0Mn-≤0.045P-≤0.015S-≤16.5-18.5Cr-≤0.1-0.2N-≤6.0-8.0Ni	
1.4319	X3CrNiN17-8	Fe-≤0.05C-≤1.00Si-≤2.00Mn-≤0.045P-≤0.015S-≤16.00-18.00Cr-≤0.04-0.08N-≤7.00-8.00Ni	
1.4330	X 2 CrNi 25 20	Fe-≤0.03C-≤1.0Si-≤1.5Mn-≤0.045P-≤0.035S-≤18.0-22.0Cr-≤23.0-27.0Ni	
1.4333	X 5 NiCr 32 21	Fe-≤0.07C-≤1.40Si-≤2.40Mn-≤0.045P-≤0.030S-≤19.00-22.00Cr-≤30.00-34.00Ni	S33200 (UNS) S33200 (SAE) B 710 (N 08330) (ASTM)
1.4335	X1CrNi25-21	Fe-≤0.02C-≤0.25Si-≤2.0Mn-≤0.025P-≤0.010S-≤24.0-26.0Cr-≤0.20Mo-≤0.110N-≤20.0-22.0Ni	

German Standard		Materials Compositions	US-Standard
Mat.-No.	DIN-Design	Percent in Weight	SAE/ASTM/UNS
1.4340	GX40CrNi27-4	Fe-≤0.30-0.50C-≤2.00Si-≤1.50Mn-≤0.045P-≤0.030S-≤26.0-28.0Cr-≤3.5-5.5Ni	A 743
1.4347	GX6CrNiN26-7	Fe-≤0.08C-≤1.50Si-≤1.50Mn-≤0.035P-≤0.020S-25.00-27.00Cr-0.100-0.200N-5.50-7.50Ni	
1.4361	X1CrNiSi18-15-4	Fe-≤0.015C-≤3.70-4.50Si-≤2.00Mn-≤0.025P-≤0.010S-≤16.5-18.5Cr-≤0.20Mo-≤0.110N-≤14.0-16.0Ni	A 336
1.4362	X2CrNiN23-4	Fe-≤0.03C-≤1.00Si-≤2.00Mn-≤0.035P-≤0.015S-≤22.0-24.0Cr-≤0.10-0.60Cu-≤0.10-0.60Mo-≤0.050-0.200N-≤3.5-5.5Ni	
1.4371	X2CrMnNiN17-7-5	Fe-≤0.30C-≤1.0Si-≤6.0-8.0Mn-≤0.045P-≤0.015S-≤16.0-17.0Cr-≤0.15-0.20N-≤3.5-5.5Ni	SAE 202
1.4401	X5CrNiMo17-12-2	Fe-≤0.07C-≤1.0Si-≤2.0Mn-≤0.045P-≤0.015S-≤16.5-18.5Cr-≤2.0-2.5Mo-≤0.110N-≤10.0-13.0Ni	SAE 316
1.4404	X2CrNiMo17-12-2; X2 CrNi-Mo 17 12 2	Fe-≤0.03C-≤1.0Si-≤2.0Mn-≤0.045P-≤0.015S-≤16.5-18.5Cr-≤2.0-2.5Mo-≤10.0-13.0Ni≤0.110N	SAE 316 L
1.4405	X 5 CrNiMo 16 5; GX4CrNiMo16-5-1	Fe-≤0.07C-≤1.0Si-≤1.0Mn-≤0.035P-≤0.025S-≤15.0-16.5Cr-≤0.5-2.0Mo-≤4.5-6.0Ni	
1.4406	X2CrNiMoN17-11-2; X2 CrNi-MoN 17 12 2	Fe-≤0.03C-≤1.0Si-≤2.0Mn-≤0.045P-≤0.015S-≤16.5-18.5Cr-≤2.0-2.5Mo-≤0.12-0.22N-≤10.0-12.0Ni	SAE 316 LN
1.4408	GX5CrNiMo19-11-2	Fe-≤0.07C-≤1.50Si-≤1.50Mn-≤0.04P-≤0.030S-≤18.0-20.0Cr-≤2.0-2.5Mo-≤9.0-12.0Ni	CF-8M
1.4410	X2CrNiMoN25-7-4	Fe-≤0.03C-≤1.0Si-≤2.0Mn-≤0.035P-0.015S-≤24.0-26.0Cr-≤3.0-4.5Mo-≤0.200-0.350N-≤6.0-8.0Ni	2507; A 182
1.4413	X3CrNiMo13-4	Fe-≤0.05C-≤0.30-0.60Si-≤0.50-1.00Mn-≤0.03P-≤0.02S-≤12.00-14.00Cr-≤0.30-0.70Mo-≤3.50-4.00Ni	UNS S42400 A 988 (S 41500) SAE S41500
1.4417	GX2CrNiMoN25-7-3	Fe-≤0.03C-≤1.0Si-≤1.5Mn-≤0.030P-≤0.020S-≤24.0-26.0Cr-≤1.0Cu-≤3.0-4.0Mo-≤0.150-0.250N-≤6.0-8.5Ni-≤1.0W	3RE60; A 789
1.4418	X4CrNiMo16-5-1	Fe-≤0.06C-≤0.7Si-≤1.5Mn-≤0.040P-≤0.015S-≤15.0-17.0Cr-≤0.8-1.5Mo-≤0.020N-≤4.0-6.0Ni	

German Standard		Materials Compositions	US-Standard
Mat.-No.	DIN-Design	Percent in Weight	SAE/ASTM/UNS
1.4424	X2CrNiMoSi18-5-3	Fe-≤0.03C-1.40-2.00Si-1.20-2.00Mn-≤0.035P-≤0.015S-18.00-19.00Cr-2.50-3.00Mo-0.050-1.00N-4.50-5.20Ni	
1.4427	X12CrNiMoS18-11	Fe-≤0.12C-≤1.0Si-≤2.0Mn-≤0.06P-≤0.15-0.35S-≤16.5-18.5Cr-≤2-2.5Mo-≤10.5-13.5Ni	
1.4429	X2CrNiMoN17-13-3	Fe-≤0.03C-≤1.0Si-≤2.0Mn-≤0.045P-≤0.015S-≤16.5-18.5Cr-≤2.5-3.0Mo-≤11.0-14.0Ni-≤0.12-0.22N	SAE 316 LN
1.4430	X2CrNiMo19-12	Fe-≤0.02C-≤1.40Si-≤1.90Mn-≤0.025P-≤0.015S-17.20-19.80Cr-2.50-3.00Mo-10.70-13.30Ni	S 31688 UNS S 31683 UNS
1.4434	X2CrNiMoN18-12-4	Fe-≤0.03C-≤1.00Si-≤2.00Mn-≤0.045P-≤0.015S-≤16.50-19.50Cr-≤3.00-4.00Mo-≤0.10-0.20N-≤10.50-14.00Ni	
1.4435	X2CrNiMo18-14-3; X2 CrNi-Mo 18 4 3	Fe-≤0.03C-≤1.0Si-≤2.0Mn-≤0.045P-≤0.015S-≤17.0-19.0Cr-≤2.5-3.0Mo-≤0.110N-≤12.5-15.0Ni	SAE 316 L
1.4436	X3CrNiMo17-13-3; X5 CrNi-Mo 17 13 3	Fe-≤0.05C-≤1.0Si-≤2.0Mn-≤0.045P-≤0.015S-≤16.5-18.5Cr-≤2.5-3.0Mo-≤0.110N-≤10.5-13.0Ni	SAE 316
1.4438	X2CrNiMo18-15-4	Fe-≤0.03C-≤1.0Si-≤2.0Mn-≤0.045P-≤0.015S-≤17.5-19.5Cr-≤3.0-4.0Mo-≤0.110N-≤13.0-16.0Ni	SAE 317 L
1.4439	X2CrNiMoN17-13-5	Fe-≤0.030C-≤1.00Si-≤2.00Mn-≤0.045P-≤0.015S-≤16.5-18.5Cr-≤4.00-5.00Mo-≤12.5-14.5Ni-≤0.12-0.22N	
1.4440	X2CrNiMo18-16-5	Fe-≤0.03C-≤1.00Si-2.00-3.00Mn-≤0.025P-≤0.025S-17.00-20.00Cr-4.00-5.00Mo-16.00-19.00Ni	SAE 317 L
1.4441	X2CrNiMo18-15-3	Fe-≤0.03C-≤1.00Si-≤2.00Mn-≤0.025P-≤0.01S-≤17.00-19.00Cr-≤0.50Cu-≤2.50-3.20Mo-≤0.10N-≤13.00-15.50Ni	
1.4442	X2CrNiMoN18-15-4	Fe-≤0.03C-1.0Si-2.0Mn-≤0.025P-≤0.010S-17.0-18.5Cr-3.7-4.2Mo-0.1-0.2N-14.0-16.0Ni	SAE 317 LN
1.4447	X5CrNiMo18-13	Fe-≤0.06C-1.5Si-2.0Mn-≤0.025P-≤0.020S-17.0-19.0Cr-4.0-5.0Mo-12.5-15.5Ni	
1.4449	X3CrNiMo18-12-3	Fe-≤0.035C-≤1.00Si-≤2.00Mn-≤0.045P-≤0.015S-≤17.0-18.2Cr-≤1.0Cu-≤2.25-2.75Mo-≤0.08N-≤11.5-12.5Ni	SAE 317

German Standard		Materials Compositions	US-Standard
Mat.-No.	DIN-Design	Percent in Weight	SAE/ASTM/UNS
1.4457	X8CrNiMo17-5-3	Fe-≤0.07-0.11C-≤0.50Si-≤0.50-1.25Mn-≤16.00-17.00Cr-≤2.50-3.25Mo-≤4.00-5.00Ni	
1.4460	X3CrNiMoN27-5-2	Fe-≤0.05C-≤1.0Si-≤2.0Mn-≤0.035P-≤0.015S-≤25.0-28.0Cr-≤1.3-2.0Mo-≤0.05-0.2N-≤4.5-6.5Ni	SAE 329
1.4462	X2CrNiMoN22-5-3	Fe-≤0.03C-≤1.00Si-≤2.00Mn-≤0.035P-≤0.015S-≤21.0-23.0Cr-≤2.50-3.50Mo-≤4.50-6.50Ni-≤0.1-0.22N	2205; A 182
1.4463	GX6CrNiMo24-8-2	Fe-≤0.07C-1.5Si-1.5Mn-≤0.045P-≤0.030S-23.0-25.0Cr-2.0-2.5Mo-7.0-8.5Ni	
1.4464	GX40CrNiMo27-5	Fe-0.30-0.50C-≤2.00Si-≤1.50Mn-≤0.045P-≤0.030S-26.00-28.00Cr-2.00-2.50Mo-4.00-6.00Ni	
1.4465	X1CrNiMoN25-25-2	Fe-≤0.02C-≤0.70Si-≤2.0Mn-≤0.020P-≤0.015S-≤24.0-26.0Cr-≤2.0-2.5Mo-≤22.0-25.0Ni-≤0.08-0.16N	SAE 310 MoLN
1.4466	X1CrNiMoN25-22-2	Fe-≤0.02C-≤0.7Si-≤2.0Mn-≤0.025P-≤0.010S-≤24.0-26.0Cr-≤2.0-2.5Mo-≤21.0-23.0Ni-≤0.1-0.16N	
1.4467	X2CrMnNiMoN26-5-4	Fe-≤0.03C-≤0.8Si-≤4.0-6.0Mn-≤0.03P-≤0.015S-≤24.5-26.5Cr-≤2.0-3.0Mo-≤0.3-0.45N-≤3.5-4.5Ni	
1.4469	GX2CrNiMoN26-7-4	Fe-≤0.03C-1.0Si-1.0Mn-≤0.035P-≤0.025S-25.0-27.0Cr-≤1.30Cu-3.0-5.0Mo-0.12-0.22N-6.0-8.0Ni	UNS J93404
1.4492	X 8 CrNiMoN 17 5	Fe-≤0.07-0.11C-≤0.5Si-≤0.5-1.25Mn-≤0.04P-≤0.03S-≤16.0-17.0Cr-≤2.5-3.25Mo-≤4.0-5.0Ni	
1.4500	GX7NiCrMoCuNb25-20	Fe-≤0.08C-≤1.50Si-≤2.0Mn-≤0.045P-≤0.03S-≤19.0-21.0Cr-≤1.5-2.5Cu-≤2.5-3.5Mo-≤24.0-26.0Ni	A 351
1.4501	X2CrNiMoCuWN25-7-4	Fe-≤0.03C-≤1.0Si-≤1.0Mn-≤0.035P-≤0.015S-≤24.0-26.0Cr-≤0.5-1.0Cu-≤3.0-4.0Mo-≤0.2-0.3N-≤6.0-8.0Ni-≤0.5-1.0W	UNS S32760
1.4502	X8CrTi18	Fe-≤0.09C-≤1.40Si-≤1.40Mn-≤0.030P-≤0.020S-16.70-18.30Cr-≤0.060N-0.35-0.65Ti	
1.4503	X3NiCrCuMoTi27-23	Fe-≤0.04C-0.75Si-0.75Mn-≤0.030P-≤0.015S-22.0-24.0Cr-2.5-3.5Cu-2.5-3.0Mo-26.0-28.0Ni-0.4-0.7Ti	
1.4504		Fe-≤0.09C-≤0.50Si-≤1.00Mn-≤0.025P-≤0.025S-≤0.75-1.25Al-≤16.00-17.25Cr-≤6.50-7.75Ni	UNS S17780 SAE S17700 A 705 (631)

German Standard		Materials Compositions	US-Standard
MatNo.	DIN-Design	Percent in Weight	SAE/ASTM/UNS
1.4505	X4NiCrMoCuNb20-18-2	Fe-≤0.05C-≤1.0Si-≤2.0Mn-≤0.045P-≤0.015S-≤16.5-18.5Cr-≤2.0-2.5Mo-≤19.0-21.0Ni-≤1.8-2.2Cu	
1.4506	X5NiCrMoCuTi20-18	Fe-≤0.07C-≤1.0Si-≤2.0Mn-≤0.045P-≤0.030S-≤16.5-18.5Cr-≤2.0-2.5Mo-≤19.0-21.0Ni-≤1.8-2.2Cu	
1.4507	X2CrNiMoCuN25-6-3	Fe-≤0.03C-≤0.7Si-≤2.0Mn-≤0.035P-≤0.015S-≤24.0-26.0Cr-≤1.0-2.5Cu-≤0.15-0.30N-≤5.5-7.5Ni-≤2.7-4.0Mo	UNS S43940
1.4509	X2CrTiNb18	Fe-≤0.03C-≤1.0Si-≤1.0Mn-≤0.040P-≤0.015S-≤17.5-18.5Cr-≤0.10-0.60Ti	
1.4510	X6CrTi17	Fe-≤0.05C-≤1.0Si-≤1.0Mn-≤0.040P-≤0.015S-≤16.0-18.0Cr	SAE 430 Ti
1.4511	X3CrNb17; X6CrNb17	Fe-≤0.05C-≤1.0Si-≤1.0Mn-≤0.040P-≤0.015S-≤16.0-18.0Cr	
1.4512	X2CrTi12	Fe-≤0.03C-≤1.0Si-≤1.0Mn-≤0.040P-≤0.015S-≤10.5-12.5Cr	SAE 409
1.4515	GX2CrNiMoCuN26-6-3	Fe-≤0.03C-≤1.0Si-≤2.0Mn-≤0.030P-≤0.020S-≤0.12-0.25N-≤24.5-26.5Cr-≤2.5-3.5Mo-≤5.5-7.0Ni-≤0.8-1.3Cu	
1.4517	GX2CrNiMoCuN25-6-3-3	Fe-max. 0.03C-1.0Si-1.5Mn-0.035P-0.025S-24.5-26.5Cr-2.75-3.50Cu-2.50-3.50Mo-0.12-0.22N-5.0-7.0Ni	UNS J93372
1.4519	X2CrNiMoCu20-25	Fe-≤0.02C-≤1.40Si-≤2.10-4.90Mn-≤0.025P-≤0.02S-≤19.20-21.80Cr-≤0.90-1.90Cu-≤4.10-5.90Mo-≤24.30-26.70Ni	
1.4520	X2CrTi17	Fe-≤0.025C-≤0.5Si-≤0.5Mn-≤0.04P-≤0.015S-≤16.0-18.0Cr-≤0.015N-≤0.3-0.6Ti	
1.4521	X2CrMoTi18-2	Fe-≤0.025C-≤1.0Si-≤0.040P-≤0.015S-≤1.0Mn-≤17.0-20.0Cr-≤1.8-2.5Mo-≤0.030N	SAE 444
1.4522	X2CrMoNb18-2	Fe-≤0.025C-≤1.0Si-≤1.0Mn-≤0.040P-≤0.015S-≤17.0-19.0Cr-≤1.8-2.3Mo-≤0.25Ni	SAE 443
1.4523	X2CrMoTiS18-2	Fe-≤0.03C-≤1.0Si-≤0.5Mn-≤0.040P-≤0.15-0.35S-≤17.5-19.0Cr-≤2.0-2.5Mo-≤0.30-0.80Ti	
1.4525	GX5CrNiCu16-4; GX4CrNi-CuNb16-4	Fe-≤0.07C-≤0.80Si-≤1.00Mn-≤0.035P-≤0.025S-≤15.00-17.00Cr-≤2.50-4.00Cu-≤0.80Mo-≤0.050N-≤0.350Nb-≤3.50-5.50Ni	
1.4527	GX4NiCrCuMo30-20-4	Fe-≤0.06C-≤1.50Si-≤1.50Mn-≤0.040P-≤0.030S-19.00-22.00Cr-3.00-4.00Cu-2.00-3.00Mo-27.50-30.50Ni	

German Standard		Materials Compositions	US-Standard
MatNo.	DIN-Design	Percent in Weight	SAE/ASTM/UNS
1.4528	X105CrCoMo18-2	Fe-≤1.0-1.1C-≤1.0Si-≤1.0Mn-≤0.045P-≤0.030S-≤16.5-18.5Cr-≤1.0-1.5Mo-≤0.30-0.80Ti-≤1.3-1.8Co≤0.07-0.12V	
1.4529	X1NiCrMoCuN25-20-7	Fe-≤0.02C-≤0.50Si-≤1.00Mn-≤0.030P-≤0.010S-≤19.0-21.0Cr-≤6.0-7.0Mo-≤0.5-1.5Cu-≤0.15-0.25N-≤24.0-26-0Ni	A 249; ASTM N08926
1.4530	X1CrNiMoAlTi12-9	Fe-≤0.015C-0.10Si-0.10Mn-≤0.020P-0.6-1.0Al-11.0-12.5Cr-1.5-2.5Mo-8.5-10.5Ni-0.25-0.40Ti	
1.4532	X8CrNiMoAl15-7-2	Fe-≤0.1C≤0.7Si≤1.2Mn≤0.04P≤0.015S-0.7-1.5Al-14.0-16.0Cr-2.0-3.0Mo-6.5-7.8Ni	UNS S15700
1.4533	X6CrNiTi18-10S	Fe-≤0.06C-≤1.00Si-≤2.00Mn-≤0.035P-≤0.015S-≤0.20Co-≤17.00-19.00Cr-≤9.00-12.00Ni	
1.4534	X3CrNiMoAl13-8-2	Fe-≤0.05C-≤0.10Si-≤0.10Mn-≤0.010P-≤0.008S-≤0.90-1.20Al-≤12.25-13.25Cr-≤2.0-2.50Mo-≤0.01N-≤7.50-8.50Ni	S 13800
1.4536	GX2NiCrMoCuN25-20	Fe-≤0.03C≤1.0Si≤1.0Mn≤0.035P≤0.01S-19.0-21.0Cr-1.5-2.0Cu-2.5-3.5Mo-24.0-26.0Ni	UNS J94650
1.4537	X1CrNiMoCuN25-25-5	Fe-≤0.02C-≤0.70Si-≤2.00Mn-≤0.030P-≤0.010S-24.00-26.00Cr-1.00-2.00Cu-4.70-5.70Mo-0.170-0.250N-24.00-27.00Ni	
1.4539	X1NiCrMoCu25-20-5	Fe-≤0.02C-≤0.70Si-≤2.0Mn-≤0.030P-≤0.010S-≤19.0-21.0Cr-≤4.0-5.0Mo-≤24.0-26.0Ni-≤1.2-2.0Cu-≤0.150N	SAE 904 L
1.4540		Fe-≤0.06C-1.0Si-1.0Mn-15.0-17.0Cr-2.5-4.0Cu-0.050N-0.15-0.40Nb-3.5-5.0Ni	UNS J92130
1.4541	X6CrNiTi18-10	Fe-≤0.08C-≤1.0Si-≤2.0Mn-≤0.045P-≤0.015S-≤17.0-19.0Cr-≤9.0-12.0Ni	SAE 321
1.4542	X5CrNiCuNb16-4	Fe-≤0.07C-≤0.70Si-≤1.5Mn-≤0.040P-≤0.015S-≤15.0-17.0Cr-≤3.0-5.0Ni-≤3.0-5.0Cu-≤0.60Mo	SAE 630; 17-4 PH
1.4544	X 10 CrNiMnTi 18 10	Fe-≤0.08C-≤1.0Si-≤2.0Mn-≤0.035P-≤0.025S-≤17.0-19.0Cr-≤9.0-11.5Ni	SAE 321; UNS J92630
1.4545		Fe-≤0.07C-≤1.00Si-≤1.00Mn-≤0.03P-≤0.015S-≤14.00-15.50Cr-≤2.50-4.50Cu-≤0.50Mo-≤3.50-5.50Ni	UNS S15500 SAE S15500 A 705 (XM-12)
1.4546	X5CrNiNb18-10	Fe-≤0.08C-≤1.0Si-≤2.0Mn-≤0.045P-≤0.030S-≤17.0-19.0Cr-≤9.0-11.5Ni	SAE 347

German Standard		Materials Compositions	US-Standard
MatNo.	DIN-Design	Percent in Weight	SAE/ASTM/UNS
1.4547	X1CrNiMoCuN20-18-7	Fe-≤0.02C-≤0.70Si-≤1.0Mn-≤0.030P-≤0.010S-≤19.5-20.5Cr-≤0.5-1.0Cu-≤6.00-7.00Mo-≤0.18-0.25N-≤17.5-18.5Ni	254 SMO; A 182
1.4548	X5CrNiCuNb17-4-4	Fe-≤0.07C-≤1.0Si-≤1.0Mn-≤0.025P-≤0.025S-≤15.0-17.5Cr-≤3.0-5.0Cu-≤0.15-0.45Nb-≤3.00-5.00Ni	17-4 PH; SAE 630
1.4550	X6CrNiNb18-10	Fe-≤0.08C-≤1.0Si-≤2.0Mn-≤0.045P-≤0.015S-≤17.0-19.0Cr≤9.0-12.0Ni	SAE 347
1.4551	X5CrNiNb19-9	Fe-≤0.06C-≤1.40Si-≤1.90Mn-≤0.025P-≤0.015S-≤18.20-19.80Cr-≤8.20-9.80Ni	UNS S34780S34781S34788
1.4552	GX5CrNiNb19-11	Fe-≤0.07C≤1.5Si≤1.5Mn≤0.04P≤0.03S-18.0-20.0Cr-9.0-12.0Ni	UNS J92710
1.4557	GX2CrNiMoCuN20-18-6	Fe-≤0.025C-≤1.00Si-≤1.20Mn-≤0.03P-≤0.01S-≤19.50-20.50Cr-≤0.50-1.00Cu-≤6.00-7.00Mo-≤0.18-0.24N-≤17.50-19.50Ni	
1.4558	X2NiCrAlTi32-20	Fe-≤0.03C-≤0.70Si-≤1.0Mn-≤0.020P-≤0.015S-≤0.15-0.45Al≤20.0-23.0Cr-≤32.0-35.0Ni	
1.4561	X1CrNiMoTi18-13-2	Fe-≤0.02C-≤0.50Si-≤2.0Mn-≤0.035P-≤0.015S-≤17.0-18.5Cr-≤2.0-2.5Mo-≤11.5-13.5Ni-≤0.4-0.6Ti	
1.4562	X1NiCrMoCu32-28-7	Fe-≤0.015C-≤0.30Si-≤2.0Mn-≤0.020P-≤0.010S-≤26.0-28.0Cr-≤1.0-1.4Cu-≤6.0-7.0Mo-≤0.15-0.25N-≤30.0-32.0Ni	Alloy 31
1.4563	X1NiCrMoCu31-27-4	Fe-≤0.02C-≤0.70Si-≤2.0Mn-≤0.030P-≤0.010S-≤26.0-28.0Cr-≤3.0-4.0Mo-≤30.0-32.0Ni-≤0.70-1.5Cu-≤0.11N	B 668
1.4564		Fe-≤0.09C-≤1.00Si-≤1.00Mn-≤0.04P-≤0.03S-≤0.75-1.50Al-≤16.00-18.00Cr-≤0.50Cu-≤6.50-7.75Ni	UNS 17700 SAE S17700 A 705 (631)
1.4565	X2CrNiMnMoNbN25-18-5-4	Fe-≤0.03C-≤1.0Si-≤3.5-6.5Mn-≤0.030P-≤0.015S-≤23.0-26.0Cr-≤3.0-5.0Mo-≤0.3-0.5N-≤0.15Nb-≤16.0-19.0Ni	UNS S34565
1.4566	X3CrNiMnMoCuNbN 23-17-5-3	Fe-≤0.04C≤1.0Si-4.5-6.5Mn≤0.03P-≤0.015S-21.0-25.0Cr-0.3-1.0Cu-3.0-4.5Mo-15.0-18.0Ni-0.1-0.3Nb	
1.4567	X3CrNiCu18-9-4; X 3 CrNiCu 18 9	Fe-≤0.04C-≤1.00Si-≤2.00Mn-≤0.045P-≤0.015S-≤17.00-19.00Cr-≤3.00-4.00Cu-≤0.110N-≤8.50-10.50Ni	304 Cu (SAE) S 30430 (UNS) A 493 (S 30430) (ASTM)
1.4568	X7CrNiMoAl17-7	Fe-≤0.09C-≤0.70Si-≤1.0Mn-≤0.040P-0.015S-≤16.0-18.0Cr-≤6.5-7.80Ni-≤0.70-1.5Al	17-7 PH; SAE 631

German Standard		Materials Compositions	US-Standard
Mat.-No.	DIN-Design	Percent in Weight	SAE/ASTM/UNS
1.4571	X6CrNiMoTi17-12-2	Fe-≤0.08C-≤1.0Si-≤2.0Mn-≤0.045P-0.015S-≤16.5-18.5Cr-≤2.0-2.5Mo-≤10.5-13.5Ni	SAE 316 Ti
1.4573	GX3CrNiMoCuN24-6-5	Fe-≤0.40C-≤1.0Si-≤1.0Mn-≤0.030P-≤0.020S-≤22.0-25.0Cr-≤1.5-2.5Cu-≤4.5-6.0Mo-≤0.15-0.25N-≤4.5-6.5Ni	SAE 316 Ti
1.4574		Fe-≤0.09C-≤1.00Si-≤1.00Mn-≤0.040P-≤0.030S-≤0.75-1.50Al-≤14.00-16.00Cr-≤2.00-3.00Mo-≤6.50-7.75Ni	S 15700 (SAE) A 579 (63) (ASTM) S 15700 (UNS)
1.4575	X1CrNiMoNb28-4-2	Fe-≤0.015C-≤1.0Si-≤1.0Mn-≤0.025P-≤0.015S-≤26.0-30.0Cr-≤1.8-2.5Mo-≤0.035N-≤3.0-4.5Ni	25-4-4; A 176
1.4577	X3CrNiMoTi25-25	Fe-≤0.04C-≤0.50Si-≤2.0Mn-≤0.030P-≤0.015S-≤24.0-26.0Cr-≤2.0-2.5Mo-≤24.0-26.0Ni	
1.4580	X6CrNiMoNb17-12-2	Fe-≤0.08C-≤1.0Si-≤2.0Mn-≤0.045P-≤0.015S-≤16.5-18.5Cr-≤2.0-2.5Mo-≤10.5-13.5Ni	SAE 316 Cb UNS J92971
1.4581	GX5CrNiMoNb19-11-2	Fe-≤0.07C≤1.5Si≤1.5Mn≤0.04P≤0.03S-18.0-20.0Cr≤2.0-2.5Mo-9.0-12.0Ni	
1.4582	X4CrNiMoNb25-7	Fe-≤0.06C-≤1.00Si-≤2.00Mn-≤0.045P-≤0.030S-≤24.00-26.00Cr-≤1.30-2.00Mo-≤6.50-7.50Ni	
1.4583	X10CrNiMoNb18-12	Fe-≤0.10C-≤1.00Si-≤2.00Mn-≤0.045P-≤0.030S-≤16.5-18.5Cr-≤2.5-3.0Mo-≤12.0-14.5Ni	318 (Spec)
1.4585	GX7CrNiMoCuNb1818	Fe-≤0.080C-≤1.50Si-≤2.0Mn-≤0.045P-≤0.030S-≤16.5-18.5Cr-≤2.0-2.5Mo-≤19.0-21.0Ni-≤1.8-2.4Cu	
1.4586	X5NiCrMoCuNb22-18	Fe-≤0.07C-1.0Si-2.0Mn-≤0.045P-≤0.030S-16.5-18.5Cr-1.5-2.0Cu-3.0-3.5Mo-21.5-23.5Ni	
1.4589	X5CrNiMoTi15-2	Fe-≤0.080C-≤1.0Si-≤1.0Mn-≤0.045P-≤0.030S-≤13.0-15.5Cr-≤0.2-1.2Mo-≤1.0-2.5Ni-≤0.3-0.5Ti	UNS S42035
1.4591	X1CrNiMoCuN33-32-1	Fe-≤0.015C-≤0.5Si-≤2.0Mn-≤0.020P-≤0.010S-≤31.0-35.0Cr-≤0.3-1.2Cu-≤0.5-2.0Mo-≤0.35-0.6N-≤30.0-33.0Ni	
1.4592	X1CrMoTi29-4	Fe-≤0.025C-1.0Si-1.0Mn-≤0.030P-≤0.010S-28.0-30.0Cr-3.5-4.5Mo-0.045N	
1.4593	GX3CrNiMoCuN24-6-2-3	Fe-≤0.04C-1.5Si-1.5Mn-≤0.030P-≤0.020S-23.0-26.0Cr-2.75-3.5Cu-2.0-3.0Mo-0.1-0.2N-5.0-8.0Ni	

German Standard		Materials Compositions	US-Standard
MatNo.	DIN-Design	Percent in Weight	SAE/ASTM/UNS
1.4603	X1CrTi17	Fe-≤0.02C-≤1.00Si-≤1.00Mn-≤0.040P-≤0.015S-≤16.00-18.00Cr	
1.4604	X2CrTi20	Fe-≤0.03C-≤1.00Si-≤1.00Mn-≤0.040P-≤0.015S-≤19.00-21.00Cr-≤0.40-0.80Ti	
1.4652	X1CrNiMoCuN24-22-8	Fe-≤0.02C-≤0.5Si-≤2.0-4.0Mn-≤0.03P-≤0.005S-≤23.0-25.0Cr-≤0.3-0.6Cu-≤7.0-8.0Mo-≤0.45-0.55N-≤21.0-23.0Ni	
1.4712	X10CrSi6	Fe-≤0.12C-≤2.00-2.50Si-≤1.00Mn-≤0.045P-≤0.030S-≤5.50-6.50Cr	
1.4713	X10CrAl7; X10CrAlSi7	Fe-≤0.12C-≤0.50-1.00Si-≤1.00Mn-≤0.040P-≤0.015S-≤0.5-1.0Al-≤6.00-8.00Cr	
1.4718	X45CrSi9-3	Fe-0.4-0.5C-2.7-3.3Si-≤0.6Mn-≤0.04P-≤0.03S-8.0-10.0Cr-≤0.5Ni	S65007 (UNS)
1.4720	X7CrTi12	Fe-≤0.08C-1.0Si-1.0Mn-≤0.040P-≤0.030S-10.5-12.5Cr	SAE 409
1.4722	X 10 CrSi 13	Fe-≤0.12C-≤1.90-2.40Si-≤1.00Mn-≤0.045P-≤0.030S-≤12.0-14.0Cr	
1.4724	X10CrAl13; X10CrAlSi13	Fe-≤0.12C-≤0.70-1.40Si-≤1.00Mn-≤0.040P-≤0.015S-≤0.70-1.20Al-≤12.0-14.0Cr	
1.4725	CrAl 14 4; (X8CrAl14-4)	Fe-≤0.1C-≤0.5Si-≤1.0Mn-≤0.045P-≤0.03S-3.5-5.0Al-13.0-15.0Cr	K91670 (UNS)
1.4742	X10CrAlSi18; X10CrAl18	Fe-≤0.12C-≤0.70-1.40Si-≤1.00Mn-≤0.040P-≤0.015S-≤0.70-1.20Al-≤17.00-19.00Cr	
1.4749	X18CrN28	Fe-0.15-0.20C-1.0Si-1.0Mn-≤0.040P-≤0.015S-26.0-29.0Cr-0.15-0.25N	
1.4762	X10 CrAl 24; X10CrAlSi25	Fe-≤0.12C-≤0.70-1.40Si-≤1.00Mn-≤0.040P-≤0.015S-≤1.20-1.70Al-≤23.0-26.0Cr	SAE 446
1.4765	CrAl 25 5; (X8CrAl25-5)	Fe-≤0.10C-≤1.00Si-≤0.60Mn-≤0.045P-≤0.030S-≤4.50-6.00Al-≤22.00-25.00Cr	
1.4773	X8Cr30	Fe-≤0.09C-≤1.90Si-≤1.40Mn-≤0.030P-≤0.025S-≤28.80-31.20Cr-≤2.00Ni	
1.4776	GX40CrSi28	Fe-0.30-0.50C-1.0-2.5Si-1.0Mn-≤0.040P-≤0.030S-27.0-30.0Cr-0.50Mo-1.0Ni	UNS J92605
1.4777	GX130CrSi29	Fe-1.20-1.40C-1.0-2.5Si-0.5-1.0Mn-≤0.035P-≤0.030S-27.0-30.0Cr-0.50Mo-1.0Ni	
1.4821	X15CrNiSi25-4; X20 CrNiSi 25 4	Fe-0.1-0.2C-0.8-1.5Si-≤2.0Mn-≤0.04P-≤0.015S-24.5-26.5Cr-≤0.11N-3.5-5.5Ni	
1.4828	X15CrNiSi20-12	Fe-≤0.20C-≤1.50-2.50Si-≤2.0Mn-≤0.045P-≤0.015S-≤19.0-21.0Cr-≤0.11N-≤11.0-13.0Ni	SAE 309

German Standard		Materials Compositions	US-Standard
Mat.-No.	DIN-Design	Percent in Weight	SAE/ASTM/UNS
1.4829	X12CrNi22-12	Fe-≤0.14C-0.90-1.90Si-1.90Mn-≤0.025P-≤0.015S-20.8-23.2Cr-10.2-12.8Ni	UNS S30980
1.4833	X7CrNi23 14; X12CrNi23-14	Fe-≤0.15C-≤1.00Si-≤2.00Mn-≤0.045P-≤0.015S-≤22.0-24.0Cr-≤0.11N-≤12.0-14.0Ni	SAE 309 S
1.4835	X9CrNiSiNCe21-11-2	Fe-≤0.05-0.12C-≤1.4-2.5Si-≤1.0Mn-≤0.045P-≤0.015S-≤0.030-0.080Ce-≤20.0-22.0Cr-≤0.12-0.20N-≤10.0-12.0Ni	253 MA; A 182
1.4841	X15CrNiSi25-20; X15CrNi-Si25-21	Fe-≤0.20C-≤1.50-2.50Si-≤2.00Mn-≤0.045P-≤0.015S-≤24.0-26.0Cr-≤0.11N-≤19.0-22.0Ni	3RE60; SAE 310; SAE 314
1.4845	X8CrNi25-21; X12CrNi25-21	Fe-≤0.15C-≤1.50Si-≤2.00Mn-≤0.045P-≤0.015S-≤24.0-26.0Cr-≤0.11N-≤19.0-22.0Ni	SAE 310 S
1.4847	X8CrNiAlTi20-20	Fe-≤0.08C-≤1.0Si-≤1.0Mn-≤0.030P-≤0.015S-≤0.6Al-18.0-22.0Cr-18.0-22.0Ni-0.6Ti	334 (SAE)
1.4848	GX40CrNiSi25-20	Fe-≤0.30-0.50C-≤1.00-2.50Si-≤1.50Mn-≤0.035P-≤0.030S-≤24.0-26.0Cr-≤19.0-21.0Ni	A 297 (HK)
1.4856	GX40NiCrSiNbTi35-25	Fe-≤0.35-0.45C-≤1.00-1.50Si-≤0.5-1.50Mn-≤0.035P-≤0.030S-≤23.0-27.0Cr-≤0.9-1.5Nb-≤33.0-37.0Ni-≤0.10-0.25Ti	
1.4857	GX40NiCrSi35-25	Fe-≤0.30-0.50C-≤1.00-2.50Si-≤1.50Mn-≤0.035P-≤0.030S-≤24.0-26.0Cr-≤34.0-36.0Ni	A 297 (HP)
1.4862	X8NiCrSi38-18	Fe-≤0.1C-1.5-2.5Si-0.8-1.5Mn-≤0.03P-≤0.03S-17.0-19.0Cr-≤0.5Cu-35.0-39.0Ni-≤0.2Ti	
1.4864	X12NiCrSi35-16; X12 NiCrSi 36 16	Fe-≤0.15C-≤1.0-2.0Si-≤2.0Mn-≤0.045P-≤0.015S-≤15.0-17.0Cr-≤0.11N-≤33.0-37.0Ni	SAE 330
1.4871	X53CrMnNiN21-9	Fe-≤0.48-0.58C-≤0.25Si-≤8.00-10.00Mn-≤0.045P-≤0.030S-≤20.00-22.00Cr-≤0.35-0.50N-≤3.25-4.50Ni	S 63008 (UNS) EV 8 (SAE)
1.4873	X45CrNiW18-9	Fe-≤0.40-0.50C-≤2.00-3.00Si-≤0.80-1.50Mn-≤0.045P-≤0.030S-≤17.00-19.00Cr-≤8.00-10.00Ni-≤0.80-1.20W	
1.4875	X55CrMnNiN20-8	Fe-≤0.50-0.60C-≤0.25Si-≤7.00-10.00Mn-≤0.045P-≤0.030S-≤19.50-21.50Cr-≤0.20-0.40N-≤1.50-2.75Ni	S 63012 (UNS) EV 12 (SAE)
1.4876	X10NiCrAlTi32-21; X10 Ni-CrAlTi 32 20	Fe-≤0.12C-≤1.00Si-≤2.00Mn-≤0.030P-≤0.015S-≤0.15-0.60Al-≤19.00-23.00Cr-≤30.00-34.00Ni-≤0.15-060Ti	N 08332 (UNS) B 366 (N08332) (ASTM) N 8810 (SAE)
1.4877	X6NiCrNbCe32-27	Fe-0.04-0.08C-≤0.3Si-≤1.0Mn-≤0.02P-≤0.01S-≤0.025Al-0.05-0.1Ce-26.0-28.0Cr-0.11N-0.6-1.0Nb-31.0-33.0Ni	S33228 (UNS)

German Standard		Materials Compositions	US-Standard
Mat.-No.	DIN-Design	Percent in Weight	SAE/ASTM/UNS
1.4878	X10CrNiTi18-10; X12 CrNiTi 18 9	Fe-≤0.10C-≤1.0Si-≤2.0Mn-≤0.045P-0.015S-≤17.0-19.0Cr-≤9.0-12.0Ni	
1.4903	X10CrMoVNb9-1	Fe-≤0.08-0.12C-≤0.20-0.50Si-≤0.30-0.60Mn-≤0.020P-≤0.010S-≤0.04Al-≤8.0-9.5Cr-≤0.85-1.05Mo-≤0.030-0.070N-≤0.06-0.1Nb-≤0.4Ni-≤0.18-0.25V	A 182
1.4913	X19CrMoNbVN11-1	Fe-0.17-0.23C-0.50Si-0.40-0.90Mn-≤0.025P-≤0.015S-≤0.02Al-≤0.0015B-10.0-11.5Cr-0.5-0.8Mo-0.05-0.1N-0.25-0.55Nb-0.20-0.60Ni-0.10-0.30V	
1.4919	X6CrNiMo17-13	Fe-≤0.04-0.08C-≤0.75Si-≤2.0Mn-≤0.035P-≤0.015S-≤0.0015-0.0050B-≤16.0-18.0Cr-≤2.0-2.5Mo-≤0.11N-≤12.0-14.0Ni	SAE 316 H
1.4922	X20CrMoV11-1	Fe-≤0.17-0.23C-≤0.50Si-≤1.00Mn-≤0.030P-≤0.030S-≤10.0-12.5Cr-≤0.80-1.20Mo-≤0.30-0.80Ni-≤0.25-0.35V	
1.4943	X4NiCrTi25-15	Fe-≤0.06C-≤1.0Si-≤2.0Mn-≤0.025P-≤0.015S-≤0.35Al-≤0.003-0.01B-≤13.5-16.0Cr-≤1.0-1.50Mo-≤24.00-27.00Ni-≤1.70-2.00Ti-≤0.10-050V	SAE HEV 7 UNS S66545 ASI S 66286 A 891
1.4944		Fe-≤0.08C-≤1.0Si-≤2.0Mn-≤0.025P-≤0.015S-≤0.35Al-≤0.003-0.01B-≤13.50-16.0Cr-≤1.0-1.50Mo-≤24.00-27.00Ni-≤1.90-2.30Ti-≤0.10-0.50V	UNS S66286; ASI 660; A 638
1.4947		Fe-≤0.07C-0.50-1.2Si-1.5-2.0Mn-≤0.035P-≤0.025S-22.0-23.0Cr-≤0.3Cu-≤0.75Mo-9.5-10.5Ni	UNS J93001
1.4948	X6CrNi18-10	Fe-≤0.04-0.08C-≤0.75Si-≤2.0Mn-≤0.035P-≤0.015S-≤17.0-19.0Cr-≤10.0-12.0Ni	SAE 304 H
1.4958	X5NiCrAlTi31-20	Fe-0.03-0.08C-≤0.7Si-≤1.5Mn-0.015P-≤0.01S-0.2-0.5Al-0.5Co-19.0-22.0Cr-≤0.5Cu-≤0.03N-≤0.1Nb-30.0-32.5Ni-0.2-0.5Ti	N08810 (UNS)
1.4959	X8NiCrAlTi32-21	Fe-0.05-0.1C-≤0.7Si-≤1.5Mn-≤0.015P-≤0.01S-0.2-0.65Al-≤0.5Co-19.0-22.0Cr-≤0.5Cu-≤0.03N-30.0-34.0Ni-0.25-0.65Ti	N08811 (UNS)
1.4961	X8CrNiNb16-13	Fe-≤0.04-0.1C-≤0.3-0.6Si-≤1.5Mn-≤0.035P-≤0.015S-≤15.0-17.0Cr-≤12.0-14.0Ni	
1.4970	X 10 NiCrMoTiB 15 15	Fe-0.08-0.12C-0.25-0.45Si-1.6-2.0Mn-≤0.03P-≤0.015S-0.003-0.006B-14.5-15.5Cr-1.05-1.25Mo-15.0-16.0Ni-0.35-0.55Ti	

German Standard		Materials Compositions	US-Standard
Mat.-No.	DIN-Design	Percent in Weight	SAE/ASTM/UNS
1.4971	X12CrCoNi21-20	Fe-≤0.08-0.16C-≤1.00Si-≤2.00Mn-≤0.035P-≤0.015S-≤18.50-21.00Co-≤20.00-22.50Cr-≤2.50-3.50Mo-≤0.10-0.20N-≤0.75-1.25Nb-≤19.00-21.00Ni-≤2.00-3.00W	HEV 1 (SAE) 661 (SAE) R 30155 (UNS)
1.4977	X 40 CoCrNi 20 20	Fe-0.35-0.45C-≤1.00Si-≤1.50Mn-≤0.045P-≤0.030S-19.00-21.00Co-19.00-21.00Cr-3.50-4.50Mo-3.50-4.50Nb-19.00-21.00Ni-3.50-4.50W	R 30590 UNS
1.4980	X6NiCrTiMoVB25-15-2; X5NiCrTi26-15	Fe-0.03-0.08C-≤1.0Si-1.0-2.0Mn-≤0.025P-≤0.015S-≤0.35Al-0.003-0.01B-13.5-16.0Cr-1.0-1.5Mo-24.0-27.0Ni-1.9-2.3Ti-0.1-0.5V	663 (SAE)
1.4981	X8CrNiMoNb16-16	Fe-≤0.04-0.10C-≤0.30-0.60Si-≤1.50Mn-≤0.035P-≤0.015S-≤15.5-17.5Cr-≤1.60-2.00Mo-≤15.5-17.5Ni	
1.4982	X10CrNiMoMnNbVB15-10-1	Fe-≤0.07-0.13C-≤1.00Si-≤5.50-7.00Mn-≤0.040P-≤0.030S-≤0.003-0.009B-≤14.00-16.00Cr-≤0.80-1.20Mo-≤0.110N-≤0.75-1.25Nb-≤9.00-11.00Ni-≤0.15-0.40V	
1.4986	X8CrNiMoBNb16-16	Fe-≤0.04-0.1C-≤0.3-0.6Si-≤1.5Mn-≤0.045P-≤0.030S-≤0.05-0.1B-≤15.5-17.5Cr-≤1.6-2.0Mo-≤15.5-17.5Ni	
1.4988	X8CrNiMoVNb16-13	Fe-≤0.04-0.1C-≤0.3-0.6Si-≤1.5Mn-≤0.035P-≤0.015S-≤15.5-17.5Cr-≤1.1-1.5Mo-≤12.5-14.5Ni-≤0.60-0.85V-≤0.06-0.14N	
1.5069	36Mn7	Fe-≤0.35C≤0.5Si≤1.6Mn≤0.025P≤0.025S	UNS H13400
1.5094	38MnS6	Fe-≤0.35-0.40C-≤0.20-0.65Si-≤1.30-1.60Mn-≤0.045P-≤0.045-0.065S-≤0.01-0.05Al-≤0.10-0.20Cr-≤0.015-0.020N	
1.5122	37MnSi5	Fe-≤0.33-0.41C-≤1.1-1.4Si-≤1.1-1.4Mn-≤0.035P-≤0.035S	
1.5219	41MnV5	Fe-0.38-0.44C-0.1-0.4Si-1.1-1.3Mn≤0.035P≤0.035S-0.1-0.15V	
1.5415	15 Mo 3; 16Mo3	Fe-≤0.12-0.2C-≤0.35Si-≤0.4-0.9Mn-≤0.030P-≤0.025S-≤0.30Cr-≤0.30Cu-≤0.25-0.35Mo-≤0.30Ni	A 204 (A)
1.5431	G12MnMo7-4	Fe-0.08-0.15C-≤0.6Si-1.5-1.8Mn≤0.02P≤0.015S-≤0.2Cr-0.3-0.4Mo≤0.05Nb≤0.1V	
1.5511	35B2	Fe-≤0.32-0.39C-≤0.4Si-≤0.5-0.8Mn-≤0.035P-≤0.035S-≤0.02Al-≤0.0008-0.005B	
1.5662	X8Ni9	Fe-≤0.10C-0.35Si-0.30-0.80Mn-≤0.020P-≤0.010S-≤0.10Mo-8.5-10.0Ni-0.05V	UNS K71340

German Standard		Materials Compositions	US-Standard
MatNo.	DIN-Design	Percent in Weight	SAE/ASTM/UNS
1.5680	X12Ni5; 12 Ni 19	Fe-≤0.15C-≤0.35Si-0.3-0.8Mn-≤0.02P-≤0.01S-4.75-5.25Ni-≤0.05V-≤0.5Cr-≤0.5Mo-≤0.5Cu	A 2515 (SAE)
1.5736	36NiCr10	Fe-max. 0.32-0.40C-0.15-0.35Si-0.40-0.80Mn-0.035P-0.035S-0.55-0.95Cr-2.25-2.75Ni	SAE 3435
1.6354		Fe-≤0.03C-≤0.10Si-≤0.10Mn-≤0.01P-≤0.01S-≤0.05-0.15Al-≤8.00-9.50Co-≤4.60-5.20Mo-≤17.00-19.00Ni-≤0.60-0.90Ti	UNS J93150
1.6511	36CrNiMo4	Fe-≤0.32-0.40C-≤0.4Si-≤0.5-0.8Mn-≤0.035P-≤0.035S-≤0.9-1.2Cr-≤0.15-0.3Mo-≤0.9-1.2Ni	SAE 9840
1.6545	30NiCrMo2-2	Fe-≤0.27-0.34C-≤0.15-0.4Si-≤0.7-1.0Mn-≤0.035P-≤0.035S-≤0.4-0.6Cr-≤0.15-0.3Mo-≤0.4-0.7Ni	SAE 8630
1.6562	40 NiCrMo 8 4	Fe-≤0.37-0.44C-≤0.20-0.35Si-≤0.70-0.90Mn-≤0.02P-≤0.015S-≤0.005-0.05Al-≤0.70-0.95Cr-≤0.30-0.40Mo-≤1.65-2.00Ni	UNS G43406 SAE 4340 UNS H 43406 A 829 SAE E 4340 H
1.6565	40NiCrMo6	Fe-0.35-0.45C-≤0.15-0.35Si-≤0.50-0.70Mn-≤0.035P-≤0.035S-≤0.90-1.4Cr-≤0.20-0.30Mo-≤1.4-1.7Ni	SAE 4340
1.6580	30CrNiMo8	Fe-≤0.26-0.34C-≤0.4Si-≤0.3-0.6Mn-≤0.035P-≤0.035S-≤1.8-2.2Cr-≤0.3-0.5Mo-≤1.8-2.2Ni	
1.6582	34CrNiMo6	Fe-≤0.3-0.38C-≤0.4Si-≤0.5-0.8Mn-≤0.035P-≤0.035S-≤1.3-1.7Cr-≤0.15-0.3Mo-≤1.3-1.7Ni	
1.6751	22NiMoCr3-7	Fe-≤0.17-0.25C-≤0.35Si-≤0.5-1.0Mn-≤0.02P-≤0.02S-≤0.05Al-≤0.3-0.5Cr-≤0.18Cu-≤0.5-0.8Mo-≤0.6-1.2Ni-≤0.03V	A 508
1.6900	X 12 CrNi 18 9	Fe-≤0.12C-≤1.00Si-≤2.00Mn-≤0.045P-≤0.030S-≤17.00-19.00Cr-≤0.5Mo-≤8.00-10.00Ni	UNS J92801
1.6903	X 10 CrNiTi 18 10	Fe-≤0.10C-≤1.00Si-≤2.00Mn-≤0.045P-≤0.030S-≤17.0-19.0Cr-≤0.5Mo-≤10.0-12.0Ni	
1.6906	X 5 CrNi 18 10	Fe-≤0.07C-≤1.0Si-≤2.0Mn-≤0.045P-≤0.030S-≤17.0-19.0Cr-≤0.50Mo-≤9.0-11.5Ni	
1.6932	28NiCrMoV8-5	Fe-0.24-0.32C-≤0.4Si-0.15-0.4Mn-≤0.035P-≤0.035S-1.0-1.5Cr-0.35-0.55Mo-1.8-2.1Ni-0.05-0.15V	
1.6944		Fe-≤0.35-0.40C-≤0.15-0.35Si-≤0.50-0.80Mn-≤0.015P-≤0.01S-0.65-0.90Cr-≤0.30-0.40Mo-≤1.65-2.00Ni-≤0.08-0.15V	

Key to materials compositions

German Standard		Materials Compositions	US-Standard
Mat.-No.	DIN-Design	Percent in Weight	SAE/ASTM/UNS
1.6948	27NiCrMoV11-6; 26NiCrMoV11-5	Fe-≤0.22-0.32C-≤0.15Si-≤0.15-0.40Mn-≤0.010P-≤0.007S-≤1.20-1.80Cr-≤0.25-0.45Mo-≤2.40-3.10Ni-≤0.05-0.15V	
1.6952	24NiCrMoV14-6	Fe-≤0.20-0.28C-≤0.15-0.40Si-≤0.30-0.60Mn-≤0.035P-≤0.035S-≤1.20-1.80Cr-≤0.35-0.55Mo-≤3.00-3.80Ni-≤0.04-0.12V	K 42885 (UNS) A 649 (6, 7, 8) (ASTM) A 470 (5, 6, 7) (ASTM)
1.6956	33NiCrMoV14-5; 33NiCrMo14-5	Fe-≤0.28-0.38C-≤0.40Si-≤0.15-0.40Mn-≤0.035P-≤0.035S-≤1.00-1.70Cr-≤0.30-0.60Mo-≤2.90-3.80Ni-≤0.08-0.25V	
1.6957	27NiCrMoV15-6	Fe-0.22-0.32C≤0.15Si-0.15-0.4Mn≤0.01P≤0.007S-1.2-1.8Cr-0.25-0.45Mo-3.4-4.0Ni-0.05-0.15V	ASTM A 470
1.7005	45Cr2	Fe-≤0.42-0.48C-≤0.15-0.40Si-≤0.50-0.80Mn-≤0.025P-≤0.035S-≤0.40-0.60Cr	
1.7033	34Cr4	Fe-≤0.3-0.37C-≤0.4Si-≤0.6-0.9Mn-≤0.035P-≤0.035S-≤0.9-1.2Cr	UNS G51320
1.7035	41Cr4	Fe-≤0.38-0.45C-≤0.4Si-≤0.6-0.9Mn-≤0.035P-≤0.035S-≤0.9-1.2Cr	SAE 5140; UNS H51400
1.7120		Fe-≤0.1C-≤0.25Si-≤0.45Mn-≤0.16Cu-≤0.07Ni-≤0.05Cr-≤0.035P-≤0.035S	
1.7131	16MnCr5	Fe-0.14-0.19C-≤0.40Si-1.00-1.30Mn-≤0.035P-≤0.035S-0.80-1.10Cr	G 51170 UNS A 711 (5115) ASTM
1.7147	20MnCr5	Fe-≤0.17-0.22C-≤0.40Si-≤1.10-1.40Mn-≤0.035P-≤0.035S-≤1.00-1.30Cr	UNS H51200 A 752 (5120) SAE 5120H
1.7214		Fe-≤0.22-0.29C-≤0.15-0.35Si-≤0.5-0.8Mn-≤0.02P-0.015S-≤0.90-1.20Cr-≤0.15-0.20Mo-0.30Ni	
1.7218	25CrMo4	Fe-≤0.22-0.29C-≤0.40Si-≤0.60-0.90Mn-≤0.035P-≤0.035S-≤0.90-1.20Cr-≤0.15-0.30Mo	SAE 4130
1.7219	26 CrMo 4; 26CrMo4-2	Fe-≤0.22-0.29C-≤0.35Si-≤0.5-0.8Mn-≤0.03P-≤0.025S-≤0.9-1.2Cr-≤0.15-0.30Mo	A 372
1.7220	34CrMo4	Fe-≤0.3-0.37C-≤0.4Si-≤0.6-0.9Mn-≤0.035P-≤0.035S-≤0.9-1.2Cr-≤0.15-0.30Mo	SAE 4130
1.7225	42CrMo4	Fe-≤0.38-0.45C-≤0.40Si-≤0.60-0.90Mn-≤0.035P-≤0.035S-≤0.90-1.20Cr-≤0.15-0.30Mo	UNS G41400 A 866 (4140) SAE 4140 RH
1.7242	16CrMo4	Fe-≤0.13-0.20C-≤0.15-0.35Si-≤0.50-0.80Mn-≤0.035P-≤0.035S-≤0.90-1.20Cr-≤0.20-0.30Mo-≤0.40Ni	

German Standard		Materials Compositions	US-Standard
MatNo.	DIN-Design	Percent in Weight	SAE/ASTM/UNS
1.7259	26CrMo7	Fe-≤0.22-0.30C-≤0.15-0.35Si-≤0.50-0.70Mn-≤0.035P-≤0.035S-≤1.50-1.80Cr-≤0.20-0.25Mo	
1.7273	24CrMo10	Fe-≤0.20-0.28C-≤0.15-0.35Si-≤0.50-0.80Mn-≤0.035P-≤0.035S-≤2.30-2.60Cr-≤0.20-0.30Mo-≤0.80Ni	
1.7276	10CrMo11	Fe-≤0.08-0.12C-≤0.15-0.35Si-≤0.30-0.50Mn-≤0.035P-≤0.035S-≤2.70-3.00Cr-≤0.20-0.30Mo	
1.7279	17 MnCrMo 3 3	Fe-≤0.20C-0.50-0.90Si-0.70-1.10Mn-≤0.035P-≤0.035S-0.60-1.00Cr-0.20-0.60Mo-0.06-0.12V-0.06-0.12Zr	
1.7281	16CrMo9-3	Fe-≤0.12-0.20C-≤0.15-0.35Si-≤0.30-0.50Mn-≤0.035P-≤0.035S-≤2.00-2.50Cr-≤0.30-0.40Mo	
1.7335	13 CrMo 4 4; 13CrMo4-5	Fe-≤0.08-0.18C-≤0.35Si-≤0.4-1.0Mn-≤0.030P-≤0.025S-≤0.7-1.15Cr-≤0.3Cu-≤0.4-0.6Mo	A 182
1.7357	G17CrMo5-5	Fe-≤0.15-0.20C-≤0.60Si-≤0.50-1.0Mn-≤0.020P-≤0.020S-≤1.00-1.50Cr-≤0.45-0.65Mo	A 217; UNS J11872
1.7362	X12CrMo5	Fe-≤0.08-0.15C-≤0.50Si-≤0.30-0.60Mn-≤0.025P-≤0.020S-≤4.00-6.00Cr-≤0.45-0.65Mo	SAE 501
1.7375	12CrMo9-10	Fe-≤0.10-0.15C-≤0.30Si-≤0.30-0.80Mn-≤0.015P-≤0.010S-≤0.01-0.04Al-≤2.00-2.50Cr-≤0.20Cu-≤0.9-1.10Mo-≤0.012N-≤0.30Ni	UNS K21590
1.7380	10CrMo9-10	Fe-≤0.08-0.14C-≤0.50Si-≤0.40-0.80Mn-≤0.030P-≤0.025S-≤2.00-2.50Cr-≤0.30Cu-≤0.90-1.10Mo	A 182 (F22); UNS J21890
1.7383	11CrMo9-10	Fe-≤0.08-0.15C-≤0.50Si-≤0.40-0.80Mn-≤0.030P-≤0.025S-≤2.00-2.50Cr-≤0.30Cu-≤0.90-1.10Mo	
1.7386	X12CrMo9-1	Fe-≤0.07-0.15C-≤0.25-1.0Si-≤0.30-0.60Mn-≤0.025P-≤0.020S-≤8.0-10.0Cr-≤0.90-1.1Mo	SAE 504; UNS S50488
1.7388	X7CrMo9-1	Fe-≤0.04-0.09C-≤0.45-0.75Si-≤0.43-0.72Mn-≤0.015P-≤0.015S-≤8.60-9.90Cr-≤0.90-1.10Mo	S 50480 (UNS)
1.7707	30CrMoV9	Fe-0.26-0.34C-≤0.4Si-0.4-0.7Mn-≤0.035P-≤0.035S-2.3-2.7Cr-≤0.25Mo-≤0.6Ni-0.1-0.2V	G43406 (UNS)

German Standard		Materials Compositions	US-Standard
MatNo.	DIN-Design	Percent in Weight	SAE/ASTM/UNS
1.7711	40CrMoV4-6; 40CrMoV4-7	Fe-≤0.36-0.44C-≤0.40Si-≤0.45-0.85Mn-≤0.03P-≤0.03S-≤0.015Al-≤0.90-1.20Cr-≤0.50-0.65Mo-≤0.25-0.35V	A 437 (B4D)
1.7715	14MoV6-3	Fe-≤0.1-0.18C-≤0.1-0.35Si-≤0.4-0.7Mn-≤0.035P-≤0.035S-≤0.3-0.6Cr-≤0.5-0.7Mo-≤0.22-0.32V	UNS K11591
1.7734		Fe-≤0.12-0.18C-≤0.20Si-≤0.80-1.10Mn-≤0.02P-≤0.015S-≤1.25-1.50Cr-≤0.80-1.00Mo-≤0.20-0.30V	
1.7766	17CrMoV10	Fe-≤0.15-0.20C-≤0.15-0.35Si-≤0.30-0.50Mn-≤0.035P-≤0.035S-≤2.70-3.00Cr-≤0.20-0.30Mo-≤0.10-0.20V	
1.7779	20 CrMoV 13 5; 20CrMoV13-5-5	Fe-≤0.17-0.23C-≤0.15-0.35Si-≤0.30-0.50Mn-≤0.025P-≤0.020S-≤3.00-3.30Cr-≤0.50-0.60Mo-≤0.45-0.55V	
1.7783	X41CrMoV5-1	Fe-≤0.38-0.43C-≤0.80-1.0Si-≤0.20-0.40Mn-≤0.015P-≤0.010S-≤4.75-5.25Cr-≤1.2-1.4Mo-≤0.40-0.60V	SAE 610
1.8070	21CrMoV5-11	Fe-≤0.17-0.25C-≤0.30-0.60Si-≤0.30-0.60Mn-≤0.035P-≤0.035S-≤1.20-1.50Cr-≤1.00-1.20Mo-≤0.60Ni-≤0.25-0.35V	
1.8075	10CrSiMoV7	Fe-≤0.12C-≤0.9-1.2Si-≤0.35-0.75Mn-≤0.035P-≤0.035S-≤1.6-2Cr-≤0.25-0.35Mo-≤0.25-0.35V	
1.8159	51CrV4; 50 CrV 4	Fe-≤0.47-0.55C-≤0.40Si-≤0.70-1.10Mn-≤0.035P-≤0.035S-≤0.90-1.20Cr-≤0.10-0.25V	UNS G61500 A 866 (6150) SAE 6150H
1.8719	15MnCrMo3-2	Fe-0.10-0.20C-0.15-0.35Si-0.60-1.00Mn-≤0.025P-≤0.025S-0.0005B-0.40-0.65Cr-0.15-0.50Cu-0.40-0.60Mo-0.70-1.00Ni-0.03-0.08V	
1.8812	18MnMoV5-2	Fe-≤0.20C-0.20-0.50Si-1.00-1.50Mn-≤0.030P-≤0.025S-0.10-0.30Mo-≤0.02N-0.05-0.10V	A 202 (A) (ASTM) A 202 (B) (ASTM) A 302 (A) (ASTM)
1.8850	S460MLH	Fe-≤0.16C-≤0.60Si-≤1.70Mn-≤0.030P-≤0.025S-0.02Al-≤0.20Mo-≤0.025N-≤0.050Nb-≤0.30Ni-≤0.05Ti-≤0.12V	A 514 (F) (ASTM) A 517 (F) (ASTM) A 592 (F) (ASTM)
1.8901	S460N	Fe-≤0.2C-≤0.6Si-1.0-1.7Mn-≤0.035P-≤0.03S-≤0.3Cr-≤0.7Cu-≤0.1Mo-≤0.05Nb-≤0.8Ni-≤0.03Ti-≤0.2V	ASTM A 572
1.8905	P460N; StE 460	Fe-≤0.20C-≤0.60Si-≤1.00-1.70Mn-≤0.030P-≤0.025S-≤0.02Al-≤0.30Cr-≤0.70Cu-≤0.10Mo-≤0.025N-≤0.050Nb-≤0.80Ni-≤0.03Ti-≤0.20V	A 225 (C), A 633 (E)

German Standard		Materials Compositions	US-Standard
Mat.-No.	DIN-Design	Percent in Weight	SAE/ASTM/UNS
1.8907	StE 500	Fe-≤0.21C-≤0.1-0.6Si-≤1.0-1.7Mn-≤0.035P-≤0.03S-≤0.02Al-≤0.30Cr-≤0.70Cu-≤0.10Mo-≤0.020N-≤0.05Nb-≤1.0Ni-≤0.2Ti-≤0.22V	6386 B; UNS K02001
1.8912	S420NL; TStE 420	Fe-≤0.2C-≤0.6Si-≤1.0-1.7Mn-≤0.03P-≤0.025S-≤0.02Al-≤0.3Cr-≤0.7Cu-≤0.1Mo-≤0.025N-≤0.050Nb-≤0.8Ni-≤0.03Ti-≤0.2V	A 737: UNS K02002
1.8924	S500Q; StE 500V	Fe-≤0.2C-≤0.8Si-≤1.7Mn-≤0.025P-≤0.015S-≤1.5Cr-≤0.5Cu-≤0.7Mo-≤0.06Nb-≤2.0Ni-≤0.05Ti-≤0.15Zr	
1.8931	S690Q; StE 690V	Fe-≤0.2C-≤0.8Si-≤1.7Mn-≤0.025P-≤0.015S-≤1.5Cr-≤0.5Cu-≤0.7Mo-≤0.06Nb-≤2.0Ni-≤0.05Ti-≤0.15Zr	
1.8935	WstE 460; P460NH	Fe-≤0.20C-≤0.60Si-≤1.0-1.70Mn-≤0.030P-≤0.025S-≤0.02Al-≤0.30Cr-≤0.70Cu-≤0.10Mo-≤0.025N-≤0.050Nb-≤0.8Ni-≤0.03Ti-≤0.2V	A 350; UNS K02900
1.8940	S890Q	Fe-≤0.2C-≤0.8Si-≤1.7Mn-≤0.025P-≤0.015S-≤1.5Cr-≤0.5Cu-≤0.7Mo-≤0.06Nb-≤2.0Ni-≤0.05Ti-≤0.15Zr	
1.8946	S355J2WP	Fe-≤0.12C-≤0.75Si-≤1.0Mn-0.06-0.15P-≤0.035S-0.30-1.25Cr-0.25-0.55Cu-≤0.009N-≤0.65Ni	K02601 (UNS)
1.8952	L450QB	Fe-≤0.16C-≤0.45Si-≤1.60Mn-≤0.025P-≤0.020S-≤0.015-0.06Al-≤0.30Cr-≤0.25Cu-≤0.10Mo-≤0.012N-≤0.05Nb-≤0.30Ni-≤0.06Ti-≤0.09V	
1.8961	S235J2W; WTSt 37-3	Fe-≤0.13C-≤0.4Si-≤0.2-0.6Mn-≤0.040P-≤0.035S-≤0.02Al-≤0.4-0.8Cr-≤0.25-0.55Cu-≤0.015-0.060Nb-≤0.65Ni-≤0.02-0.10Ti-≤0.02-0.10V	
1.8962	9CrNiCuP3-2-4	Fe-≤0.12C-≤0.25-0.75Si-≤0.2-0.5Mn-≤0.07-0.15P-≤0.035S-≤0.5-1.25Cr-≤0.25-0.55Cu-≤0.65Ni	A 242; UNS K11430
1.8963	S355J2G1W; WTSt 52-3	Fe-≤0.16C-≤0.50Si-≤0.50-1.5Mn-≤0.035P-≤0.035S-≤0.02Al-≤0.40-0.80Cr-≤0.25-0.55Cu-≤0.3Mo-≤0.015-0.060Nb-≤0.65Ni-≤0.02-0.10Ti-≤0.02-0.12V-≤0.15Zr	A 588 (A)
1.8972	L415NB	Fe-≤0.21C-≤0.45Si-≤1.60Mn-≤0.025P-≤0.020S-≤0.015-0.060Al-≤0.30Cr-≤0.25Cu-≤0.10Mo-≤0.012N-≤0.050Nb-≤0.30Ni-≤0.04Ti-≤0.15V	API 5LX 60 (API)

German Standard		Materials Compositions	US-Standard
MatNo.	DIN-Design	Percent in Weight	SAE/ASTM/UNS
1.8975	L450MB; StE 445.7	Fe-≤0.16C-≤0.45Si-≤1.6Mn-≤0.025P-≤0.02S-0.015-0.06Al-≤0.3Cr-≤0.25Cu-≤0.1Mo-≤0.05Nb-≤0.3Ni-≤0.06Ti	API 5LX65
1.8977	L485MB; StE 480.7	Fe-≤0.16C-≤0.45Si-≤1.70Mn-≤0.025P-≤0.020S-≤0.015-0.06Al-≤0.30Cr-≤0.25Cu-≤0.10Mo-≤0.012N-≤0.06Nb-≤0.30Ni-≤0.06Ti-≤0.10V	API 5LX70
2.4060	Ni 99,6	≤99.60Ni-≤0.08C-≤0.15Si-≤0.35Mn-≤0.005S-≤0.15Cu-≤0.25Fe-≤0.15Mg-≤0.10Ti	UNS N02200
2.4061	LC-Ni 99,6	Fe-≤0.02C-0.15Si-0.35Mn-≤0.005S-≤0.15Cu-≤0.25Fe-≤0.15Mg-99.6Ni-≤0.10Ti	UNS N02201
2.4066	Ni 99,2; S-Ni 99,2	≤99.20Ni-≤0.10C-≤0.25Si-≤0.35Mn-≤0.005S-≤0.25Cu-≤0.40Fe-≤0.15Mg-≤0.10Ti	UNS N02200
2.4068	LC-Ni 99	≤99.0Ni-≤0.02C-≤0.25Si-≤0.35Mn-≤0.005S-≤0.25Cu-≤0.40Fe-≤0.15Mg-≤0.10Ti	UNS N02201
2.4360	NiCu 30 Fe	≤63.0Ni-≤0.15C-≤0.50Si-≤2.0Mn-≤0.020S-≤0.5Al-≤28.0-34.0Cu-≤1.0-2.5Fe-≤0.3Ti	UNS N04400
2.4361	LC-NiCu 30 Fe	Fe-≤0.04C-≤0.3Si-≤2.0Mn-≤0.02S-≤0.5Al-28.0-34.0Cu-1.0-2.5Fe-63.0Ni-≤0.3Ti	N04402 (UNS)
2.4363	NiCu30Fe5	≤0.30C-≤0.50Si-≤2.0Mn-0.025-0.60S-≤0.50Al-28.00-34.00Cu-≤2.50Fe-63.00-70.00Ni-≤0.30Ti	
2.4365	G-NiCu 30 Nb	Ni-≤0.15C-0.5-1.5Si-0.5-1.5Mn-≤0.5Al-26.0-33.0Cu-1.0-2.5-1.0-1.5Nb	UNS N24130
2.4366	EL-NiCu 30 Mn	≤62.0Ni-≤0.15C-≤1.0Si-≤1.0-4.0Mn-≤0.030P-≤0.015S-≤0.5Al-≤27.0-34.0Cu-≤0.5-2.5Fe-≤1.0Nb≤1.0Ti	B 127-98
2.4368	G-NiCu 30Si4	Fe-≤0.25C-3.5-4.5Si-0.5-1.5Mn-27.0-31.0Cu-1.0-2.5Fe-60.0-68.0Ni	UNS N10665
2.4374	NiCu30Al	≤0.25C-≤1.00Si-≤1.50Mn-≤0.010S-2.00-4.00Al-27.00-34.00Cu-≤2.00Fe-≥63.00Ni-0.25-1.00Ti	
2.4375	NiCu 30 Al	Ni-≤0.20C-≤0.50Si-≤1.5Mn-≤0.015S-≤2.2-3.5Al-≤27.0-34.0Cu-≤0.5-2.0Fe-≤63.0Ni-≤0.3-1.0Ti	UNS N05500
2.4566	ACN 17	Ni-≤0.12C-≤10Si-≤1.2Mn-≤4Co-≤3Cu	
2.4600	NiMo29Cr	Ni-≤0.01C-≤0.1Si-≤3.0Mn-≤0.025P-≤0.015S-≤0.1-0.5Al-≤3.0Co-≤0.5-3.0Cr-≤0.5Cu-≤1.0-6.0Fe-≤26.0-32.0Mo-≤0.4Nb-≤0.2Ti-≤0.2V-≤3.0W	

German Standard		Materials Compositions	US-Standard
MatNo.	DIN-Design	Percent in Weight	SAE/ASTM/UNS
2.4602	NiCr21Mo14W	Ni-≤0.01C-≤0.08Si-≤0.5Mn-≤0.025P-≤0.010S-≤2.5Co-≤20.0-22.5Cr-≤2.0-6.0Fe-≤12.5-14.5Mo-≤0.35V-≤2.5-3.5W	UNS N06022
2.4603	NiCr30FeMo	Ni-≤0.03C-≤0.08Si-≤2.0Mn-≤0.04P-≤0.02S-≤5.0Co-≤28.0-31.5Cr-≤1.0-2.4Cu-13.0-17.0Fe-≤4.0-6.0Mo-≤0.3-1.5Nb-≤1.5-4.0W	UNS N06002
2.4605	NiCr23Mo16Al	Ni-≤0.01C-≤0.10Si-≤0.5Mn-≤0.025P-≤0.015S-≤0.1-0.4Al-≤0.3Co-≤22.0-24.0Cr-≤0.5Cu-≤1.5Fe-≤15.0-16.5Mo	UNS N06059
2.4606	NiCr21Mo16W	Ni-≤0.01C-≤0.08Si-≤0.75Mn-≤0.025P-≤0.015S-≤0.5Al-≤1.0Co-≤19.0-23.0Cr-≤2.0Fe-≤15.0-17.0Mo-≤0.02-0.25Ti-≤0.2V-≤3.0-4.0W	UNS N06686
2.4607	SG-NiCr23Mo16	Ni-≤0.015C-≤0.08Si-≤0.50Mn-≤0.02P-≤0.015S-≤0.1-0.4Al-≤0.3Co-≤22.0-24.0Cr-≤1.5Fe-≤15.0-16.5Mo	UNS N06059
2.4608	NiCr26MoW	Fe-0.03-0.08C-0.7-1.5Si-≤2.0Mn-≤0.03P-≤0.015S-2.0-4.0Cu-24.0-26.0Cr-≤0.5Cu-2.5-4.0Mo-44.0-47.0Ni-2.5-4.0W	N06333 (UNS)
2.4610	NiMo16Cr16Ti	Ni-≤0.01C-≤0.08Si-≤1.0Mn-≤0.025P-≤0.015S-≤2.0Co-≤14.0-18.0Cr-≤0.5Cu-≤3.0Fe-≤14.0-18.0Mo-≤0.7Ti	UNS N06455
2.4612	EL-NiMo15Cr15Ti	≤0.02C-≤0.20Si-≤1.00Mn-≤0.015S-≤2.00Co-≤14.00-18.00Cr-≤3.00Fe-≤14.00-17.00Mo-at least56Ni	
2.4615	SG-NiMo27	Fe-≤0.02C-≤0.10Si-≤1.0Mn-≤0.015S-≤1.0Cr-≤2.0Fe-26.0-30.0Mo-64.0Ni	UNS N10665
2.4617	NiMo28	Ni-≤0.01C-≤0.08Si-≤1.0Mn-≤0.025P-≤0.015S-≤1.0Co-≤1.0Cr-≤0.5Cu-≤2.0Fe-≤26.0-30.0Mo	UNS N10665
2.4618	NiCr22Mo6Cu	Ni-≤0.05C-≤1.0Si-≤1.0-2.0Mn-≤0.025P-≤0.015S-≤2.5Co-≤21.0-23.5Cr-≤1.5-2.5Cu-≤18.0-21.0Fe-≤5.5-7.5Mo-≤1.75-2.5Nb-≤1.0W	UNS N06007
2.4619	NiCr22Mo7Cu	Ni-≤0.015C-≤1.0Si-≤1.0Mn-≤0.025P-≤0.015S-≤5.0Co-≤21.0-23.5Cr-≤1.5-2.5Cu-≤18.0-21.0Fe-≤6.0-8.0Mo-≤0.5Nb-≤1.5W	UNS N06985
2.4621	EL-NiCr20Mo9Nb	Ni-≤0.1C≤0.8Si≤2.0Mn≤0.4Al-20.0-23.0Cr≤0.5Cu≤6.0Fe-8.0-10.0Mo-2.0-4.0Nb≤0.4Ti	
2.4623	EL-NiCr23Mo7Cu	Ni-≤0.02C≤1.0Si≤1.0Mn≤0.04P≤0.03≤5.0Co-21.0-23.5Cr-1.5-2.5Cu-18.0-21.0Fe-6.0-8.0Mo≤0.5Nb≤0.5Ta≤1.5W	

Key to materials compositions

German Standard		Materials Compositions	US-Standard
Mat.-No.	DIN-Design	Percent in Weight	SAE/ASTM/UNS
2.4627	SG-NiCr22Co12Mo	Fe-≤0.1C-≤0.5Si-≤1.0Mn-≤0.015S-0.8-1.5Al-10.0-14.0Co-20.0-24.0Cr-≤0.5Cu-≤1.0Fe-8.0-10.0Mo-50.0Ni-≤0.6Ti	N06617 (UNS)
2.4630	NiCr20Ti	Fe-0.08-0.15C-≤1.0Si-≤1.0Mn-18.0-21.0Cr-≤0.5Cu-≤5.0Fe-0.2-0.6Ti	N06075 (UNS)
2.4631	NiCr20TiAl	Ni-0.04-0.1C≤1.0Si≤1.0Mn≤0.03P≤0.015S-1.0-1.8Al≤2.0Co-18.0-21.0Cr≤0.2Cu≤1.5Fe-1.8-2.7Ti	UNS N07080
2.4632	NiCr20Co18Ti	Ni-≤0.13C≤1.0Si≤1.0Mn≤0.02P≤0.015S-1.0-2.0Al-15.0-21.0Co-18.0-21.0Cr≤0.2Cu≤1.5Fe-2.0-3.0Ti	UNS N07090
2.4633	NiCr25FeAlY	Fe-0.15-0.25C-≤0.5Si-≤0.5Mn-≤0.02P-≤0.01S-1.8-2.4Al-24.0-26.0Cr-≤0.1Cu-8.0-11.0Fe-0.1-0.2Ti-0.05-0.12Y-0.01-0.1Zr	
2.4634	NiCo20Cr15MoAlTi	Ni-0.12-0.17C-≤1.0Si-≤1.0Mn-≤0.045P-≤0.015S-4.5-4.9Al-18.0-22.0Co-14.0-15.7Cr-≤0.2Cu-≤1.0Fe-4.5-.5.5Mo-0.9-1.5Ti	UNS N13021
2.4636	NiCo15Cr15MoAlTi	Ni-0.12-0.2C≤1.0Si≤1.0Mn≤0.045P≤0.03S-4.5-5.5Al-13.0-17.0Co-14.0-16.0Cr≤0.2Cu≤1.0Fe-3.0-.5.0Mo-3.5-4.5Ti	NIMONIC alloy 115
2.4641	NiCr21Mo6Cu	Fe-≤0.025C-≤0.50Si-≤1.0Mn-≤0.025P-≤0.015S-≤0.2Al-≤1.0Co-20.0-23.0Cr-1.5-3.0Cu-5.5-7.0Mo-39.0-46.0Ni-0.6-1.0Ti	UNS N08042
2.4642	NiCr29Fe	≤58.0Ni-≤0.05C-≤0.5Si-≤0.5Mn-≤0.020P-≤0.015S-≤0.5Al-≤27.0-31.0Cr-≤0.5Cu-≤7.0-11.0Fe-≤0.5Ti	UNS N06690
2.4650	NiCo20Cr20MoTi	Ni-≤0.04-0.08C-≤0.4Si-≤0.6Mn-≤0.007S-≤0.3-0.6Al-≤0.005B-≤19.0-21.0Co-≤19.0-21.0Cr-≤0.2Cu-≤0.7Fe-≤5.6-6.1Mo-≤1.9-2.4Ti	UNS N07263
2.4652	EL-NiCr26Mo	≤37.0-42.0Ni-≤0.03C-≤0.7Si-≤1.0-3.0Mn-≤0.015S-≤0.1Al-≤23.0-27.0Cr-≤1.5-3.0Cu-≤30.0Fe-≤3.5-7.5Mo-≤37.0-42.00Ni-≤1.0Ti	UNS S32654
2.4654	NiCr20Co13Mo4Ti3Al; NiCr19Co14Mo4Ti	Fe-0.02-0.2C-≤0.15Si-≤0.1Mn-≤0.015P-≤0.015S-1.2-1.6Al-0.003-0.01B-12.0-15.0Co-18.0-21.0Cr-≤0.1Cu-≤2.0Fe-3.5-5.0Mo-2.8-3.3Ti-0.02-0.08Zr	N07001 (UNS)
2.4658	NiCr7030; NiCr 70 30	Fe-≤0.1C-0.5-2.0Si-≤1.0Mn-≤0.02P-≤0.15S-≤0.3Al-≤1.0Co-≤29.0-32.0Cr-≤0.5Cu-≤5.0Fe-60.0Ni	N06008 (UNS)

German Standard		Materials Compositions	US-Standard
MatNo.	DIN-Design	Percent in Weight	SAE/ASTM/UNS
2.4660	NiCr20CuMo	≤32.0-38.0Ni-≤0.07C-≤1.0Si-≤2.0Mn-≤0.025P-≤0.015S-≤19.0-21.0Cr-≤3.0-4.0Cu-≤2.0-3.0Mo	UNS N08020
2.4662	NiCr13Mo6Ti3	Fe-0.02-0.06C-≤0.40Si-≤0.50Mn-≤0.020P-≤0.020S-≤0.35Al-0.01-0.02B-≤1.0Co-11.0-14.0Cr-≤0.04Cu-5.0-6.5Mo-40.0-45.0Ni-2.8-3.1Ti	UNS N09901
2.4663	NiCr23Co12Mo	Ni-≤0.05-0.10C-≤0.2Si-≤0.2Mn-≤0.01P-≤0.01S-≤0.7-1.4Al≤0.006B-≤11.0-14.0Co-≤20.0-23.0Cr-≤0.5Cu-≤2.0Fe-≤8.5-10.0Mo-≤0.2-0.6Ti	UNS N06617
2.4665	NiCr22Fe18Mo	Fe-0.05-0.15C-≤1.0Si-≤1Mn-≤0.02P-≤0.015S-≤0.5Al-0.01-0.1B-0.5-2.5Co-20.5-23.0Cr-≤0.5Cu-17.0-20.0Fe-8.0-10.0Mo-0.2-1.0W	680 (SAE)
2.4667	SG-NiCr19NbMoTi	Fe-≤0.08C-≤0.40Si-≤0.40Mn-≤0.015S-0.2-0.8Al-≤0.006B-17.0-21.0Cr-≤0.30Cu-≤22.0Fe-2.8-3.3Mo-4.8-5.5Nb-50.0Ni-0.60-1.20Ti	
2.4668	NiCr19Fe19Nb5Mo3	Fe-0.02-0.08C-≤0.35Si-≤0.35Mn-≤0.015P-≤0.015S-0.3-0.7Al-0.006B-≤1.0Co-17.0-21.0Cr-≤0.30Cu-2.8-3.3Mo-4.7-5.5Nb-50.0-55.0Ni-0.60-1.20Ti	UNS N07718
2.4669	NiCr15Fe7TiAl; NiCr15Fe7-Ti2Al	Ni-≤0.08C-≤0.5Si-≤1.0Mn-≤0.02P-≤0.015S-≤0.4-1.0Al≤1.0Co-≤14.0-17.0Cr-≤0.5Cu-≤5.0-9.0Fe-≤0.7-1.2Nb-≤2.25-2.75Ti	UNS N07750
2.4670	G-NiCr13Al6MoNb	Ni-≤0.08-0.20C-≤0.50Si-≤0.25Mn-≤0.015P-≤0.015S-≤5.50-6.50Al-≤0.005-0.15B-≤1.00Co-≤12.00-14.00Cr-≤0.50Cu-≤3.80-5.20Mo-≤1.50-2.50Nb-≤0.40-1.00Ti-≤0.05-0.15Zr	UNS N07713
2.4675	NiCr23Mo16Cu	Ni-≤0.01C-≤0.08Si-≤0.5Mn-≤0.025P-≤0.015S-≤0.5Al≤2.0Co-≤22.0-24.0Cr-≤1.3-1.9Cu-≤3.0Fe-≤15.0-17.0Mo	
2.4679	G-NiCr35	Ni-≤0.10C-≤1.00Si-≤0.30Mn-≤34.00-36.00Cr-≤1.00Fe-≤0.30N	
2.4680	G-NiCr50Nb	Ni-≤0.10C-≤1.00Si-≤0.50Mn-≤0.02P-≤0.02S-≤48.00-52.00Cr-≤1.00Fe-≤0.50Mo-≤0.16N-≤1.00-1.80Nb	
2.4681	CoCr26Ni9Mo5W	Ni-≤1.0C-≤1.0Si-≤1.5Mn-≤23.5-27.5Cr-≤1.0-3.0Fe-≤4.0-6.0Mo-≤0.12N-≤7.0-11.0Ni-≤1.0-3.0W	

Key to materials compositions

German Standard		Materials Compositions	US-Standard
Mat.-No.	DIN-Design	Percent in Weight	SAE/ASTM/UNS
2.4683	CoCr22NiW	Fe-0.05-0.15C-0.2-0.5Si-≤1.25Mn-≤0.02P-≤0.015S-20.0-24.0Cr-≤3.0Fe-0.02-0.12La-20.0-24.0Ni-13.0-16.0W	R30188 (UNS)
2.4686	G-NiMo 17 Cr	Ni-≤0.03C-≤0.5Si≤1.0Mn-≤2.5Co-15.5-17.5Cr-≤7.0Fe-16.0-18.0Mo	
2.4694	NiCr16Fe7TiAl	Fe-≤0.08C-≤0.5Si-≤0.5Mn-≤0.015P-≤0.01S-0.8-1.6Al-14.0-17.0Cr-≤0.5Cu-5.0-9.0Fe-0.7-1.2Nb-70.0Ni-2.0-2.6Ti	N07031 (UNS)
2.4800	S-NiMo 30	−≤60.0Ni-≤0.05C-≤1.0Si-≤1.0Mn-≤0.045P-≤0.025S-≤2.5Co-≤1.0Cr-≤4.0-7.0Fe-≤26.0-30.0Mo-≤0.2-0.4V	UNS N10001
2.4810	NiMo 30	≤62.0Ni-≤0.05C-≤0.5Si-≤1.0Mn-≤0.030P-≤0.015S-≤2.5Co-≤1.0Cr-≤0.5Cu-≤4.0-7.0Fe-≤26.0-30.0Mo-≤0.6V	UNS N10001
2.4811		Fe-≤0.03C-≤0.05Si-≤0.80Mn-≤0.030P-≤0.015S-19.0-21.0Cr-≤0.50Cu-≤2.5Fe-14.0-17.0Mo-58.0Ni-0.35V	
2.4816	NiCr15Fe	≤72.0Ni-≤0.05-0.1C-≤0.5Si-≤1.0Mn-≤0.020P-≤0.015S-≤0.3Al-≤0.0060B-≤1.0Co-≤14.0-17.0Cr-≤0.5Cu-≤6.0-10.0Fe-≤0.3Ti	UNS N06600
2.4817	LC-NiCr15Fe	≤72.0Ni-≤0.025C-≤0.50Si-≤1.0Mn-≤0.020P-≤0.015S-≤0.3Al-≤0.0060B-≤1.0Co-≤14.0-17.0-≤0.5Cu-≤6.0-10.0Fe-≤0.3Ti	
2.4819	NiMo16Cr15W	Ni-≤0.01C-≤0.08Si-≤1.0Mn-≤0.025P-≤0.015S-≤2.5Co-≤14.5-16.5Cr-≤0.5Cu-≤4.0-7.0Fe-≤15.0-17.0Mo-≤0.35V-≤3.0-4.5W	UNS N10276
2.4831	SG-NiCr21Mo9Nb	Fe-≤0.10C-≤0.50Si-≤0.50Mn-≤0.015S-≤0.4Al-20.0-23.0Cr-≤0.50Cu-≤5.0Fe-8.0-10.0Mo-3.0-4.5Nb-60.0Ni-≤0.40Ti	UNS N06625
2.4851	NiCr23Fe	Fe-0.03-0.1C-≤0.5Si-≤1.0Mn-≤0.02P-≤0.015S-1.0-1.7Al-≤0.006B-21.0-25.0Cr-≤0.5Cu-≤18.0Fe-58.0-63.0Ni-≤0.5Ti	N06601 (UNS)
2.4856	NiCr22Mo9Nb	Ni-≤0.03-0.10C-≤0.5Si-≤0.5Mn-≤0.020P-≤0.015S-≤0.4Al-≤1.0Co-≤20.0-23.0Cr-≤0.5Cu-≤5.0Fe-≤8.0-10.0Mo-≤3.15-4.15Nb-≤0.4Ti	UNS N06625
2.4858	NiCr21Mo	≤38.0-46.0Ni-≤0.025C-≤0.50Si-≤1.0Mn-≤0.025P-≤0.015S-≤0.20Al-≤1.0Co-≤19.5-23.5Cr-≤1.5-3.0Cu-≤2.5-3.5Mo-≤0.6-1.2Ti	B 163 UNS N08825

German Standard		Materials Compositions	US-Standard
MatNo.	DIN-Design	Percent in Weight	SAE/ASTM/UNS
2.4869	NiCr80-20	Fe-≤0.15C-0.50-2.0Si-≤1.0Mn-≤0.020P-≤0.015S-≤0.3Al-≤1.0Co-19.0-21.0Cr-≤0.50Cu-≤1.0Fe-75.0Ni	UNS N06003
2.4882		Fe-≤0.12C-0.50-2.0Si-≤1.0Mn-≤0.020P-≤0.015S-≤0.3Al-≤1.0Co-19.0-21.0Cr-≤0.50Cu-≤1.0Fe-75.0Ni	UNS N10001
2.4883		Fe-≤0.03C-≤0.50Si-≤1.0Mn-≤2.50Co-15.50-17.50Cr-≤7.0Fe-16.0-18.0Mo	UNS N10002
2.4886	SG-NiMo16Cr16W; UP-Ni-Mo16Cr16W	≤50.0Ni-≤0.02C-≤0.08Si-≤1.0Mn-≤0.015S-≤14.5-16.5Cr-≤4.0-7.0Fe-≤15.0-17.0Mo-≤0.4V-≤3-4.5W	UNS N10276
2.4887	EL-NiMo15Cr15W	≤50.0Ni-≤0.02C-≤0.20Si-≤1.0Mn-≤0.015S-≤14.5-16.5Cr-≤4.0-7.0Fe-≤15.0-17.0Mo-≤0.4V-≤3-4.5W	
2.4951	NiCr20Ti	Ni-≤0.08-0.15C-≤1.0Si-≤1.0Mn-≤0.020P-≤0.015S-≤0.3Al-≤0.0060B-≤5.0Co-≤18.0-21.0Cr-≤0.5Cu-≤5.0Fe-≤0.2-0.6Ti	UNS N06075
2.4952	NiCr20FeMo3TiCuAl	Fe-≤0.03C-≤0.5Si-≤1.0Mn-≤0.03P-≤0.03S-0.1-0.5Al-19.5-22.5Cr-1.5-3.0Cu-≤22.0Fe-2.5-3.5Mo-≤0.5Nb-42.0-46.0Ni-1.9-2.4Ti	
2.4964	CoCr20W15Ni	≤9.0-11.0Ni-≤0.05-0.15C-≤0.4Si-≤2.0Mn-≤0.020P-≤0.015S-≤19.0-21.0Cr-≤3.0Fe-≤9.0-11.0Ni-≤14.0-16.0W	UNS R30605
2.4973	NiCr19CoMo	Ni-≤0.12C-≤0.50Si-≤0.10Mn-≤1.40-1.80Al-≤10.00-12.00Co-≤18.00-20.00Cr-≤5.00Fe-≤9.00-10.50Mo-≤2.80-3.30Ti	N 07041 (UNS) 683 (SAE)
2.4975	NiFeCr12Mo	Fe-≤0.10C-≤0.60Si-≤2.00Mn-≤0.020P-≤0.010S-≤0.350Al-≤1.00Co-11.00-14.00Cr-5.00-7.00Mo-40.00-45.00Ni-2.35-3.10Ti	
2.4976	NiCr20Mo	Ni-≤0.10C-≤1.00Si-≤1.00Mn-≤0.020P-≤0.010S-0.50-1.80Al-≤2.00Co-18.00-21.00Cr-≤5.00Fe-4.00-5.00Mo-1.80-2.70Ti	
2.4983	NiCr18Co	Ni-≤0.15C-≤0.50Si-≤1.00Mn-≤0.02P-≤0.01S-≤2.50-3.20Al-≤17.00-20.00Co-≤17.00-20.00Cr-≤4.00Fe-≤3.00-5.00Mo-≤2.50-3.20Ti	UNS N07500 ASTM B637 (N07500)(864)
2.4999	MP35N	35.0Ni-≤0.01C-≤20.0Cr-≤9.5Mo	

Table 2: Chemical compositions of different American, CIS, Bulgarian and other steels

Steel	Materials Compositions, Percent in Weight	Note
000Ch16N13M2	Fe-≤0.07C-≤1.0Si-≤2.0Mn-16.5-18.5Cr-2-2.5Mo-10-13Ni-≤0.045P-≤0.015S-≤0.11N	CIS, formerly USSR, identical with SAE 316
000Ch16N13M3	Fe-≤0.07C-≤1.0Si-≤2.0Mn-16.5-18.5Cr-2-2.5Mo-10-13Ni-≤0.045P-≤0.015S-≤0.11N	CIS, formerly USSR, identical with SAE 316
000Ch16N16M4	Fe-≤0.07C-≤1.0Si-≤2.0Mn-16.5-18.5Cr-2-2.5Mo-10-13Ni-≤0.045P-≤0.015S-≤0.11N	CIS, formerly USSR, identical with SAE 316
000Ch18N10	Fe-≤0.03C-≤0.8Si-≤2.0Mn-17-19Cr-≤0.3Mo-9-11Ni	CIS, formerly USSR/Bulg.
000Ch18N11	Fe-≤0.03C-≤1.0Si-≤2.0Mn-18-20Cr-10-12Ni-≤0.045P-≤0.015S-≤0.11N	Bulg., comparable with 1.4306
000Ch20N20	Fe-≤0.03C-≤18.57Cr-≤19.40Ni-≤0.71Mn-≤0.26Si	CIS, formerly USSR
000Ch21N6M2	Fe-≤0.036C-≤21.1Cr-6.5Ni-2.4Mo	CIS, formerly USSR
000Ch21N10M2	Fe-≤0.02C-≤19.8Cr-≤10.5Ni-≤2.1Mo	CIS, formerly USSR
000Ch21N21M4B	Fe-≤0.03C-≤20-22Cr-20-21Ni-3.4-3.7Mo-≤0.6Mn-≤0.6Si-≤0.03P-≤0.02S-0.45-0.8Nb	CIS, formerly USSR
005Ch25B	Fe-0.005C-0.007N-25Cr	CIS, formerly USSR
00Ch18G8N2T	Fe-≤0.08C-≤0.8Si-7-9Mn-17-19Cr-≤0.3Mo-1.8-2.8Ni-≤0.2W-≤0.3Cu-≤0.2Ti-0.2-0.5Al	CIS, formerly USSR
00Ch18N10	Fe-≤0.015C-≤0.7Si-≤1.7Mn-≤17.3Cr-≤10.4Ni	CIS, formerly USSR
0Ch20N6M2T	not available (n.a.)	CrNiMoTi 20 6 2
0Ch21N6M2T	n. a.	CrNiMoTi 21 6 2
1Ch21N5T	n. a.	–
1H18N9	n. a.	cf. 1.4541; SAE 321
2Ch18N9	n. a.	Fe-0.2C-18Cr-9Ni; cf. 1.4310, UNS 30200
02Ch12N10S5	Fe-0.02C-12Cr-10Ni-5Si, Nb-stabilized	CIS, formerly USSR
02Ch12N10S5B	Fe-0.02C-12Cr-10Ni-5Si, Nb-stabilized	CIS, formerly USSR
02Ch12N10S5T	Fe-0.02C-12Cr-10Ni-5Si, Nb-stabilized	CIS, formerly USSR
02Ch17NS6	Fe-0.02C-4-6.5Si-0.43-0.52Mn-16.3-18Cr-10.5-18.2Ni-0.005-0.008S-0.012-0.014P	CIS, formerly USSR
02Ch8N22S6	Fe-≤0.02C-≤5.4-6.7Si-0.6Mn-≤0.030P-≤0.020S-7.5-10Cr-0.3Mo-21-23Ni-≤0.2Ti-≤0.20W	CIS, formerly USSR
02Ch8N22S6B	Fe-0.02C-8Cr-122Ni-6Si, Nb-stabilized	CIS, formerly USSR
02Ch8N22T	Fe-0.02C-8Cr-122Ni-6Si, Ti-stabilized	CIS, formerly USSR
03Ch16N15M3	Fe-≤0.03C-15.0-17.0Cr-14.0-16.0Ni-2.5-3.0Mo-Ti	CIS, formerly USSR

Steel	Materials Compositions, Percent in Weight	Note
03Ch18N11	Fe-0.03C-≤0.80Si-≤0.70-2.0Mn-≤0.035P-0.020S-17.0-19.0Cr-≤0.10Mo-10.5-12.5Ni-≤0.20W-0.30Cu-≤0.50Ti	CIS, formerly USSR, comparable with DIN-Mat.No. 1.4306
03Ch18N14	Fe-0.03C-18Cr-14Ni	CIS, formerly USSR
03Ch21N21M4B	n. a.	CrNiMoB 21 21 4
03Ch21N21M4GB	Fe-≤0.03C-≤0.60Si-1.8-2.5Mn-≤0.030P-≤0.020S-20-22Cr-3.4-3.7Mo-20-22Ni-≤0.2W-≤0.3Cu-≤0.2Ti, Nb 15 x C-0.80	CIS, formerly USSR
03Ch23N6	Fe-≤0.03C-≤0.40Si-1.0-2.0Mn-≤0.035P-≤0.020S-22-24Cr-≤5.3-6.3Ni	CIS, formerly USSR
03Ch25	Fe-about 0.03C-25Cr-0.6Ni	CIS, formerly USSR
03ChN28MDT	Fe-≤0.03C-≤0.80Si-≤0.80Mn-≤0.035P-≤0.020S-22.0-25.0Cr-2.5-3.0Mo-26.0-29.0Ni-0.50-0.90Ti-2.5-3.5Cu	CIS, formerly USSR
04Ch18N10	Fe-≤0.04C-≤0.8Si-≤2.0Mn-17-19Cr-≤0.030P-≤0.02S-≤0.3Mo-9-11Ni-≤0.2W-≤0.3Cu-≤0.2Ti	CIS, formerly USSR
04Ch18N10T	Fe-≤0.04C-≤0.8Si-≤2.0Mn-17-19Cr-≤0.3Mo-9-11Ni-≤0.2W-≤0.3Cu-≤0.2Ti	CIS, formerly USSR
05Ch16N15M3	Fe-0.05C-16Cr-15Ni-3Mo	CIS, formerly USSR
06Ch17G15NAB	Fe-0.05C-18.36Cr-16.5Mn-1.6Ni-0.31Nb-á.12Si-0.01Ce-0.017P-0.014S	CIS, formerly USSR
06Ch23N28M3D3T	Fe-≤0.06C-≤0.8Si-≤2.0Mn-22-25Cr-≤2.4-3Mo-26-29Ni-0.5-0.9Ti-2.5-3.5Cu	CIS, formerly USSR/Bulg.
06Ch28MDT	Fe-≤0.06C-≤0.8Si-≤0.80Mn-22.0-25.0Cr-2.5-3.0Mo-26.0-29.0Ni-0.50-0.90Ti-2.5-3.5Cu	CIS, formerly USSR
06ChN28MDT	Fe-≤0.06C-≤0.8Si-≤0.8Mn-≤0.035P-≤0.02S-22-25Cr-2.5-3.0Mo-26-29Ni-0.5-0.9Ti-2.5-3.5Cu	CIS, formerly USSR
06ChN40B	Fe-0.055C-17.01Cr-39.04Ni-1.99Mn-0.50Nb-0.60Si-0.013S-0.022P	CIS, formerly USSR
06XH28M?T	n. a.	
07Ch13AG20	Fe-≤0.07C-≤0.60Si-≤19-22Mn-≤0.035P-≤0.025S-≤0.0030B-≤0.1Ca-≤0.1Ce -12.-14.8Cr-≤0.30Cu ≤0.1Mg-≤0.30Mo-≤0.08-0.18N-≤1.0Ni-≤0.20W≤0.20Ti	CIS, formerly USSR
07Ch16N4B	Fe-0.05-0.10C-≤0.60Si-≤0.2-0.5Mn-≤0.025P-≤0.020S-15-16.5Cr-≤0.30Cu-≤0.30Mo-0.2-0.4Nb-3.5-4.5Ni-≤0.20W	CIS, formerly USSR
07Ch17G15NAB	Fe-0.05C-18.4Cr-16.5Mn-1.6Ni-0.01Ce-0.005B-0.32N	CIS, formerly USSR
07Ch17G17DAMB	Fe-0.06C-17.6Cr-15.2Mn-0.43Mo-0.3Nb-0.005B-0.38N	CIS, formerly USSR
08Ch17N5M3	Fe-0.06C-0.10C-≤0.80Si-≤0.80Mn-≤0.035P-≤0.020S-16.0-17.5Cr-3.0-3.5Mo-4.5-5.5Ni-≤0.20W-≤0.30Cu-≤0.20Ti	CIS, formerly USSR

Steel	Materials Compositions, Percent in Weight	Note
08Ch17N15M3B	Fe-≤0.08C-16.0-18.0Cr-14.0-16.0Ni-3.0-4.0Mo-Ti	CIS, formerly USSR
08Ch17N15M3T	Fe-≤0.08C-≤0.80Si-≤2.0Mn-≤0.35P-≤0.020S-16.0-18.0Cr-3.00-4.00Mo-14.0-16.0Ni≤0.20W-0.30Cu-0.30-0.60Ti	CIS, formerly USSR
08Ch17T	Fe-≤0.08C-≤0.80Si-≤0.80Mn-≤0.035P-≤0.025S-16.0-18.0Cr-≤0.6Ni-≤0.30Cu	CIS, formerly USSR
08Ch18G8N2M2T	Fe-0.08C-18.2Cr-3.42Ni-8.9Mn-2.32Mo-0.22Ti	CIS, formerly USSR
08Ch18G8N2T	Fe-≤0.08C-≤0.80Si-7.0-9.0Mn-≤0.035P-17.0-19.0Cr-≤0.30Mo-1.80-2.80Ni-≤0.30Cu-≤0.035P-≤0.025S-≤0.20-0.50Ti-≤0.20W	CIS, formerly USSR
08Ch18N10	Fe-≤0.08C-≤0.8Si-≤2.0Mn-≤0.035P-≤0.020S-17-19Cr-0.3Mo-9.0-11.0Ni-≤0.2W-≤0.3Cu	CIS, formerly USSR
08Ch18N10T	Fe-≤0.08C-≤0.8Si-≤2.0Mn-≤0.035P-≤0.020S-17-19Cr-0.5Mo-9.0-11.0Ni-0.5Ti-≤0.2W-≤0.3Cu	CIS, formerly USSR
08Ch21N6M2T	Fe-≤0.08C-≤0.8Si-≤0.8Mn-≤0.035P-≤0.025S-20-22Cr-1.8-2.5Mo-5.5-6.5Ni-0.2-0.4Ti-≤0.2W-≤0.3Cu	CIS, formerly USSR
08Ch22N6M2T	Fe-≤0.08C-≤0.80Si-≤0.80Mn-20.0-22.0Cr-1.80-2.50Mo-5.50-6.50Ni-≤50.20W-≤0.30Cu-≤0.035P-≤0.025S-0.20-0.40Ti	CIS, formerly USSR
08Ch22N6T	Fe-≤0.08C-≤0.8Si-≤0.8Mn-≤0.035P-≤0.025S-21-23Cr-≤0.3Mo-5.3-6.3Ni-≤0.2W-≤0.3Cu, 5x% C max. 0.65Ti	CIS, formerly USSR
08ChP	Fe-0.25Cr-0.25Ni-0.25Mo	CIS, formerly USSR
08-KP	n. a.	cf. 1.0335; UNS G 10060
08X21H6M2T	n. a.	–
08X22H6T	n. a.	–
09Ch16N15M3B	Fe-≤0.09C-≤0.80Si-≤0.80Mn-≤0.035P-≤0.020S-15.0-17.0Cr-≤0.30Cu-2.5-3.0Mo-0.6-0.9Nb-14.0-16.0Ni-≤0.20Ti-≤0.20W	CIS, formerly USSR
09G2S	Fe-≤0.12C-0.5-0.8Si-1.3-1.7Mn-≤0.035P-≤0.035S-≤0.3Cr-≤0.3Ni-≤0.3Cu	Bulg.
0Ch17N16M3T	Fe-≤0.080C-≤0.8Si-≤2.00Mn-16.0-18.0Cr-3.00-4.00Mo-14.0-16.0Ni-≤ 0.035P-≤0.025S-0.30-0.60Ti	CIS, formerty USSR
0Ch18G8N3M2T	Fe-about 18 Cr-8Mn-3Ni-2Mo, Ti	CIS, formerly USSR
0Ch18N10T	Fe-≤0.08C-≤1.0Si-≤2.0Mn-≤0.045P-≤0.015S-17-19Cr-9-12Ni, Ti 5xC-0.70	CIS, formerly USSR/Bulg.
0Ch18N12B	Fe-≤0.08C-≤1.0Si-≤2.0Mn-≤0.045P-≤0.015S-17-19Cr-≤9.0-12Ni, Nb 10xC-1.00	CIS, formerly USSR/Bulg., comparable with DIN-Mat. No. 1.4550
0Ch20N14S2	Fe- about 20Cr-14Ni-2Si	CIS, formerly USSR

Steel	Materials Compositions, Percent in Weight	Note
0Ch21N5T	Fe-≤0.08Cr-21Cr-5Ni, Ti	CIS, formerly USSR s. text HNO3
0Ch23N18	Fe-≤0.20C-≤1.00Si-1.50Mn-22.0-25.0Cr-≤0.30Mo-17.0-20.0Ni-≤0.035P-≤0.025S	CIS, formerly USSR
0Ch23N28M3D3T	Fe-≤0.06C-≤0.8Si-≤2Mn-22-25Cr-2.4-3Mo-26-29Ni-0.5-0.9Ti-2.5-3.5Cu	CIS, formerly USSR/Bulg.
0Ch25T	Fe-≤0.01C-25Cr, Ti-stabilized	CIS, formerly USSR
0H17N12M2T	Fe-≤0.05C-≤1.0Si-≤2.0Mn-16-18Cr-2-3Mo-11-14Ni-≤0.045P-≤0.030S, Ti 5xC-0.60	Poland
10Ch13 (1Ch13)	Fe-0.08-0.15C-≤1.0Si-≤1.5Mn-≤0.040P-≤0.015S-11.5-13.5Cr-≤0.75Ni	CIS, formerly USSR/Bulg.
10Ch14AG15	Fe-≤0.10C-≤0.80Si-14.5-16.5Mn-≤0.045P-≤0.030S-13.0-15.0Cr-<0.60Ni-≤0.60Cu-≤0.20Ti-0.15-0.25N	CIS, formerly USSR
10Ch14G14N4T	Fe-≤0.10C-≤0.80Si-13.0-15.0Mn-13.0-15.0Cr-≤0.30Mo-2.80-4.50Ni-≤0.20W-≤0.30Cu-≤0.035P-≤0.020S-5x% C max. 0.60Ti	CIS, formerly USSR
10Ch17	Fe-0.10C-17Cr	CIS, formerly USSR
10Ch17N13M2T	Fe-≤0.10C-≤0.80Si-≤2.0Mn-≤0.035P-≤0.020S-16.0-18.0Cr-≤0.30Cu-2.0-3.0Mo-12.0-14.0Ni-≤0.20W, Ti>-5x% C	USA, comparable with DIN-Mat. No. 1.4571
10Ch17N13M3T	Fe-≤0.10C-≤0.80Si-≤2.0Mn-≤0.035P-≤0.020S-16.0-18.0Cr-≤0.30Cu-3.0-4.0Mo-12.0-14.0Ni-≤0.20W-≤0.7Ti	USA, comparable with DIN-Mat. No. 1.4573
10Ch18N9T(Ch18N9T)	Fe-0.08C-1.0Si-≤2Mn-≤0.045P-≤0.015S-17-19Cr-≤0.3Mo-9-12Ni-Ti5xC-0.70	CIS, formerly USSR/Bulg.
10Ch18N10M2T	Fe-≤0.10C-18Cr-10Ni-2Mo, Ti stabilized	CIS, formerly USSR
10Ch18N10T	Fe-≤0.10C-≤0.80Si-≤1.0-2.0Mn-≤0.035P-≤0.020S-≤17.0-19.0Cr-≤10.0-11.0Ni	CIS, formerly USSR/Bulg.
12Ch13G18D	Fe-0.12C-13Cr-18Mn-Cu	CIS, formerly USSR
12Ch17G9AN4	Fe-≤0.12C-≤0.80Si-8.0-10.5Mn-≤0.035P-≤0.020S-16.0-18.0Cr-≤0.30Mo-3.5-4.5Ni-≤0.20W-≤0.30Cu-≤0.20Ti-0.15-0.25N	CIS, formerly USSR
12Ch18N9T	Fe-≤0.12C-≤0.80Si-≤2.0Mn-≤0.035P-≤0.020S-17.0-19.0Cr-≤0.50Mo-8.0-9.5Ni-≤0.20W-≤0.30Cu-Ti = 5x % C	CIS, formerly USSR
12Ch18N10T	Fe-≤0.12C-≤0.8Si-≤2.0Mn-≤0.025P-≤0.020S-17.0-19.0Cr-≤0.30Cu-≤0.50Mo-9.0-11.0Ni-≤0.20W-≤0.70Ti	CIS, formerly USSR; comparable with DIN-Mat. No. 1.4878
12Ch2M1	Fe-0.12C-2Cr-1Mo	CIS, formerly USSR/Bulg., comparable with DIN-Mat. No. 1.7380, A 182, F 22, B.S. 1501-622

Steel	Materials Compositions, Percent in Weight	Note
12Ch2N4A	Fe-0.09-0.15C-0.17-0.37Si-0.30-0.60Mn-≤0.025P-≤0.025S-1.25-1.65Cr-3.25-3.65Ni-≤0.30Cu-≤0.15Mo-≤0.03Ti-≤0.05V-≤0.12W	CIS, formerly USSR
12Ch21N5T	Fe-0.09-0.14C-≤0.80Si-≤0.80Mn-≤0.035P-≤0.025S-20.0-22.0Cr-≤0.30Mo-4.80-5.80Ni-≤0.20W-≤0.30Cu-0.25-0.50Ti-≤0.08Al	CIS, formerly USSR
12ChN2	Fe-0.09-0.16C-0.17-0.37Si-0.30-0.60Mn-≤0.035P-≤0.035S-0.60-0.90Cr-1.50-1.90Ni-≤0.30Cu-≤0.15Mo-≤0.03Ti-≤0.05V-≤0.20W	CIS, formerly USSR/Bulg.
13-4-1	Fe-0.043C-12.7Cr-3.9Ni-1.5Mo-0.68Mn-0.39Si-0.009P-0.013S-0.030N	CIS, formerly USSR
14Ch17N2	Fe-≤0.11-0.17C-≤0.80Si-≤0.80Mn-16.0-18.0Cr-≤0.30Mo-1.50-2.50Ni-≤0.20W-≤0.30Cu-≤0.030P-≤0.025S-≤0.20Ti	CIS, formerly USSR
15Ch17N2	Fe-0.13C-0.49Si-0.52Mn-17.17Cr-1.75Ni-0.012P-0.09S	CIS, formerly USSR
15Ch25T	Fe-≤0.15C-≤1.00Si-≤0.80Mn-≤0.035P-≤0.025S-24.0-27.0Cr-≤0.30Cu-≤1.00Ni-0.09Ti	CIS, formerly USSR
15Ch28	Fe-≤0.15C-≤1.00Si-≤0.80Mn-27.0-30.0Cr-0.60Ni-≤0.30Cu-≤0.035P-≤0.025S-≤1.0Ni-≤0.20Ti	CIS, formerly USSR
15Ch2M2FBS	Fe-about 0.15C-2Cr-2Mo-V-Nb-Si	CIS, formerly USSR
15Ch5M	Fe-≤0.15C-≤0.50Si-≤0.50Mn-4.50-6.00Cr-0.40-0.60Mo-≤0.60Ni-≤0.03Ti-≤0.030P-≤0.025S-≤0.20Cu-≤0.05V-≤0.30W	CIS, formerly USSR
16GS	Fe-≤0.12-0.18C-≤0.40-0.70Si-0.90-1.20Mn-≤0.30Cr-≤0.30Ni-≤0.30Cu-≤0.035P-≤0.040S-≤0.05Al-≤0.08As-≤0.012N-≤0.03Ti	CIS, formerly USSR/Bulg., comparable with DIN-Mat. No. 1.0481, A 414, A 515, A 516
18/8-CrNi-steel	Fe-≤0.12C-1.0Si-≤2.0Mn-≤0.045P-≤0.030S-≤17.0-19.0Cr≤-8.0-10.0Ni	–
1815-LCSi	Fe-0.006C-18.3Cr-15.1Ni-1.5Mn-4.1Si-0.005S-0.010P-0.010N	UNS S30600, comparable with DIN-Mat. No. 1.4361
18-18-2	Fe-≤0.08C-1.5-2.5Si-≤2.0Mn-≤0.030P-≤0.030S-17.0-19.0Cr-17.5-18.5Ni	USA
18G2A	Fe-≤0.20C-≤0.50Si-≤0.9-1.7Mn-≤0.025P-≤0.020S-≤0.0200Al-≤0.30Cr-≤0.50Ni-≤0.30Cu.-≤0.08Mo-≤0.020N-≤0.050Nb-≤0.03Ti-≤0.1V	Poland
20ChGS2	Fe-≤0.25C-≤1.0Si-≤1.5Mn-≤0.04P-≤0.015S-≤14.0Cr	Russia
20Ch13 (2Ch13)	Fe-0.16-0.25C-≤0.8Si-≤0.8Mn-12-14Cr-≤0.6Ni-≤0.030P-≤0.025S-≤0.30Cu-≤0.20Ti	CIS, formerly USSR/Bulg., comparable with DIN-Mat. No. 1.4021, SAE 420, 420 S 29

Steel	Materials Compositions, Percent in Weight	Note
20Ch23N18	Fe-≤0.2C-≤1.0Si-≤2.0Mn-22-25Cr-≤0.3Mo-17-20Ni-≤0.2W-≤0.3Cu-≤0.2Ti-≤0.035P-≤0.025S	CIS, formerly USSR
20Ch2G2SR	Fe-0.16-0.26C-0.75-1.55Si-1.4-1.8Mn-≤0.040P-≤0.040S-1.4-1.8Cr-≤0.30Ni-≤0.30Cu-0.02-0.08Ti-0.015-0.050Al-0.001-0.007B	CIS, formerly USSR
23Ch2G2T	Fe-0.19-0.26C-0.40-0.70Si-1.4-1.7Mn-≤0.045P-≤0.045S-1.35-1.70Cr-≤0.30Ni-≤0.30Cu-0.02-0.08Ti-0.015-0.05Al	CIS, formerly USSR
36NChTJu	Fe-≤0.05C-≤0.3-0.7Si-≤0.8-1.2Mn-≤0.020P-≤0.020S-11.5-13Cr-35-37Ni-0.9-1.2Al-2.7-3.2Ti	CIS, formerly USSR
40Ch13	Fe-0.36-0.45C-≤0.80Si-≤0.80Mn-≤0.030P-≤0.025S-12.0-14.0Cr-≤0.30Cu-≤0.60Ni-≤0.20Ti	CIS, formerly USSR, comparable with DIN-Mat. No. 1.4031
45 G2	Fe-0.41-0.49C-0.17-0.37Si-1.4-1.8Mn-0.035P-≤0.035S-≤0.30Cr-≤0.30Cu-≤0.15Mo-≤0.30Ni-≤0.03Ti-≤0.05V-≤0.20W	CIS, formerly USSR, comparable with DIN-Mat. No. 1.0912
50Ch	Fe-0.46-0.54C-0. 17-0.37Si-0.50-0.80Mn-≤0.035P-≤0.035S-0.80-1.10Cr-≤0.30Ni-≤0.30Cu-≤0.15Mo-≤0.03Ti-≤0.05V-≤0.20W	CIS, formerly USSR
70G	Fe-0.67-0.75C-0.17-0.37Si-0.90-1.20Mn-≤0.035P-≤0.035S-≤0.25Cr-≤0.25Ni- ≤0.20Cu	CIS, formerly USSR
80S	Fe-0.74-0.82C-0.60-1.10Si-0.50-0.90Mn-≤0.040P-≤0.045S-≤0.30Cr-≤0.30Ni-≤0.30Cu-0.015-0.040Ti	CIS, formerly USSR
2320	Fe-≤0.08-1.0Si-1.0Mn-≤0.040P-≤0.030S-16.0-18.0Cr-≤1.0Ni	Sweden, comparable with DIN-Mat. No. 1.4016; SAE 430, 10Ch17T
ASTM A-159	Fe-3.1-3.4C-1.9-2.3Si-0.6-0.9Mn-0.15S-0.15P	USA
ASTM A-516 Gr. 70	Fe-0.27C-0.13-0.45Si-0.79-1.30Mn-≤0.035P-≤0.040S	USA, comparable with DIN-Mat. No. 1.0050 and No. 1.0481
ASTM XM-27	Fe-≤0.01C-≤0.40Si-≤0.40Mn-≤0.020P-≤0.020S-25.0-27.5Cr-≤0.20Cu-≤0.75-1.50Mo-≤0.015N-≤0.050-0.2Nb-0.50Ni, Ni+Cu≤0.50	USA, comparable with SAE XM-27
C 1204	Fe-0.20C-0.35Si-0.50Mn-0.050P-0.050S-0.30Cr	Yugoslavia, comparable with DIN-Mat. No. 1.0425, B.S. 1501 Gr. 161-400, 164-350, 164-400; 16 K
C 90	Fe-0.85-0.94C-≤-0.35Si-≤0.35Mn-≤0.03P-≤0.03S	Italy
Carpenter 20 Cb-3	Fe-≤0.06C-≤1.00Si-≤2.00Mn-19.0-21.0Cr-2.0-3.0Mo-32.5-35Ni-3.0-4.0Cu-≤0.035P-≤0.035S	USA
Ch12M	Fe-1.45-1.65C-0.15-0.35Si-0.15-0.4Mn-11-12.5Cr-0.4-0.6Mo-≤0.35Ni-15-0.3V-≤0.2W≤0.3Cu-≤0.03Ti	CIS, formerly USSR
Ch14N40SB	Fe-0.034C-4.0Si-0.05Mn-14.4Cr-38.9Ni-0.63Nb	CIS, formerly USSR

Key to materials compositions | 453

Steel	Materials Compositions, Percent in Weight	Note
Ch15T	Fe-≤0.1C-≤0.8Si-≤0.8Mn-14-16Cr-≤0.3Mo-≤0.6Ni, 5x%C≤Ti≤0.8	CIS, formerly USSR/Bulg.
Ch16N15M3	n. a.	–
Ch17	Fe-≤0.08C-≤1Si-≤1Mn-≤0.040P-≤0.015S-16.0-18.0Cr	Bulg., comparable with DIN-Mat. No. 1.4016, SAE 430, 12Ch17T, X6Cr17
Ch17M2TL	n. a.	–
Ch17N2	Fe-17Cr-2Ni	CIS, formerly USSR
Ch17N5M3	n. a.	–
Ch17N12M3T	Fe-≤0.12C-≤1.5Si-≤2.0Mn-16-19Cr-3-4Mo-11-13Ni-0.3-0.6Ti	CIS, formerly USSR/Bulg.
Ch17N13M2T	n. a.	cf. 1.4571; SAE 316 Ti; CrNiMoTi 17 13 2
Ch17N13M3T	n. a.	cf. 1.4571; UNS S31635; CrNiMoTi 17 13 3
Ch17N18M2T	Fe-0.09C-0.6Si-1.4Mn-16.9Cr-1.9Mo-12.3Ni, Ti stab. (p.a.)	CIS, formerly USSR/Bulg.
Ch17T	Fe-≤0.05C-≤1Si-≤1Mn-≤0.040P-≤0-0.15S-16.0-18.0Cr, Ti4x(C+N)+0.15-0.80	Bulg., comparable with DIN-Mat. No. 1.4510, 08Ch17T, X3CrTi17
Ch18AG14	Fe-18Cr-14Mg-0.5N	
Ch18N9T Ch18N10T	Fe-≤0.08C-≤1Si-≤2Mn-≤0.045P-≤0.015S-17.0-19.0Cr-9.0-12.0Ni, Ti 5xC-0.70	CIS, formerly USSR/Bulg., comparable with DIN-Mat. No. 1.4541, X6CrNiTi18-10
Ch18N10	Fe-0.08C-18.4Cr-10.2Ni-1.08Mn-0.3Si-0.005P-0.014S-0.005N	
Ch18N12M2T	Fe-≤0.15C-≤5 1.5Si-≤2Mn-17-19Cr-2-2.5Mo-11-13Ni, 4x%C≤Ti≤0.8	CIS, formerly USSR/Bulg.
Ch18N12T	Fe-0.08C-1.0Si-≤2.0Mn-≤0.045P-0.015S-17.0-19.0Cr-9.0-12.0Ni-Ti = 5x%C≥0.70	CIS, formerly USSR/Bulg.
Ch18N14	Fe-0.035C-18.8Cr-14.6Ni-0.35Mn-0.75Si-0.005P-0.03S-0.004N	
Ch18N40T	Fe-<0.08C-<1.0Si-<2.0Mn-<0.045P-<0.015S-17-19Cr-9-12Ni-<0.7Ti	Comparable with DIN-Mat. No. 1.4541, SAE 321
Ch20N20	Fe-0.004-0.015C-19.4-21.8Cr-19.3-20.8Ni-0.05-5.40Si-0.002-0.1P	CIS, formerly USSR/Bulg.
Ch21N5	n. a.	–
Ch21N6M2	n. a.	–
Ch21N6M2T	n. a.	–
Ch22N5	Fe-0.07C-21.54Cr-5.73Ni	CIS, formerly USSR

Steel	Materials Compositions, Percent in Weight	Note
Ch23N18	Fe-≤0.20C-≤1.00Si-≤1.50Mn-22.0-25.0Cr-≤0.30Mo-17.0-20.0Ni-≤0.035P-≤0.025S	CIS, formerly USSR/Bulg.
Ch23N27M2T	Fe-27Ni-23Cr-2Mo-Ti	CIS, formerly USSR/Bulg.
Ch23N28M3D3T	Fe-28Ni-23Cr-3Mo-3Cu-Ti	CIS, formerly USSR
Ch25T	Fe-≤0.15C-≤1.00Si-≤0.80Mn-24.0-27.0Cr-≤0.30Mo-≤0.60Ni-≤0.035P-≤025S-5x%C≤Ti≤0.9	CIS, formerly USSR/Bulg.
Ch28N18	Fe-0.16C-1.7Mn-1.1Si-22.6Cr-18Ni-0.4Ti (p.a.)	CIS, formerly USSR/Bulg.
ChN28MDT	Fe-0.03-0.046C-22.2-23.5Cr-26.55-27.88Ni-2.55-3.06Mo-2.68-3.38Cu-0.54-0.76Ti-0.15-0.30Mn-0.39-0.69Si-0.021-0.43P-0.008-0.017S	CIS, formerly USS
ChN40B	Fe-0.032C-18.2Cr-40.4Ni-0.08Si-0.05Mn-0.49Nb	CIS, formerly USSR
ChN40S	Fe-0.031C-20.0Cr-38.9Ni-4.2Si-0.05Mn-0.13Nb	CIS, formerly USSR
ChN40SB	Fe-0.04C-18.8Cr-39.4Ni-4.3Si-0.06Mn-0.63Nb	CIS, formerly USSR
ChN58W	Ni-0.03C-14.5-16.5Cr-15-17Mo-3.0-4.5W-1.5Fe-1.0Mn-<0.12Si-0.02S-0.025P	CIS, formerly USSR
ChN60V	Fe-0.01-0.02C-0.09N-max.0.05Zr-0.1Ti-0.015Ce-max.1.7Nb-max0.009B	CIS, formerly USSR
ChN77TJuR	Fe-≤0.07C-≤0.60Si-≤0.40Mn-≤0.015P-≤0.007S-0.6-1.0Al-≤0.01B-≤0.02Ce-19.0-22.0Cr-≤1.0Fe-≤0.001Pb-2.4-2.8Ti	CIS, formerly USSR
ChN78T	Fe-≤0.25C-≤1.0Si-≤1.5Mn-≤0.04P-≤0.015S-≤14.0Cr	Russia
FC 20	Fe-3.92C-1.12Si-0.63Mn-0.072P-0.012S	Japan
JS 700	Fe-0.04C-≤1.0Si-≤2.0Mn-≤0.040P-≤0.030S-19-23Cr-4.3-5.0Mo-24-26Ni, Nb≥8xC≤0.40	USA
OZL-17u	Fe-0.04C-0.32Si-1.5Mn-23.2Cr-0.2Mo-29.4Ni-0.01P-0.01S-0.1Ti	CIS, formerly USSR
SAE 1008	Fe-0.10C-0.30-0.50Mn-≤0.030P-≤0.050S	USA, comparable with DIN-Mat. No. 1.0204
SAE 1018	Fe-0.15-0.20C-0.60-0.90Mn-≤0.030P-≤0.050S	USA
SAE C-1018	Fe-0.20C-0.25Si-0.58Mn-0.16Cr-0.04Mo-0.012-0.014P-0.02S	USA
S35C	Fe-0.32-0.38C-0.15-0.35Si-0.60-0.90Mn-≤0.030P-≤0.035S-≤0.20Cr-≤0.30Cu-≤0.20Ni	Japan, comparable with DIN-Mat. No. 1.0501
SIS 2333	Fe-≤0.05C-1.0Si-2.0Mn-≤0.045P-≤0.030S-17.0-19.0Cr-8.0-11.0Ni	Sweden, comparable with DIN-Mat. No. 1.4303; SAE 304, 03Ch18N11
SKH 2	Fe-0.73-0.83C-0.45Si-0.40Mn-≤0.030P-≤0.030S-3.8-4.5Cr-≤0.25Cu-17.0-19.0W-1.0-1.2V	Japan, comparable with DIN 1.3355
SKH-4A	Fe-0.80C-0.29Si-0.31Mn-4.16Cr-17.64W-9.3Co-1.1V	Japan
SKH-9	Fe-0.85C-0.16Si-0.31Mn-4.14Cr-4.97Mo-6.03W-1.88V	Japan

Steel	Materials Compositions, Percent in Weight	Note
SS41	Fe-≤0.050P-≤0.050S	Japan, comparable with DIN-Mat. No. 1.0040
St35b-2	Fe-≤0.16C-0.17-0.40Si-0.35-0.65Mn-≤0.30Cr-≤0.30Ni-≤0.30Cu-0.050P-0.050S	Germany, formerly GDR, comparable with DIN-Mat. No. 1.0309
St35hb	Fe-≤0.18C-≤0.17Si-0.35-0.65Mn-0.050P-0.050S	Germany, formerly GDR
St38	Fe-≤0.20C-≤0.080-≤0.060S	Germany, formerly GDR, comparable with DIN-Mat. No. 1.0037
St38b-2	Fe-0.12-0.20C-0.17-0.37Si-0.40-0.65Mn-≤0.045P-≤0.050S-Cr+Cu+Ni≤0.70	Germany, formerly GDR, comparable with DIN-Mat. No. 1.0038, BS 4360-40C and A 570 Gr. 36
St5	Fe-≤0.045P-≤0.045S-≤0.009N	Poland, comparable with DIN-Mat. No. 1.0050
SUS 304	Fe-≤0.08C-≤1.0Si-≤2.0Mn-≤0.045P-≤0.030S-18.0-20.0Cr-8-10.5Ni	Japan, comparable with DIN-Mat. No. 1.4301
SUS 430	Fe-≤0.12C-≤0.75Si-≤1.00Mn-16.0-18.0Cr-≤0.60Ni-≤0.040P-≤0.030S	Japan, comparable with DIN-Mat. No. 1.4016, SAE 430, 430 S 15, 12Ch17
Sv-08	Fe-≤0.10C-≤0.03Si-0.35-0.60Mn-≤0.040P-≤0.040S-≤0.015Cr-≤0.30Ni-≤0.01Al	CIS, formerly USSR
TsL-17	Fe-0.1C-5Cr-1Mo	CIS, formerly USSR
TsL-9	Fe-0.07C-1.00Si-2.3Mn-24.1Cr-12.9Ni-1.1Nb-0.03P-0.02S	CIS, formerly USSR
U7A	Fe-0.65-0.75C-0.10-0.30Si-0.10-0.40Mn-≤0.030P-≤0.030S	CIS, formerly USSR, comparable with DIN-Mat. No. 1.1520
U8A	Fe-0.75-0.85C-0.10-0.30Si-0.10-0.40Mn-≤0.030P-≤0.030S	CIS, formerly USSR, comparable with DIN-Mat. No. 1.1525
U10A	Fe-0.96-1.03C-0.17-0.33Si-0.17-0.28Mn-≤0.025P-≤0.018S-≤0.20Cr-≤0.20Ni-≤0.20Cu	CIS, formerly USSR, comparable with DIN-Mat. No. 1.1545, SAE W 1 10
X5CrNiMoCuTi18-18	Fe-≤0.07C-≤0.80Si-≤2.0Mn-≤0.045P-≤0.030S-16.5-18.5Cr-2.0-2.5Mo-19.0-21.0Ni-1.8-2.2Cu-Ti≥7x%C	Germany, formerly GDR, comparable with DIN-Mat. No. 1.4506
X8CrNiTi18-10	Fe-≤0.10C-≤1.0Si-≤2.0Mn-≤0.045P-≤0.015S-17.0-19.0Cr-9.0-12.0Ni-Ti≥5 x C≤0.80	Germany, formerly GDR, comparable with DIN-Mat. No. 1.4541

Index of materials

1
1,2-polybutadiene 360
1.0032 59
1.0346 82
1.0356 82
1.0503 59, 61
1.0605 77
1.0619 94
1.0984 67
1.1127 70–71
1.1151 59–60, 65–66
1.3965 131, 137, 141, 148
1.4301 131–132, 135, 153–155
1.4310 106, 108–109, 112, 131–132, 135–137
1.4401 150, 153, 155–157
1.4404 107, 143, 148
1.4435 107, 143, 148
1.4436 153, 155–157
1.4454 131, 148
1.4462 118–119, 387
1.4501 123
1.4507 108
1.4550 187, 189
1.4571 374
1.4592 107
1.4841 106
1.4845 132, 135, 137, 153, 155
1.4914 107–108
1.4922 94–95
1.4931 94
1.4943 53, 137, 185, 187, 189
1.4980 164
1.4988 94
1.5403 68
1.6310 68
1.6580 70–71
1.7218 94–95
1.7220 70–71, 73, 76
1.7259 94
1.7273 94
1.7276 94
1.7281 94
1.7335 95
1.7357 94
1.7362 95
1.7363 93
1.7375 95
1.7380 88, 93, 95
1.7381 95
1.7386 95
1.7389 93
1.7720 76–77
1.7766 94
1.7779 94–95
1.8907 67, 78
1.8924 67
1.8975 78
1.8977 78
10CrMo9-10 88, 95
10CrMo11 94
12 CrMo 19 5 95
12CrMo9-10 95
12CrMo12-10 95
13CrMo4-5 95
16CrMo9-3 94
17CrMoV10 94
1010 59
1020 59
1043 59
1074 77

2
2.0040 51
2.0060 50
2.0061 50
2.0062 50
2.0070 51
2.0076 51
2.0080 50
2.0090 51
2.4061 308
2.4066 169
2.4603 189
2.4613 189
2.4642 171
2.4654 199
2.4665 189
2.4668 24, 176–179, 182, 185, 187, 388
2.4669 172, 174–175
2.4816 171
2.4856 185, 187, 189, 388
20CrMoV13-5-5 94–95
24CrMo10 94

Index of materials

25CrMo4   94–95
26CrMo7   94
2124   36–38

3
3.7025   236, 246
3.7035   236
3.7055   236
3.7065   236
3.7115   187, 189, 277, 280
3.7165   187, 189, 261, 263–265, 267, 270–274, 277–278, 280
3.7175   270, 272, 280
30CrNiMo8   70–71
34CrMo4   70–71, 73, 76–78
38Mn6   70–71
300   372
301   106, 108–110, 112, 131, 135–137
301 A   109
301 C   109
301 H   109
304   131, 133, 153, 155
310   106
310 S   133, 135, 137, 153, 155
316   150, 153, 155–156, 158
316 L   107, 143, 148
347   188

8
8090   37–38

a
A 286   53, 137, 164, 185–187, 189
A 516   82
A 533 (B)   68
ABS   365
ACM   361, 367, 373
acrylate rubber   361, 367, 373
acrylonitrile butadiene rubber   361, 367, 373
acrylonitrile butadiene styrene copolymer   365
acrylonitrile isoprene rubber   361
Al 29-4-2   109–110, 112, 137
AL 29-4C   107
Alcryn®   366
Alloy 600   171
Alloy 690   171
Alloy 718   24, 176–178, 188, 192, 200, 388
Alloy X-750   172–175
Alsint 99,7®   353
aluminide   300
aluminide XD™   310
aluminum   20–21, 40, 385
aluminum alloy   22–23, 27, 31–32, 40, 385
aluminum borosilicate   353
aluminum inner lining   383
aluminum lithium material   36
aluminum oxide   353, 389
aluminum (pure)   20, 22

aluminum-copper material   23
aluminum-lithium alloy   39
aluminum-zinc material   31
AlZn4Mg3   32
AlZn5,5MgCu   32
AlZn7MgCu   32
Anchor-Lok® PE   371
Anchor-Lok® PP   371
Anchor-Lok® PVC   371
Anchor-Lok® PVDF   371
Araldite® LY556/HY917   380
Armco iron   59, 61
ASA   358
AU   368, 374
austenite   112
austenitic CrNiMo steel   131
austenitic steel   108, 387

b
beryllium   303–304
beryllium oxide   353
Beta I   292, 294
Beta III   292, 294–295
boron carbide   354
boron nitride   354
BR   360
brass   55
bronze   55
butadiene rubber   360
butyl rubber   361, 367, 373

c
C45   59, 61
C 75   77–78
C steel   82–83
CA   360
carbon   348
carbon graphite   374
cast steel   57, 93, 138, 385
cellulose acetate   360
cellulose hydrate   359
cellulose nitrate   360
cement   348, 389
cerium sulfide   354
chlorobutyl rubber   361, 370
chloroprene rubber   361, 367, 370, 373
chlorosulfonated polyethylene   361, 367, 373
chromium   44, 47
chromium alloy   44, 47
chromium steel   386
chromium-vanadium alloy   48–49
CIIR   361, 370
Ck22   58–61, 64–66
CMSX-2   197–198
CN   360
coatings   369, 390
cobalt alloy   44, 309
composite material   380
copper   49, 56–57, 385

copper alloy   56
copper-nickel alloy   53, 385
copper-niobium alloy   56–57
copper-tin alloy   55
copper-tin-zinc alloy   55
copper-zinc alloy   55
CR   361, 367, 370, 373
CR003A   50
CR004A   50
CR005A   50
CR006A   50
CR007A   51
CR008A   51
CR009A   51
CR020A   51
CR021A   51
CR022A   51
CR023A   51
CR024A   51
CR025A   51
crystal lattice   103
CSM   361, 367, 370, 373
CuCr17   57
Cu-DHP   51
Cu-DLP   51
Cu-DXP   51
Cu-ETP   50
Cu-ETP1   50
Cu-FRHC   50
Cu-FRTP   50
Cu-HCP   51
CuNb20   56
Cu-OF   51
Cu-OF1   51
Cu-OFE   51
Cu-PHC   51
Cu-PHCE   51
CuZn30   55

d
Derakane®   380
diallyl phthalate resins   366
duplex steel   108, 110–112, 114–118, 120–121, 123–124
duroplastic   366

e
E1-Cu58   50
E2-Cu58   50
E-BRITE 26-1   107
EC   360
EC Duro-Bond®   370
EC Duro-Bond® ETFE Lining   371
EC Duro-Bond® FEP Lining   371
EC Duro-Bond® MFA Lining   371
EC Duro-Bond® PFA Lining   371
EC Duro-Bond® PVDF Lining   371
ECTFE   364, 370
E-Cu57   50
Edlon®   370

elastomer   366–368, 390
EN AW-1199   42
EN AW-2024   24–27
EN AW-2124   36–39
EN AW-2219   23–25, 28–29, 31
EN AW-6061   22
EN AW-6061-T3   383
EN AW-7039   22, 32
EN AW-7075   22, 31–32, 35
EN AW-7178   32
EN AW-8090   36–38
EN AW-Al 99.99   42
EN AW-Al Cu4Mg1   24–25, 27, 36
EN AW-Al Cu6Mn   23, 28
EN AW-Al Li2,5Cu1,5Mg1   36
EN AW-Al Mg1SiCu   22, 383
EN AW-Al Zn4Mg3   22
EN AW-Al Zn5,5MgCu   22, 31
EN AW-AlCu6Mn   29
enamel   351
EP   360, 380
EP-CF   362
EPDM   361, 367, 373
EP-GF   362
epoxy resin   360, 380–381
ETFE   364, 371
ethyl cellulose   360
ethylene chlorotrifluoroethylene copolymer   364, 370
ethylene propylene diene rubber   361, 367, 373
ethylene tetrafluoroethylene copolymer   364, 371
ethylene vinyl alcohol copolymer   359
EU   368, 374
EVOH   359
expanded graphite   372
expansion metal   372

f
F-Cu   50
FeMnAl alloy   126–127, 129, 131, 132, 135–137
FEP   357, 364, 370
Ferralium® 255   108–112, 114–116, 137
ferritic chromium steel   99, 103, 386
ferritic stainless steel   103, 106
FFKM   367, 373
Fiberdur® Centricast II   380
Fiberdur® CSEP   380
Fiberdur® CSVE   380
Fiberdur® EP   380
Fiberdur® VE   380
fine grain structural steel   67–68, 78
FKM   361, 367, 373
fluorinated rubber   361, 367, 373
fluorosilicone rubber   368, 374
fluorothermoplastic   390
forged steel   138

Index of materials

FQ 368
FVMQ 368
FVMQ/FMQ 374

g
G12CrMo9-10 93
G17CrMo5-5 94
G41300 95
G41350 70, 76
G43400 70
gasket 372
glass 349
glass fiber reinforced plastic 380
Gore-Tex® series 300 372
Gorka® 40 348
GP240GH 94
Grade 2 244
graphite 348, 372, 389, 391
GX12CrMo5 93
GX12CrMo10 93
GX23CrMoV12-1 94
Gylon® blue 372
Gylon® Standard 372
Gylon® white 372

h
H1/S88 308–309
hafnium 247–249, 312
Halar® 370
hastelloy X alloy 187, 189
HDA-230 308
high-alloy stainless steel 103
high-purity aluminum 20
high-purity chromium 44, 46
high-temperature steel 107
high-tensile steel 81
HNBR 367, 373
HY 32 377, 379
HY 50 378–379
HY 101 379
hydrogenated nitrile rubber 367, 373
Hytrel® 366

i
IIR 361, 367, 373
IMI 115 241–242
IMI 130 241–242
IMI 155 241–242
INCONEL® alloy 625 185–187, 189, 388
INCONEL® alloy 718 176–182, 185–190, 192, 195–196, 200, 388
INCONEL® alloy 718 183–184
INCONEL®100 199
Incoloy alloy 903 159–160, 162–163
IR 360, 370
Incoloy 907 158
iron material 103
isobutene isoprene rubber 361, 367, 373
isoprene rubber 360

j
J42045 93
J82090 93
JBK-75 164

k
K02001 67
K11547 95
K12539 68
K21390 93, 95
K31545 95
KLINGER®top-chem 2000 372

l
layers 369, 390
LCP 358
liquid crystal polymer 358
low-alloy steel 57, 70, 385

m
magnesium oxide 353, 389
MAR-M 247 199, 200
martensitic steel 107
methacrylonitrile isoprene rubber 361
MFA 364
monocrystalline alloy 201
monocrystals from nickel 167
MQ 362

n
N02200 169
N07718 176, 388
N19903 159
NASA-HR-1 164
natural rubber 360, 366, 370, 373
NBR 361, 367, 372–373
Nextel® 312 353
Nextel® 440 353
Ni 99.2 169
Ni₃Al alloy 205
nickel 166–167, 169–170, 387
Nickel 201 308
nickel alloy 166, 187, 387
nickel copper alloy 55
nickel-chromium alloy 171, 176, 388
nickel-chromium-iron alloy 171, 388
nickel-chromium-molybdenum alloy 176, 388
nickel-chromium-molybdenum steel 78
nickel-copper alloy 53–54, 201, 388
NiCr15Fe 171
NiCr15Fe7TiAl 172, 174–175
NiCr19Fe19Nb5Mo 388
NiCr19Fe19Nb5Mo3 176–177, 185, 187
NiCr19NbMo 24
NiCr22Mo9Nb 185, 187, 189, 388
NiCr29Fe 171
niobium 56–57, 214–216, 218–220, 222–223, 225–227, 234–235, 312, 330, 388
NIR 361

NITRONIC® 40 131–132, 134–137, 141, 143–144, 146–150
novaflon® 200 372
novaflon® 500 372
novaphit® Super HPC 372
novaphit® Super SSTC 372
novaphit® VS 372
novapress® multi II 372
novatec® PREMIUM II 372
NR 360, 366, 370, 373

o
OF-Cu 51
OFHC copper 52–53
OT 368

p
PA6 359, 365
PA6-3-T 365
PA11 359, 365
PA12 365
PA46 365
PA66 365
PA610 365
palladium 207–208, 210
palladium alloy 210
palladium-silver alloy 207–208, 210–213
PB 356, 363
PBI 358
PBT 365
PC 358, 365
PCTFE 357
PDAP 366
PE 371
PEEK 365
PEEK-CF 362
PE-HD 356, 363
PE-HMW 363
PE-LD 356, 363, 370
PE-MD 363
PEMFC 348
perfluoro rubber 367, 373, 390
PESU 357
PET 358, 365
PE-UHMW 363
PE-X 363
PFA 364, 370
phenolic resin 348, 366, 370
PI 358
PIB 356, 363
pipe steel 78, 116, 119
platinum metal 207
PMMA 359, 365
poly(1-phenyl-1-propine) 359
poly[1-phenyl-2-[p-(triisopropylsilyl)phenyl] acetylene] 359
poly[1-phenyl-2-[p-(trimethylsilyl)phenyl] acetylene] 359
poly(1-trimethylsilyl-1-propine) 359
Polyamide 6 359, 365

Polyamide 6-3-T 365
Polyamide 11 359, 365
Polyamide 12 365
Polyamide 46 365
Polyamide 66 365
Polyamide 610 365
polybenzimidazole 358
polybenzylmethacrylate 359
polybutene 356, 363
polybutylene terephthalate 365
polycarbonate 358, 365
polychlorotrifluoroethylene 357
polycrystalline nickel 167
polydimethylsiloxane 362
poly(diphenylacetylene) 359
polyester resin 366, 381
polyether ether ketone 365, 381
polyethersulfone 357
polyethylene 356, 363, 370–371, 390
polyethylene terephthalate 358, 365, 381
polyimide 358
polyisobutylene 356, 363
polymer electrolyte membrane fuel cells 348
polymethylmethacrylate 359, 365
Poly(oxy-2,6-dimethyl-1,4-phenylene) 359
polyoxymethylene 359, 365
polyphenylene sulfide 365
polypropylene 348, 356, 363, 371, 390
polystyrene 358
polysulfone 357
polysulphide rubber 368
polytetrafluoroethylene 348, 357, 364, 372, 391
polyvinyl acetate 359
polyvinyl alcohol 359
polyvinyl benzoate 359
polyvinyl chloride 357, 363, 371, 390
polyvinylfluoride 357
polyvinylidene chloride 357, 364
polyvinylidene fluoride 348, 357, 364, 371
POM 359, 365
PP 363, 371
PP-B 363
PPE 359
PP-H 356, 363
PP-R 363
PPS 365
pressure vessel steel 68, 82–83
PS 358
PSU 357
PTFE 348, 357, 364, 372
pure aluminum 20, 22, 40
pure chromium 46–47, 49
pure copper 52
pure iron 59, 63, 103
pure nickel 166, 169–170
pure titanium 238, 241, 244–246, 259, 278, 297
pure zirconium 313

PVAC   359
PVAL   359
PVC-C   364
PVC-P   357, 363
PVC-U   357, 363, 371
PVDC   357, 364
PVDF   357, 364, 371
PVF   357
PWA 1480   196–197

q
quartz   353, 389

r
R60701   312
R60702   246, 312
R60704   312
R60705   312
R60706   312
R60801   313
R60802   313
R60804   313
R60901   313
red brass   55
refined copper   49
refractory metal   214, 230, 325
refractory product   389
Rhenoguard® Jumbo   370
RSP-AlLi-alloy   39

s
S500MC   67
S500N   67, 78
S500Q   67
S34800   187, 189
S44627   107
S50100   95
S50400   95
SAE 301   106, 108–110, 112, 131–132, 135–137
SAE 304   131–133, 153–155
SAE 310   106
SAE 310 S   132–133, 137, 153, 155
SAE 316   150, 153, 155–158
SAE 316 L   107, 143, 148
SAE 347   187–189
SAE 1010   59
SAE 1020   59
SAE 1043   59
SAE 1074   77
SAN   365
SB   358
SBR   360, 366, 370, 373
SE-Cu   51
SF-Cu   51
Sigracet® BPP   348
silica glass   349
silicon carbide   354, 390
silicon dioxide   353
silicon nitride   354

silicone rubber   361, 368, 374
silver   20
silver alloy   20
special chromium molybdenum casting   374
St 34-2   59
stainless austenitic steel   153
StE 500   67, 78
steel   99, 103, 187
styrene butadiene rubber   360, 366, 373
styrene-acrylonitrile copolymer   365
styrene-butadiene copolymer   358
super duplex steel   125
super ferrite   109, 112
super-purity aluminum   20
SW-Cu   51

t
tantalum   214–216, 218, 220, 222, 224, 312, 388
technical zirconium   312
technically pure titanium   254
tempered steel   69
tetrafluoroethylene perfluoro propyl vinyl ether copolymer   364, 370
tetrafluoroethylene-hexafluoropropylene copolymer   364, 370
tetrafluoroethylene/perfluoromethyl vinyl ether copolymer   364
thermoplastic elastomer   366
Ti 1   235–236
Ti 2   235–236
Ti 3   235–236
Ti 4   236
Ti Grade   241
Ti Grade 1   236, 238–242
Ti Grade 2   236, 238–242, 249–252, 257–258
Ti Grade 3   236, 241–243, 257–258
Ti Grade 4   236, 241–242
Ti Grade 12   249–252
Ti$_3$Al   296, 300–301
TiAl   296, 299–301, 310
TiAl1Mn1   285–286
TiAl$_2$ 296
TiAl2Mn2   285–286
TiAl$_3$ 296
TiAl3V8Cr6Mo4Zr4   290, 292
TiAl5Mo4   286–289
TiAl5Sn2,5   187, 189, 277–281
TiAl6   280–281
TiAl6Nb2Ta1Mo0,8   265, 267
TiAl6V4   24, 187, 189, 261–275, 277–281, 283–284, 297
TiAl6V4,5   259
TiAl6V6Sn2   270, 272, 280–281, 283
TiAl8Mo1V1   270, 272
TiAl14AlNb20V3   308
TiAl14AlNb20V3,2Mo2   307, 309
TiAl14Nb20V3,2Mo2   303, 304

TiAl14Nb21 308–309
TiAl24Nb11 297, 302–303, 305–306
TiAl25 299
TiAl25Nb10V3Mo1 303
TiAl33Nb5Ta1 303–304
TiAl45V3 299, 310
TiAl48 299
TiAl48Cr2 310–311
TiAl48Nb2,5Ta0,3 307, 310
TiAl48Nb2Mn2 309
TiAl48Nb2Ta0,2 297
TiAl48V1 297–298
TiAl50 297–298
TiAlSn 389
TiAlV 389
TiMo2Fe2Cr2 288, 290
TiMo7Cr11Al3 292
TiMo11,5Zr6Sn4,5 292, 294
TiMo18 290–291
TiMo30 286–287, 290, 293
titanium 235–238, 246–249, 255–256, 277–279, 312, 389
titanium alloy 235, 259, 261, 272, 283–284, 389
titanium aluminide 295–297, 299–300, 308–309
titanium carbide 354
titanium material 187, 260
titanium nitride 354
TiV13Cr11Al3 292
TiV20 295
TPE 360, 366
tungsten 325
tungsten carbide 354

u

U-0.8Ti 338
Udimet® 720 199–200
unalloyed carbon steel 85–86, 88
unalloyed steel 57
UNS J42045 93
UNS J82090 93
UNS K02001 67
UNS K12539 68
UNS K21390 93
UNS N02200 169
UNS N07718 176, 388
UNS N19903 159
UP 366
uranium 338
urethane rubber 368, 374

v

V-1Ti 331
V-5Ti 331
V-10Ti 331
V-20Ti 331
V-30Ti 331
vanadium 214, 217, 219, 221–222, 225, 228–229, 312, 325, 327, 330, 333, 336–338, 388
vanadium alloy 47, 325, 335
vanadium-chromium alloy 333–335
vanadium-niobium alloy 327, 330–331, 333
vanadium-titanium alloy 330, 332–333
VE 366, 380
vinyl ester resin 366, 380, 391
VMQ 361, 368, 374
VT 15 292, 294

w

Waspaloy® 199
Wavistrong® 380
welding material 88–89

x

X1CrMoNi29-4-2 103–106
X2CrMoTi29-4 107, 386
X2CrNiMo25-6-4 113–114
X2CrNiMoN22-5-3 387
X4NiCrTi25-15 185, 187, 189
X6CrNiNb18-10 187, 189
X 65 78
X 70 78
X8CrMnNi18-8 141
X8CrNiMoVNb16-13 94
X10CrNi18-8 106
X11CrMo9-1 95
X15CrNiSi25-21 106
X20CrMoV11-1 94–95

z

Zeron® 100 123–126
Zircaloy®-1 313
Zircaloy®-2 313, 322, 324
Zircaloy®-4 313, 315–317, 321–322, 323
zirconium 247–249, 312, 389
zirconium alloy 312–313, 322, 389
zirconium dioxide 353, 389
zirconium material 312
Zirkaloy®-4 323
Zr 701 312
Zr 702 312
Zr 704 312
Zr 705 312
Zr 706 312
Zr-2.5Nb 313–314

Subject index

a
α. 13
absorption column 381
absorption cross-section 313
absorption of hydrogen 21, 40–41, 168–169
accelerators 355
acoustic emission 313, 382
activators 355
activity coefficient 40
aerospace industry 36, 295
aging 125–126
aircraft construction 36, 238, 261, 269
alloy U-0.8Ti 341
alloy U-5.7Nb 339
alternating load test 72, 80
aluminum melt 41
American Petroleum Institute 96
ammonia converter 99
ammonia (inhibiting effect of) 76
antidegradant 355
antioxidant 355
API 96
apparatus component 348
aramid fiber 372
atomic hydrogen 21

b
battery 177
bearing technology 348
bio material 261
bipolar plate 348
blistering 42, 50
Bosch hole 86
brazed joint 51
bright annealing 50
brittle fracture 63, 72, 75, 78, 252, 254

c
carbon dioxide 64
carbon dioxide (inhibiting effect of) 76
carbon fiber 381
carbon monoxide 64
carbon monoxide (inhibiting effect of) 76
carrier gas method 18
cast steel grade 94
catalytic combustion 8
catalytic steam cracking 7
centrifugally cast pipe 380
CERT test 64, 70, 75, 78, 118, 127, 185, 288
chemical decomposition 355
chemical engineering 312
chemical plant 261
chemical protection layer 380
chloralkali electrolysis 8
cladding 178–179, 313, 314
CNSB specimen 269, 271
coal gasification 8
coking plant 8
cold compressed hydrogen 59
cold forming 88
combustion chamber 56
compact tension 18, 233
competitive adsorption 14
compressed gas cylinder 63
compressed hydrogen 22, 164, 171
compressed hydrogen damage 15, 75
compressed hydrogen resistant steels 88, 93–95
compressed pipe 78
constant extension rate tensile test 64, 127
converter gas 209–210
crack propagation facet 26
crack propagation velocity 28–29, 33
critical hydrogen concentration 101–102
crude oil fraction 7
crude oil installations 94
crystal lattice 103
CT specimen 18, 230, 244, 250, 252, 267, 269, 272, 278, 292, 322, 324
Curran test 352
cylinder store 59

d
damage type 355
DCB specimen 18, 180–181, 183
decarburization 15, 84, 88
deformation martensite 153, 155
degree of dissociation 13
delayed hydride cracking 313
deposition welding 19

deuterium   107, 113, 148
diffusion barrier   179
diffusion cleaner   208
diffusion coefficient   107
dimple   169
dissociation of hydrogen   20
dissolved hydrogen   168–169
distilled water   30–31, 45
double cantilever beam   18, 180
dry air   36, 45
dry argon   26, 31
dry helium   37–38
dry hydrogen   25–27, 30–33, 37–38, 47, 63
dry hydrogen sulfide   69
ductility   88
duroplastic   360

e
effusion time   102
elastomer   360
electric arc welding   156
electrolysis of water   8
electrolytic cell   381
electron beam welding   156
embrittlement   12, 32, 39–40, 50–55, 106, 116–117, 119, 123, 155, 166–170, 174, 179, 186, 201, 210–212, 214, 235, 240, 244, 246, 254, 259, 261, 264–265, 267, 272–273, 278, 283, 286–288, 294, 303, 330, 333
embrittlement index   117–123, 156–157, 275
embrittlement maximum   67
embrittling influence   47
engine construction   295
excess-pressure vessel   381
exhaust air duct   381
exhaust system   257–258
explosive cladding   19
external hydrogen   178–179
extrusion   39

f
fatigue behavior   26, 32
fertilizer production   312
fiber   355
fiber-reinforced composite   362, 380
filler   355
fish eye   40, 59
fitting   63
flame retardant   355
flammability   380
flat gasket   372, 391
flow additive   355
flowability   168
fluorothermoplastic   364
fluxing material   51
forged steel   137
fracture   12, 85
fracture characteristic   167

fracture mechanics   17
fracture toughness   233
fracture-mechanical test   16, 64, 80
fuel   296
fuel cell   208
fuel element   313–314
fusion reactor   99, 102, 155, 325

g
gas production   108
gas shielded welding   88
gas turbine   177
GFRP   380
gold coating   179
Gorsky effect   80
grain boundary crack   88
grain boundary embrittlement   249
grain boundary precipitate   168
graphite yarn   391

h
Haber-Bosch process   86
HCF test   37, 38
heat affected zone   90–91
heat exchanger   235
heat treatment   50
heavy water reactor   314
helium atmosphere   178
helium embrittlement   107
high-frequency alternating load   188
high-pressure autoclave   171, 173
high-pressure hydrogen   22, 265, 292, 294, 299, 307, 322
high-pressure hydrogenation   85, 96
high-pressure synthesis   86
high-purity water vapor   32
high-temperature thermoplastic   365
high-temperature water   171–172, 174, 313
high-temperature zone   157
hot compressed hydrogen   84, 90–91
hot extraction   104
hot isostatic pressing   196–197
hydride formation   7, 12, 15, 216, 245–246, 256–257, 259, 264, 269, 278, 283, 286, 299–300, 308–309
hydrogen absorption   20–21, 39
hydrogen activity   25
hydrogen atmosphere   23
hydrogen degassing   107
hydrogen diffusion   72, 208
hydrogen diffusivity   105–106
hydrogen disease   50
hydrogen electrode   208
hydrogen embrittlement   35, 40, 49–53, 55–56, 68, 76–78, 80, 99, 103, 106–107, 114, 118–120, 122–123, 126–127, 129, 131, 135, 137–139, 141, 143–145, 147–148, 150, 152–153, 155, 158, 162, 164, 174, 176–177, 185–186, 192–193, 201, 205, 209–210, 214, 244–246, 252, 256–

257, 261, 267, 272, 277, 286, 288, 290, 292, 303–305, 308–311, 321–322, 325, 327–329, 332, 387
hydrogen embrittlement index 143, 146
hydrogen gas 5, 25, 28
hydrogen plant 177
hydrogen resistance 51
hydrogen solubility 21, 40
hydrogen source 35, 99
hydrogen storage layer 383
hydrogen storage tank 278
hydrogen sulfide 69
hydrogen tank 24, 383
hydrogen trap 77
hydrogen-containing synthesis gas 78
hydrogen-containing welding material 59
hydrogen-induced brittle fracture 84
hydrogen-induced crack growth 64, 112, 118

i
ice water 120
incubation period 88
influence of alloying elements 72
influence of material strength 72
influence of strain rate 72
influence of sulfur content 168
inhibiting effect of oxygen 64
inhibitors 355
intercrystalline stress corrosion cracking 34
intermetallic compound 201
internal crack 12, 85
internal decarburization 85
internal hydrogen 178–179
ionized hydrogen 100

j
Joule-Thomson coefficient 9
Joule-Thomson inversion curve 9

l
laser treatment 192–194
lattice vacancies 12
LCF test 37–38
light water reactor 313
liquid hydrogen tank 277
liquid rocket stage 63
liquify 9
load-controlled swelling test 72
low-frequency alternating load 188
low-pressure hydrogen 307

m
magnesium content 169
manufacture of chemical equipment 363
material for pressure vessels 28
material stress 10
mechanical seal 374
methane 15, 64, 84, 88–91

methane bubble 92
microcrack formation 382
mining 348
moist air 26–27, 30–31, 36, 45
molten metal 40
monocrystalline alloy 201, 222
multilayer reactors 86

n
natural gas 78
natural gas installation 94
Nelson diagram 96–97, 99
nickel-hydrogen battery 177
nitrous dioxide 64
nuclear fusion 53
nuclear reaction 99, 102
nuclear reactor 313
nuclear waste 249
number of cycles to failure 27

o
oil production 108
organic coatings and linings 370
O-rings 373
oxygen 64

p
packing 372
palladium membrane 208–209
palladium-silver membrane 210
Paris' law 81
partial hydrogen pressure 88
penetration notch 67, 257
peristaltic pump 367–368
permeability 355
permeation 380
permeation coefficient 15, 355, 362
petrochemical process 7
pigments 355
pipe 363, 380
pipe seal 367
pipeline 63, 363
plasma 99
plastic deformation 63
plasticizers 355
plastics production 312
powder coating 370, 390
pressure chamber 57
pressure vessel 85–86
production of ammonia 99
production of hydrogen 8
production of steel 86
pulse-echo method 283

r
R value 230
radiation damage 325
rapid solidification process 39
reaction resin 366
reaction resin coating 370

Subject index

reactor construction 312
reactor technology 312
refinery 8
reforming 209
resonance method 283
rocket 56
roll cladding 19
rotor 51
RSP 39
rubber 360
rubber lining 370

s

salt melt 104, 114, 186
scrubbing tower 381
sealing band 372
sealing ring 374
seawater 108
secant method 250
semiconductor technology 8
shielding gas 106
shock test 352
single-phase material 186
slow burst test 16
slow strain rate tensile test 16, 64, 67
Snoek effect 284
soft rubber lining 370
solubility limit 12
space shuttle 177, 193
Space Station (ISS) 177
spacecraft 277, 296, 380, 389
Stahl-Eisen-Werkstoffblatt 93
static load 81
steam reforming 7
steam-embrittlement 50
storage tank 69
strain-controlled swelling test 73
stress intensity factor 17, 29, 124, 174, 181–182, 200, 233, 262, 264, 269–270, 277–278, 287, 324
stuffing box packing 372
subcritical crack growth 81
submerged arc welding 88, 90
sulfide precipitate 169
sulfur dioxide 64
sulfur dioxide (inhibiting effect of) 76
surface decarburization 85
surface roughness 67, 72
swelling 355

t

tank 383
tantalum 214
technical solubility coefficient 6
technically pure hydrogen 75
tendency to embrittlement 35
thermal conductivity 6
thermal neutron 313
thermal oxidation 256
thermal spraying 19
thermoplastic elastomer 360
thermoplastic web 370, 390
threshold temperature 88
titanium aluminide 296
titanium vessel 256
torsional stress 26–27
toughness 88
transcrystalline cleavage fracture 179
transcrystalline crack 168, 179
transport tank 63
tritium 148, 150
turbine blades 193, 198, 295

u

ultrapure hydrogen 72
ultrapure nitrogen 72
ultrasonic test 283–284
uniform elongation 78
urea 254–256
UV stabilizer 355

v

vacuum extraction 18
vacuum vessel 155
Vickers hardness test 303

w

water vapor pressure 31
wedge opening load 18, 180
weld metal 88, 91, 157
weld region 67, 88
weld seam 257, 284
weld zone 156
welded joint 155–157
wet helium 38
wet hydrogen 26–27, 32–33
Widmannstätten structure 264, 266, 283, 286, 288
WOL specimen 18, 160, 162, 180–181, 265